Two Manuals Combined into One Book

Combat Engineer

and

Engineer Operations

Department of Defense

Pentagon
Publishing

Field Manual
No. 5-10

Headquarters
Department of the Army
Washington, DC, 3 October 1995

Combat Engineer Platoon

TABLE OF CONTENTS

DISTRIBUTION RESTRICTION: Approved for public release; distribution is unlimited.

Page

PREFACE

Field Manual (FM) 5-10 is designed as a companion manual for *FM 5-34* and is structured using the following Battlefield Operating Systems (BOS) as a framework:

- Intelligence.
- Maneuver.
- Mobility and survivability.
- Fire support (FS).
- Air defense (AD).
- Combat service support (CSS).
- Command and control (C^2).

Each BOS is not discussed in detail. Instead, they are used to outline considerations for how the engineer platoon supports the total combined-arms (CA) effort. The bulk of the platoon's missions concentrates on mobility, countermobility, and survivability operations in support of the mobility and survivability BOS. However, the platoon leader and sergeant must have an understanding of the role their platoon plays in supporting all the BOS. More importantly, the platoon leader and sergeant must understand the impact of all the BOS on platoon planning, preparation, and execution.

This FM also has three appendices: *Appendix A* covers basic formations, movement techniques, and hand-and-arm signals; *Appendix B* contains checklists for security, precombat, and packing; and *Appendix C* covers case studies for mounted and dismounted breaching.

The proponent of this publication is Headquarters (HQ), United States (US) Army Engineer School. Send comments and recommendations on *Department of the Army (DA) Form 2028* directly to Commander, US Army Engineer School, ATTN: ATSE-T-PD-PM, Fort Leonard Wood, Missouri 65473-6650.

Unless this publication states otherwise, masculine nouns and pronouns do not refer exclusively to men.

CHAPTER 1
Intelligence

Intelligence is the key for preparing for the next battle. The analysis of intelligence, in detail, determines the impact of the enemy, weather, and terrain on operations. It is a continuous process that supports the planning for and execution of all operations. It is imperative that the platoon leader understand the intelligence process and his part in that process.

THREAT

The threat facing the US Army has gone through a quantum change with the breakup of the Warsaw Pact and the former Soviet Union. However, the majority of the forces we will face in mid- to high-intensity conflicts will use the "Soviet" model for their operations. *Figures 1-1 and 1-2, pages 1-2 and 1-3,* provide templates for the Soviet-style platoon and company defense as well as movement formations. A detailed analysis of threat mine-warfare capabilities and tactics can be found in *FM 90-13-1, Appendix A, and FM 20-32, Appendix C.*

RECONNAISSANCE

Reconnaissance is critical to the engineer. During offensive operations, the engineer must see the battlefield to plan breaching operations. History has proven that when a unit conducts CA rehearsals and obstacle reconnaissance it has greater success. The platoon leader should organize his platoon based on the reconnaissance mission the commander gives him. This frequently means providing expert engineers to assist the task force (TF) scouts. The platoon leader may also organize all or part of his platoon for a reconnaissance mission if there is a critical requirement for engineer-specific intelligence. If at all possible, the engineers who conduct the reconnaissance should be trained in this task as well as cross trained with the maneuver unit's scout platoon. The platoon leader must ensure that his designated reconnaissance element links up with the scout platoon.

TYPES

The general types of engineer reconnaissance missions are—

- Technical, which—
  - Is conducted to collect engineer-specific technical data.

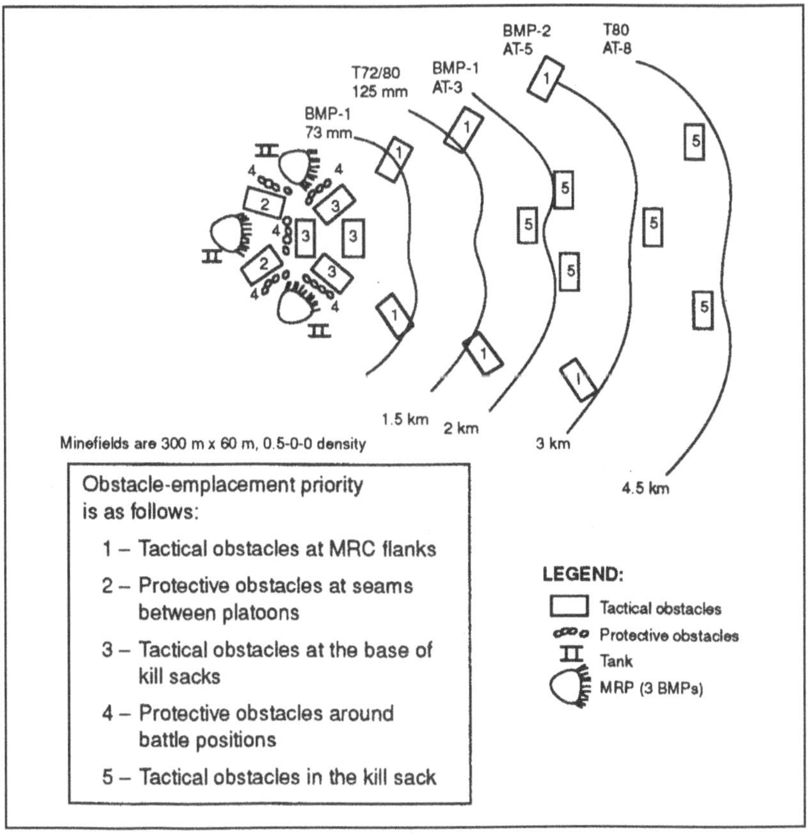

Figure 1-1. Motorized rifle company (MRC) defense template

- Is conducted to perform a specified task from higher HQ in a Level 1 threat environment.
- Involves (almost always) engineers only and is usually conducted behind the forward line of own troops/forward edge of the battle area (FLOT/FEBA).

- Tactical, which—
  - Is conducted to collect engineer-specific information requirements on an enemy point, area target, or attack axis or a route.

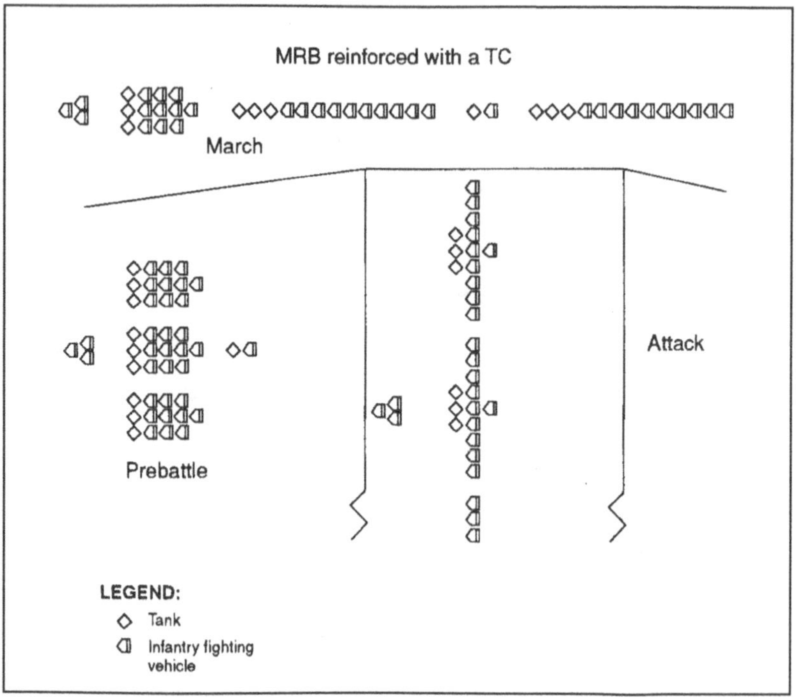

Figure 1-2. Typical threat formations

- Is conducted to support the maneuver plan as an implied task.
- Is conducted in a Level 3 threat environment.
- Requires engineer participation as part of a CA effort.
- Is usually conducted forward of the FLOT/FEBA.

PLANNING

Reconnaissance is a CA effort. The engineers either augment the existing reconnaissance element or conduct their own reconnaissance as part of the consolidated reconnaissance and surveillance (R&S) plan. Engineer reconnaissance must be focused on a specific target, and the reconnaissance element should be given specific intelligence requirements.

In heavy forces, the engineer company commander provides the reconnaissance target and does the majority of the engineer reconnaissance planning. However, in light forces, the engineer platoon leader does the engineer reconnaissance planning.

Platoon reconnaissance planning should include—

- A warning order (WO) based on brigade and TF planning, which includes—
  - Tentative target locations.
  - Type of reconnaissance or target (see *Figure 1-3).*
  - Specific information, equipment, and coordination requirements.
- The movement of reconnaissance assets to effect the linkup.
- The maximum time for leaders to prepare for the reconnaissance mission.
- The assignment of detailed reconnaissance routes.
- The actions to be taken on contact.

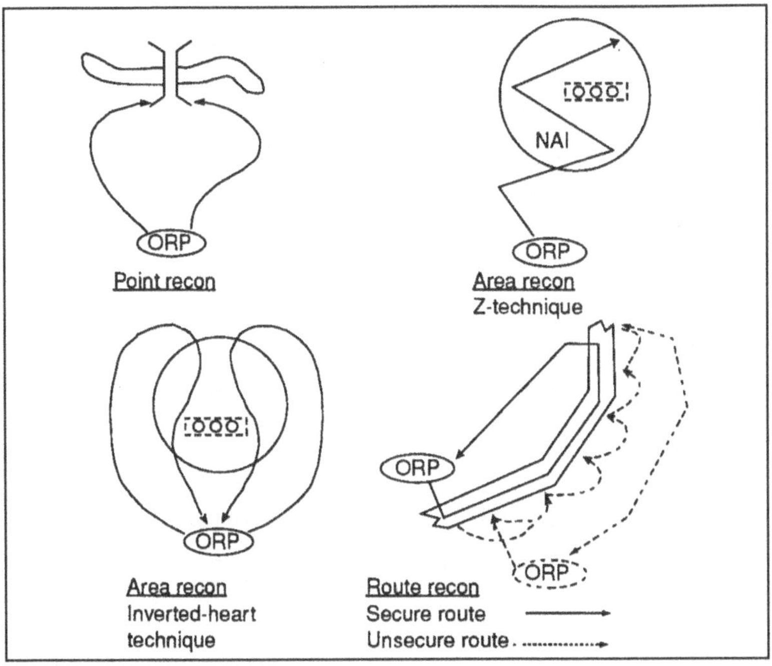

Figure 1-3. Recon techniques

- Consolidation instructions.
- Specific reporting requirements.

MOUNTED AND DISMOUNTED PATROLLING

Patrolling is a CA operation. Engineers are normally part of a larger unit's patrol. When patrolling mounted—

- Position all reconnaissance assets forward.
- Use aggressive and continuous patrols.
- Use area and zone reconnaissance.
- Allow enough time for a thorough reconnaissance.
- Use stealth, not combat power, for reconnaissance.
- Modify formations to overwatch the point element.
- Use terrain driving techniques; use draws and folds in the terrain to mask movement.
- Use night, fog, dust, and storms to conceal movement.
- Reduce speeds to avoid dust signatures.
- Travel at the highest tactical speeds, and place artillery fires on expected enemy positions to avoid long-range fires.
- Keep vehicles dispersed.

When patrolling dismounted—

- Reconnoiter all danger areas, choke points, and critical terrain features.
- Expect contact with enemy dismounted scouts.
- Use darkness, storms, and periods of limited visibility to infiltrate enemy positions.
- Use binoculars or night-vision devices at night with the illumination from ambient light (ensure that laser protection is maintained).
- Use artillery illumination and spotting rounds on reconnaissance targets to assist in land navigation and orientation.

CHAPTER 2
Maneuver

Engineer support to maneuver units requires detailed planning. Engineers may be part of the maneuver unit's formations and, if so, will provide the combat multiplier of engineer effort. *FM 90-13-1* is the primary doctrinal manual for breach planning. *FMs 20-32 and 90-7* are the primary doctrinal manuals for obstacle planning and construction. These references provide the technical information required to plan engineer support for offensive and defensive maneuvers.

The platoon leader should have an understanding of the tactical decision-making process, troop-leading procedures (TLP), and orders preparation. *Figure 2-1, page 2-2,* shows the relationship between TLP, the estimate process, and the intelligence preparation of the battlefield (IPB) process.

In addition, the platoon leader must advise the maneuver commander on the military aspects of the terrain since he is the terrain expert. Normally, the platoon leader only has a map to prepare his terrain analysis. However, he should request terrain-analysis products from his parent engineer company. In both light and armored units, the engineer company has access to terrain-analysis and terrain-visualization products or the ability to develop them. These products greatly aid platoon-level planning.

OFFENSE

The platoon leader is the key engineer expert to his supported unit. He must make several key decisions during the offense. He recommends where critical breach assets are located and which ones are best based on the situation. For example, the supporting engineer platoon leader, in an armored unit, would recommend where and how the M1 tank's plow would be used; or a light platoon leader could recommend to the TF commander a covert breach based on his terrain analysis. In both cases, the engineer platoon leader assists the maneuver commander with his planning. For more information on breaching, see *FM 90-13-1*. The following considerations are specific to the offense:

- Using a highly mobile engineer force, well forward and integrated into maneuver formations, to maintain the momentum of the attack.

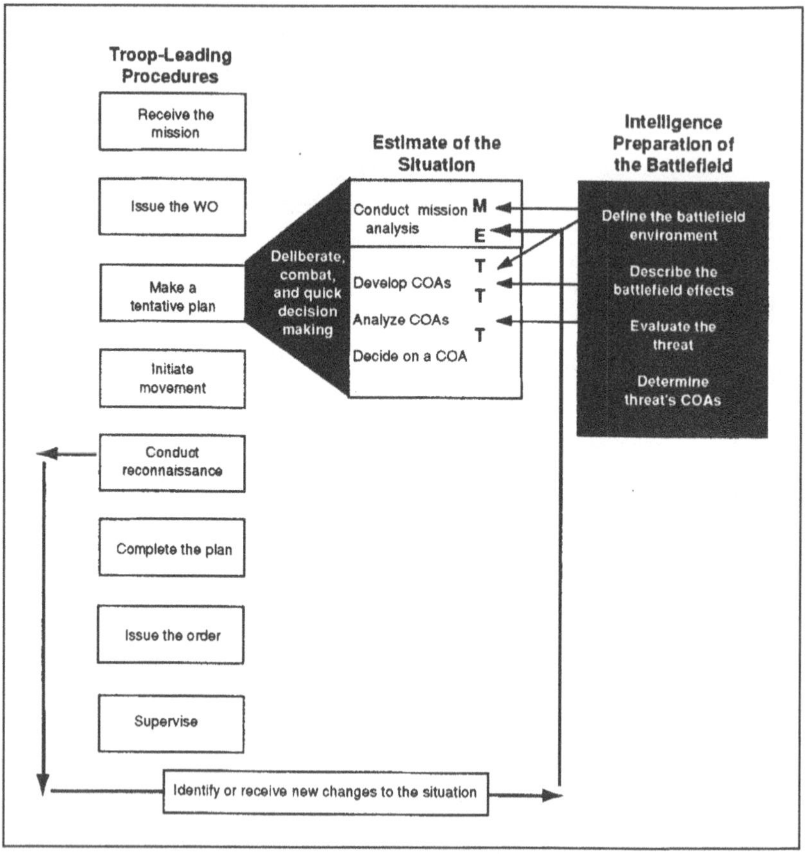

Figure 2-1. Relationship between TLP, the estimate process, and the IPB process

- Arranging for lane turnover between the forward breaching force and follow-on engineers for lane improvement and obstacle clearance.
- Ensuring that marking material and the standing operating procedure (SOP) for lane marking are coordinated between passing units.
- Ensuring that critical engineer equipment, such as armored vehicle-launched bridges (AVLBs), follow-on tactical bridging, lift capability for

line-charge reloading, and lane-marking materials to replenish marking systems, is available.

- Planning situational-obstacle capability for flank security. Scatterable mine (SCATMINE) systems work well in this application.
- Planning tactical-obstacle capability for protection against counterattacks (CATKs). Artillery- and aircraft-delivered family of scatterable mines (FASCAM) are planned for quick emplacement. Ground FASCAM systems are planned for prolonged defenses.
- Ensuring that Class IV/Class V barrier materials are available for a transition to the defense.
- Removing curbs from the AVLB if M1 tanks with rollers are to cross. The roller's dog bone should be raised.

DEFENSE

As with the offense, the platoon leader recommends to the maneuver commander where engineer assets are best used. It is critical that the platoon leader be an integral part of obstacle planning and integration. For more information on obstacle integration, see *FM 90-7*. The platoon leader should also coordinate and develop the obstacle-protection plan. It is critical that the obstacles are protected against enemy breaching attempts, especially covert attempts. The platoon leader should coordinate the inspection of the obstacles that his platoon built before the expected enemy attack. This will ensure that the enemy has not covertly breached and, if so, allows for the obstacles to be repaired before the enemy's main attack. The following considerations are specific to the defense:

- Understanding the commander's intent for obstacles.
- Ensuring that adequate transportation assets are available to haul Class IV/Class V supplies forward (see *Table 2-1, page 2-4*).
- Identifying critical engineer tasks early. Terrain preparation requires time for completion. Engineers must not remain idle while planning is in progress. If possible, mortars, air-defense artillery (ADA), fire-support-team (FIST) vehicles, CSS assets, and the tactical operations center (TOC) should be dug in while direct-fire systems are being positioned. Typically, these systems can be fortified between issuing the WO and completing the direct-fire systems' positioning.
- Practicing operations security (OPSEC) measures to prevent premature disclosure of the defense.

Table 2-1. Class IV/Class V haul capacities

Vehicle	Concertina Wire	M15 AT Mine	M21 AT Mine	M16 AP Mine	M14 AP Mine	MOPMS Mine	Flipper Mine	Volcano Mine	MICLIC Reload
HMMWV, M998 2,500 lb 215 cu ft	2	51	27	55	56	15	11	1	*1
2 1/2 -ton truck 5,000 lb 443 cu ft	4	102	55	111	113	30	23	2	1
5-ton truck 10,000 lb 488 cu ft	7	204	109	222	227	61	46	5	3
5-ton dump truck 10,000 lb ***135/291 cu ft	2/4	112/204	32/70	168/222	71/153	23/51	39/46	3/5	2/3
20-ton dump truck 40,000 lb 754 cu ft	11	628	179	888	443	132	184	20	11
HEMTT truck 20,000 lb 540 cu ft	8	408	128	444	317	94	92	10	7
12-ton S&P 24,000 lb 875 cu ft	13	489	208	533	514	148	110	12	9
40-ton lowboy 80,000 lb 1,760 cu ft	27	1,466	419	1,777	1,035	308	368	43	27
M54812,000 lb 529 cu ft	8	244	125	266	272	74	55	6	4
#Mines / wt/lb; Cube cu ft	40/ 1,180; 64	1/ 49; 1.2	4/ 91; 4.1	4/ 45; .8	90/ 44; 1.9	21/ 162; 5.7	40/ 217; 3.4	240/ 1,850; 37.6	** 2,656; 64.8

for concertina = bundles; 1 bundle = 40 rolls * Overloads vehicle **Line charge + rocket ***Without/with sideboards

- Ensuring that the engineer organization for combat allows for rapid transition to the offense. The reserve must always have a designated force of engineers. Obstacle placement must not interfere with spoiling and CATKs.
- Ensuring that engineer units are not held in reserve but remain committed and work on the commander's priority tasks.

PLATOON LEADER'S PLANNING

The platoon leader can use *Tables 2-2 and 2-3, pages 2-6 and 2-7,* as a guideline when planning and conducting offensive and defensive operations.

REHEARSAL PRINCIPLES

Regardless of the event or task to be rehearsed or the type of rehearsal used, certain principles are universal for conducting effective rehearsals. They are—

- Supporting the scheme of maneuver and the commander's intent.
- Providing clear tasks/conditions/standards (T/C/S).
- Conducting multiechelon CA rehearsals.
- Determining key participants.
- Enforcing standards/training to standard.
- Using the after-action-review (AAR) process to provide feedback to all participants.
- Complementing the preparation phase.
- Instilling confidence in the plan and in the leaders.

During the command estimate process, use the rehearsal principles. These principles produce the—

- Participant levels.
- Rehearsal technique.
- Initiation of precombat checks (PCCs), precombat inspections (PCIs), and mission-specific drills.

Proper application of the rehearsal principles enables the unit to progress through a "crawl/walk/run" process that ends with mission success (see *Figure 2-2, page 2-8*).

Table 2-2. TF-engineer offense briefing

Briefing Topic	Main Points to Brief
Engineer task organization	• How the platoon will be task-organized • Special equipment that will be task-organized to subordinate units • Times and locations for linkup
Enemy situation	• Obstacle template • Enemy's obstacle intent • Enemy's expected level of defensive preparation • Confirmed intelligence on enemy obstacles and positions (include an analysis of this intelligence) • Enemy employment of FASCAM • Impact of enemy obstacles on friendly maneuver
OBSTINTEL collection effort	• Engineer R&S plan • NAIs critical to OBSTINTEL • How the OBSTINTEL effort supports the scheme of maneuver • Scout, engineer, and maneuver OBSTINTEL responsibilities and relationships
Scheme of engineer operations	• How the platoon will support the scheme of maneuver from the objective to the LD, to include consolidation • Engineer contingency plans and pertinent decision points
TF and company/team actions at obstacles	• In-stride breaching, to include— — Areas where the unit can expect to in-stride breach — Special equipment required — Special instructions • Deliberate breaching, to include— — Transitioning from attack formations to breaching formations — How the fundamentals of breaching will be synchronized — Key events and control measures synchronizing the breach — Composition of and instructions for the support, breach, and assault forces — Scheme of fires supporting the breach — Obscuration plan — Coordinating instructions, command and signals for the breach • Assault breaching, to include— — Allocation of engineers — Special coordinating instructions
Countermobility support to offensive maneuver	• Planned locations for situational obstacles • Triggers for situational-obstacle employment
Coordinating instructions	• Lane marking, breach traffic control, order of march through the lane, specific breaching signals, and breach C^2
Service support	• What support the platoon is receiving from the supported company/team and the parent company, if different from the SOP • Medical evacuation, supplies, and vehicle recovery • Current equipment status if critical to the mission
Unresolved issues	• Issues that the plan does not adequately address

Table 2-3. TF-engineer defense briefing

Briefing Topic	Main Points to Brief
Engineer task organization	• How the platoon will be task-organized • Special equipment that will be task-organized to subordinate units • Times and locations for linkup
Enemy	• Enemy maneuver formations through the sector, to include— – Most probable enemy COA – Place and time the enemy will change formation – Enemy breaching assets and capabilities (include high-priority targets)
Countermobility support	• Scheme of obstacles, to include— – Obstacle zones and restrictions – Obstacle belts and groups and their effects – Desired effect of obstacle groups on enemy maneuver – How the scheme of obstacles support the direct- and indirect-fire plans – Triggers and decision points for situational and reserve obstacles – Location of Class IV/ClassV supply points • Obstacle status reporting, to include— – Completion of obstacles – Barrier-material status • Obstacle-protection and turnover plans
Mobility support	• Where the lanes need to be to support MSRs, company/team supply routes, and reconnaissance passages • Responsibility for lane siting and closure • Lane marking and contact points • Obstacle restrictions that support the TF's mobility • Engineer-support plan for mobility
Survivability support	• Protective-obstacle plan, to include— – Material that the unit allocates – Any engineer assistance required – Protective-obstacle reporting responsibilities • Construction plan for fighting positions, to include— – Any engineer assistance required to construct individual positions – Allocation of engineer equipment and time and the priority for constructing vehicle positions – Linkup points and times with the company/team • Equipment control plan to ensure that maneuver companies know— – How many positions will be constructed by type – When, where, and how much equipment they will receive – How long they will retain the equipment
Engineer time line	• Engineer time line, to include— – Estimated completion time – Enemy trigger to stop work – Withdrawal route or passage – Engineer consolidation plan – Class VI/V supply-point recover or destruction
Coordinating instructions	• Obstacle-siting responsibilities • Reporting times • Class IV/Class V supply-point location, security plan, and responsibility

Table 2-3. TF-engineer defense briefing (continued)

Briefing Topic	Main Points to Brief
Service support	• Critical equipment and material shortages affecting defensive preparations • Medical evacuation plan for engineers • Barrier-material configurations to support obstacle-group construction
Unresolved issues	• Issues that the plan does not adequately address

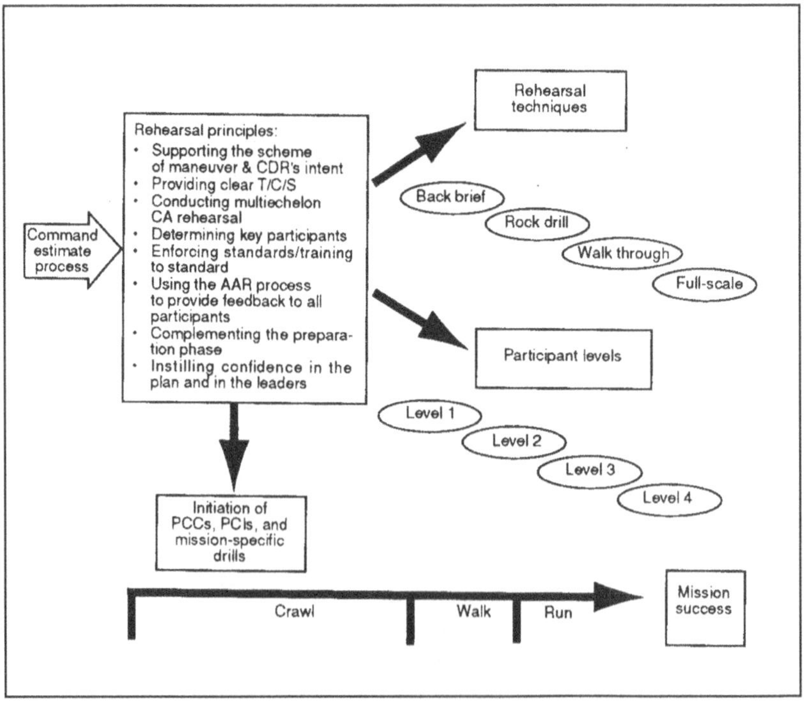

Figure 2-2. Rehearsal planning

PARTICIPANT LEVELS

The participant level details exactly who in the unit is required to attend the rehearsal. A CA-team rehearsal may have four distinct levels. The platoon leader may participate with or without his subordinates in the CA rehearsal. The three levels that apply to a platoon rehearsal follow:

- Level 1 - Squad leaders and platoon HQ.
- Level 2 - Squad leader, team leader, special teams, and vehicle commanders.
- Level 3 - Entire unit.

REHEARSAL TECHNIQUES

Rehearsal techniques fall into four different categories that follow the "crawl/walk/run" training concept. These categories are—

- Back brief.
- Rock drill.
- Walk through.
- Full-scale.

Each type of rehearsal reflects an increase in mission realism and a corresponding increase in rehearsal payoff. Each technique also increases the realism of the enemy, terrain, subunit actions, and time/distance relationships.

Back Brief

The back-brief rehearsal is an event that occurs after an operation order (OPORD) brief. Subordinate leaders repeat to the commander what he expects them to do and why. They identify all specified and implied tasks, determine their mission-essential tasks, and give their restated mission. During the back brief, they address the—

- Commander's intent.
- Concept of the operation.
- Scheme of maneuver.
- Timing to complete tasks.

During a typical back brief, each leader uses a map or a sand table and explains his mission. The back brief is the quickest of all the rehearsal techniques. It is a leader's tool and is typically the first rehearsal for the unit.

Rock Drill

A rock-drill rehearsal is the acting out of friendly and enemy actions based on the scheme of maneuver and the situation and event templates. Participants rehearse their actions by moving something that represents themselves or their unit, such as rocks or sticks. While acting out the plan, participants talk through their missions, critical tasks, actions, and decisions. Since all participants are simultaneously acting out their part of the scheme of maneuver, leaders can identify problems and disconnects in synchronization more clearly in a rock drill.

Walk Through

A walk-through rehearsal is the acting out of the scheme of maneuver using mounted or dismounted movement based on the assets employed during the attack. Participants should also communicate with the same type of equipment they will use during the action. Participants rehearse by—

- Maneuvering their vehicles (mounted movement) or themselves (dismounted movement).
- Reporting critical actions.
- Making required decisions.

Since the participants are in a more realistic environment, they rehearse the finer aspects of synchronization, C^2, and subunit actions. Aggressive portrayals of enemy actions and reactions are critical in walk-through rehearsals. This threat portrayal is the backbone of increasing the rehearsal realism. This rehearsal technique is more difficult to orchestrate than the back brief or rock drill; however, it is the optimum balance between resource constraints and realism. Successfully mastering a walk-through rehearsal should be the minimum goal for all units.

Full-Scale

During a full-scale rehearsal, participants use real-time mounted and dismounted movement over the actual or similar terrain. Typically, it is conducted at Level 3; however, a mix of participant levels is possible. At least one subunit must participate at Level 3 for a full-scale rehearsal. This type of rehearsal is obviously the most resource intensive, but it provides the most realistic training environment for the unit. It is often used to rehearse the operation plan (OPLAN) or OPORD when time is not an immediate constraint. One must plan extensively to execute full-scale rehearsals correctly and in a manner that does not waste the soldiers' time.

INITIATION OF PCCs, PCIs, AND MISSION-SPECIFIC DRILLS

The subordinate leader conducts PCCs to determine his unit's readiness to execute its assigned tactical mission. These inspections are informal. They are the leader's inspection of mission-critical tasks or areas of special interest to the commander. PCCs should cover—

- Vehicles.
- Weapons systems.

- Soldier's equipment, to include his weapon.
- Mission-critical equipment, such as explosives, detonators, and blasting machines.

The unit commander or leader conducts PCIs to determine the unit's readiness to execute its assigned tactical mission. These inspections may be formal or informal. Formal PCIs are the commander's meticulous and time-consuming inspections of all areas within the unit. They are seldom conducted during combat operations or assembly-area occupation. Informal PCIs are the commander's inspections of particular areas or activities of special interest or concern to him. He may designate the executive officer (XO) or first sergeant (1SG) to perform the PCI. PCIs should cover—

- Vehicles
- Major weapons systems.
- Soldiers.
- Communications equipment.

Mission-specific drills are rehearsals of special combat drills that subordinate units conduct. Special drills may include—

- Breach drills.
- Marking drills.
- Situational obstacles.
- Reserve demolitions.

CHAPTER 3
Mobility and Survivability

Mobility and survivability are the primary tasks of the engineer on the battlefield. The platoon leader requires a detailed understanding of this BOS to succeed. This chapter provides the tactics, techniques, and procedures (TTP) needed for the platoon to conduct mobility, countermobility, and survivability.

MOBILITY

Mobility enables the commander to maneuver tactical units into positions of advantage over the enemy. In the attack, engineers reduce enemy obstacles and fortifications that inhibit maneuver.

Expanded engineer reconnaissance capability is needed to identify routes, existing obstacles, and minefields. Open areas that are level provide good sites for aircraft landing strips; however, in most cases, the soil must be stabilized.

The application of the breaching fundamentals (suppress, obscure, secure, and reduce {SOSR}) and the organization of the force in terms of support, breach, and assault forces are standard. However, open areas may offer greater opportunity to bypass enemy obstacles because of the greater range of mobility afforded by the terrain. Exercise caution when choosing to bypass enemy obstacles, since the bypass may lead the force to the enemy's engagement area (EA). Additionally, expect the enemy to make extensive use of rapid mine-laying techniques that include SCATMINE delivery systems, as well as more conventional methods of mine emplacement.

OBSTACLE BREACHING

The platoon leader must decide where the best breach location is. Never breach where the avenue of approach crosses the obstacle. The enemy is sure to have preplanned artillery there. Pick a location away from obvious breaching sites to reduce the threat of artillery. The platoon leader should place himself where he can see the breach site and effectively command and control his platoon. He must be able to signal his breaching reserve (generally a squad) if he sees the breach failing. Observation of the breach site also reduces reporting requirements on the squad leaders to breach and mark the breach site.

Typically, there is a time lag as the maneuver force suppresses the enemy and obscures the breach site for the breach. The platoon leader should also consider dismounting an element from his lead squad to reconnoiter the best breach location during this tactical pause. This will assist the platoon leader in choosing the best breach site.

Obstacle breaching must be a CA effort. Units are more successful if they conduct—

- Obstacle reconnaissance.
- CA rehearsals.
- Breach planning.

For more information on obstacle breaching, see *FM 90-13-1*.

Breaching Fundamentals

When breaching against a defending enemy, the following fundamentals must be applied to ensure success:

- Suppress. Neutralize or destroy the enemy's weapons with our own fires. (The support force should have a 3:1 advantage over the enemy's weapons overwatching the obstacle.) Consider the effects and characteristics of the enemy's weapons. Ensure the correct focus of direct and indirect fires and the timely shift or lift of fires. Consider what ammunition is required.
- Obscure. Hamper the enemy's observation and target acquisition through the use of smoke (indirect, on board, pyrotechnic), terrain, limited visibility, and stealth.
- Secure. Eliminate enemy interference with obstacle reduction and lane usage. Occupy the obstacle with a friendly force and through the use of effective fires.
- Reduce. Create lanes through, over, or around an obstacle. Ensure that a company/team has at least one lane through an obstacle and that a battalion TF has two lanes. Consider what additional requirements are needed to guide follow-on forces through the obstacle or to mark and reduce the obstacle.

Types of Breaching Operations

When planning a breaching operation, the engineer leader should consider bypassing the obstacle as his first option. However, the enemy's intent may be to force the unit to turn or channelize into a fire sac. This is why a good reconnaissance of a possible bypass is essential. The unit conducting a bypass must have maximum security during the bypass. If bypass operations are not possible, consider the following options:

In-stride Breach. A unit conducts an in-stride breach when it can be successfully executed under the complete C^2 of a task-organized subordinate unit or when the situation is unclear and rapid movement is critical. Subordinate units serve as both the breach and assault forces. The subordinate commanders synchronize the breaching fundamentals. Doctrinally, TFs and above conduct in-stride breaching.

Deliberate Breach. A unit conducts a deliberate breach when the force-allocation ratios for support, breach, and assault forces are beyond the capability of a task-organized subordinate unit. One or more subordinate units are specifically tasked to perform the role of support, breach, and assault forces. Synchronization is critical and the principle of mass drives task organization.

Assault Breach. A unit conducts an assault breach during the penetration and destruction phases of actions on the objective to destroy an enemy that has had time to emplace protective obstacles. For the assault breach, a rule of thumb is one lane per assaulting platoon. Assault lanes are generally only footpaths until the objective is secured. Mounted assault lanes require the same effort as tactical vehicle lanes.

Covert Breach. A unit conducts a covert breach during limited visibility when surprise is essential. The covert breach relies on stealth; quiet, manual lane-reduction techniques; and dismounted maneuver. It can be employed to breach tactical obstacles or protective obstacles (more difficult). It requires the same combat ratios as other breaches for suppression and security, and these points should be considered during the planning process.

Breach Planning

The following are considered when developing a breaching plan (see *Figures 3-1 and 3-2, pages 3-4 and 3-5*):

- Reverse planning begins with actions on the objective.
- Actions on the objective drive the size of the assault force and determine the number and location of lanes to be breached.
- Lane requirements and the type of obstacle drive the allocation of mobility assets to the breach force.
- Ability of the enemy's infantry to interfere with the breach determines whether to secure the breaching site by force or by fire.
- Ability of the enemy to mass fires at the breaching site determines the amount of suppression required and the size of the support force.

The platoon leader must plan for actions after the breach is complete. Generally, the platoon is responsible for beaching, marking, and providing guides for the lane.

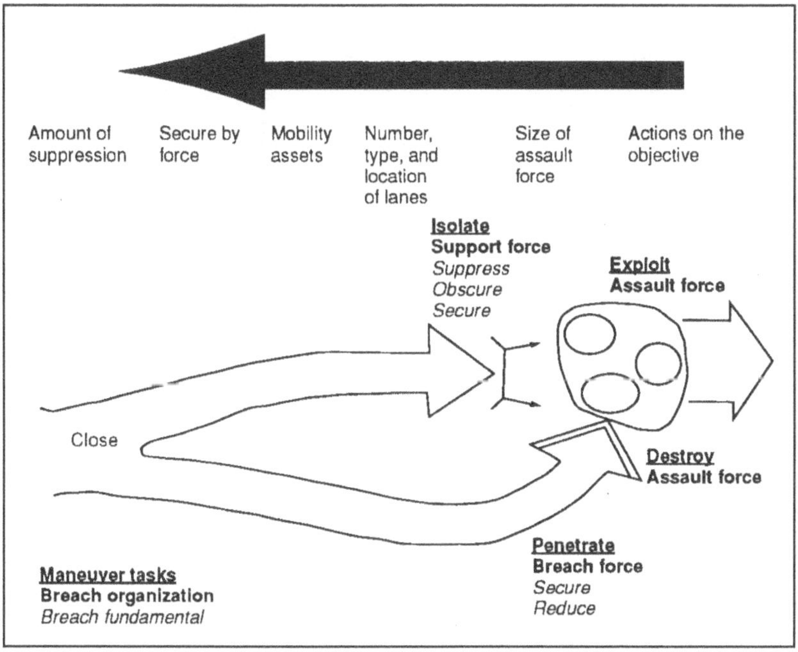

Figure 3-1. Breach planning

Normally, the platoon remains at the breach location and continues to improve the breached lane. The platoon must be prepared to reopen the lane if the enemy closes it with artillery-delivered mines. The platoon leader should also plan for vehicles that are stopped in the lane. The platoon may be forced to create an additional lane around the stopped vehicle or extract it from the breached lane.

Breach TF Organization

The breach TF is organized into three elements:

- Support force—eliminates enemy interference with the breach, isolates the objective area, and destroys enemy weapons that are able to fire on the breach force.
- Breach force—creates lanes in and through the obstacle. While creating the lane through the obstacle, it must also locally secure the breaching site and pass the assault force through the breach.

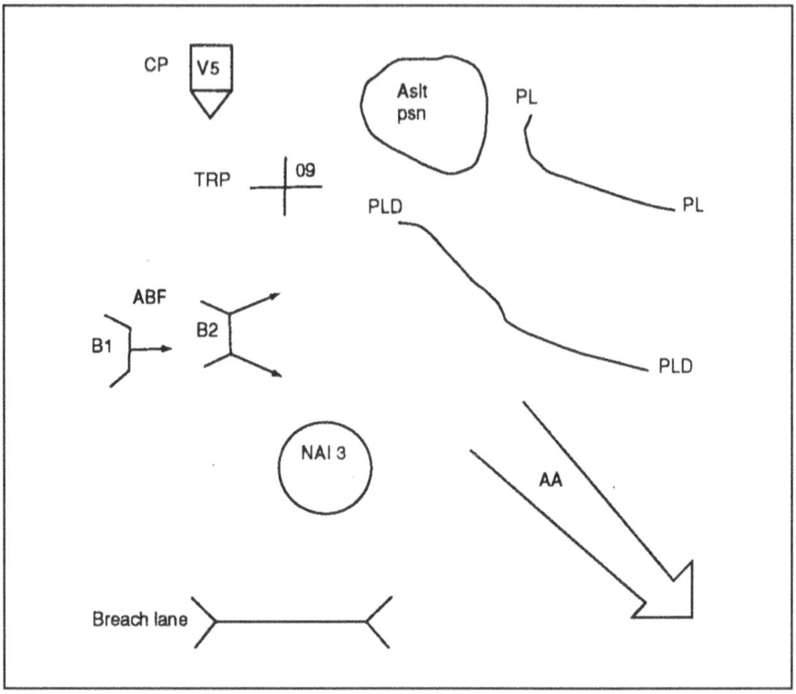

Figure 3-2. Breach control measures

- Assault force—destroys or dislodges the enemy force on the far side of the obstacle and seizes the initial foothold on the objective.

Obscurants

To obscure friendly activities and movement, units can use artillery- or mortar-delivered smoke and/or white phosphorous (WP) rounds, generators, smoke pots and M203 grenades, chemical units, and burning diesel. Obscurants—

- Should be preplanned and coordinated with the company commander and the supported TF.
- Can be used for deception.
- Can be combined with supporting fires.
- Can be used over large areas.

LANE MARKING

The critical components to any lane-marking system are the lane-marking patterns and devices. Lane-marking devices are not standardized Army-wide. *Figure 3-3* shows examples of lane-marking devices. *Table 3-1, page 3-8,* can be used as a guide to choose lane-marking devices. Lane-marking patterns have been standardized Army-wide. Standard breach lane-marking patterns are explained in *FM 90-13-1, Appendix E*. The following are lane markers and their use:

- Entrance markers—indicate the start and the width of a reduced lane. They must be visually different than handrail markers. Entrance markers are placed a minimum of 4.5 meters apart (1 meter for dismounted).
- Handrail markers—define the lane path and indicate the limits of the lane width. Mark, as a minimum, the left handrail.
- Exit markers—indicate the far-side limit of reduced lanes. Exit markers must be visually different from handrail markers but may be the same as entrance markers.
- Entrance-funnel markers—augment the entrance markers. They assist the small-unit commander in guiding the lead unit of his combat column formation.
- Final-approach marker—is a highly visible marker that augments the visual signature of entrance-funnel markers. It provides the assault-force commander with a highly visible reference point toward which to maneuver his formation.
- Far-recognition markers—are highly visible markers located between the final-approach marker and the friendly unit. They are primarily used when passing battalion-sized forces through a lane where distance, visibility, or terrain does not allow the passing force direct observation of the final-approach marker. When possible, far-recognition markers should be different from the final- approach marker.
- Traffic-control post (TCP) or guides—are a two-man team with communications means that assists the commander in controlling the movement of forces. When possible, military police (MP) should man TCPs. However, the breach force should plan to man TCPs until relieved.

The standard levels of lane marking for breach lanes and bypasses are initial, intermediate, and full (see *Figure 3-4, page 3-9*). The following are lane-marking patterns and their use:

- Initial lane-marking pattern—is emplaced by the breach force immediately after the lane is reduced and proofed. It is a signal to the assault force that the lane is ready for traffic to pass through.

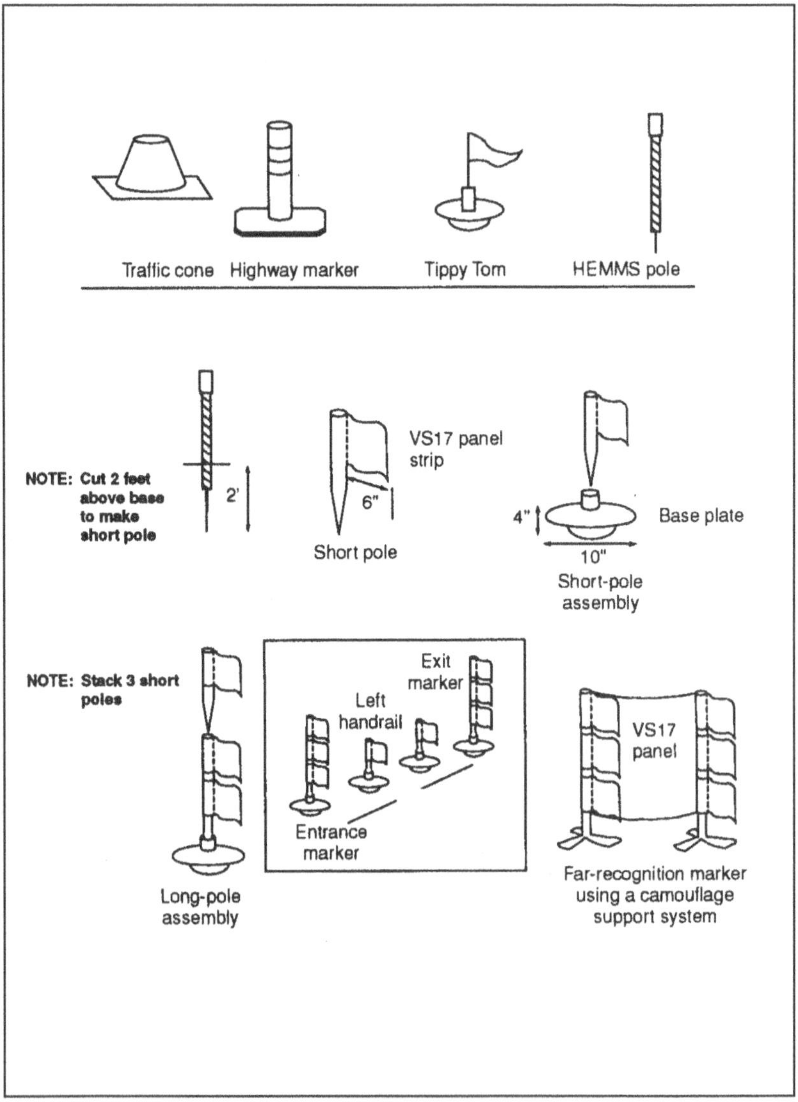

Figure 3-3. Lane-marking devices

Table 3-1. Breach lane marking

Markers	Mounted Forces	Dismounted Forces
Handrail and funnel	They— • Can be seen by the vehicle commander and driver buttoned up from 50 meters. • Are quick to emplace, minimizing soldier exposure.	They— • Are visible to a prone dismounted soldier from 50 meters. • Are lightweight and easy to emplace.
Entrance and exit	They— • Can be seen from 100 meters. • Are visually different from handrail and funnel markers. • Are easy to emplace and are man portable.	They— • Are visible from 50 meters. • Are visually different from handrail and funnel markers. • Are easy to emplace and are man portable.
Final approach and far recognition	They— • Can be seen from 500 meters. • Are visually different from each other. • Can be altered to facilitate traffic control through multiple lanes.	They— • Are visible from 100 meters. • Are visually different from each other. • Can be altered to facilitate traffic control through multiple lanes.

- Intermediate lane-marking pattern—is used for the commitment of larger combat forces who are unable to directly observe the breach or the rearward passage of sustainment traffic. It builds on the initial lane-marking pattern by adding right handrail markers, exit-funnel markers, far-recognition markers, and a far-side final-approach marker.
- Full lane-marking pattern—is usually not part of an initial breaching operation. It is a lane that will support uninterrupted two-way traffic.

Table 3-2, page 3-10, provides information as to who marks the various types of lanes, when they are to mark the lanes, and what markers they are to use. Lane marking is normally synchronized during the CA rehearsal to ensure that all key leaders understand the marking procedure and standard being employed.

COUNTERMOBILITY

Countermobility operations are conducted to augment natural terrain with obstacle systems according to the commander's concept. This adds depth to the battle in

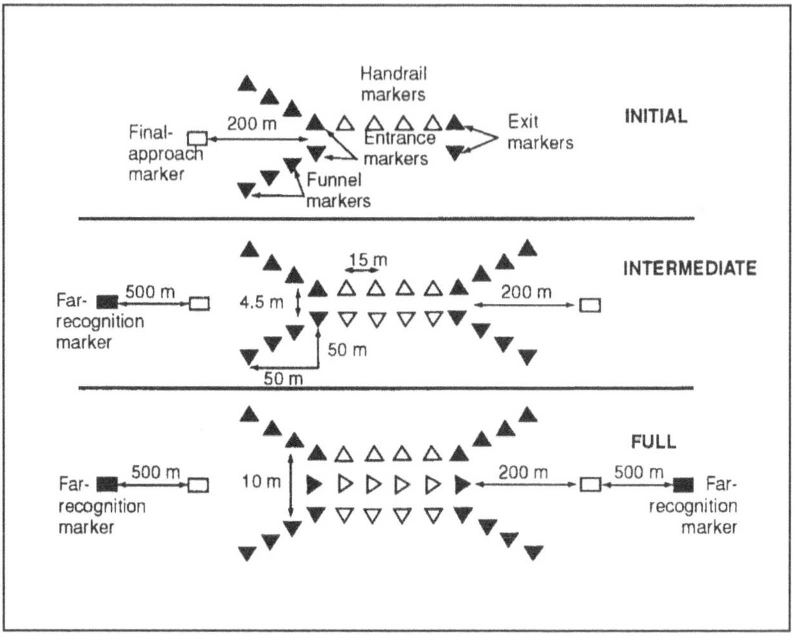

Figure 3-4. Lane-marking patterns

space and time by attacking the enemy's ability to maneuver its forces. With the enemy's movement impeded (disrupted, turned, fixed, or blocked), it is vulnerable to our forces. Engineers advise the commander on the best means to reinforce natural obstacles (terrain) and to emplace tactical obstacles that support his plan. Minefields, wire, antitank ditches (ATDs) or berms, road craters, bridge demolition, and all other traditional obstacles may be effective. For expedient obstacles, units should consider employing local materials.

ATD or berms require extensive preparation but are very effective when properly prepared. Many areas have irrigation ditches that can be used tactically. However, in sandy areas, ditches can easily be filled in. Obstacles should be planned to protect the unit's flanks during offensive operations.

All obstacles can be situational obstacles. Situational obstacles are obstacle resources held in reserve. They have a "be-prepared" mission but not an "on-order mission". Situational obstacles should be considered during planning to give the maneuver commander a flexible response to enemy or friendly maneuver.

Table 3-2. Lane-marking level—unit responsible, trigger events, and lane markers

Breach Type	Initial	Intermediate	Full
	Unit Responsible		
Deliberate	TF breach force	TF breach force	Brigade
Covert	TF breach force	TF breach force	Brigade
In-stride	Breach company/ team	TF mobility reserve	Brigade
Assault	Assault platoon	TF assault force	NA
	Trigger Events		
	When— • Lanes are reduced • Passing platoon- or company-sized forces	When passing— • Battalion- or company-sized forces • Forces that cannot see the lane • TF combat trains	When— • Passing brigade- or battalion-sized forces • Situation requires uninterrupted sustainment traffic
	Lane Markers		
	Entrance/exit Left handrail Entrance funnel Final approach	The following markers are added: • Right handrail. • Exit funnel. • Far-side final approach. • Far recognition. • Guides/TCPs.	Lane width is expanded to 10 meters. Existing markers are adjusted. Far-side-recognition markers and guides/TCPs are added.

OBSTACLE PLANNING

To assist the TF commander in the defense, the TF engineer must integrate obstacle planning throughout the TF sector and coordinate with adjacent TFs according to the overall engineer plan.

Each echelon has its own obstacle-control measures for directing the location of obstacles on the ground. The division commander and his staff use obstacle zones to control obstacles to support future operations. Within these zones, the brigades plan obstacle belts, and the battalion TF plans obstacle groups. The engineer will then emplace each obstacle by doctrinal obstacle norms to support the commander.

Protective Obstacles

They are designed to repel the enemy's assault by emplacing them around a unit's defensive position (a survivability task).

Tactical Obstacles

They are designed and emplaced to (see *Figure 3-5*)—

- Disrupt, which—
  - Breaks up C^2.
  - Frustrates the enemy's timing.
  - Interrupts formations or attacks the enemy when they are massing into formations.
  - Coerces the enemy into our lines of fire.

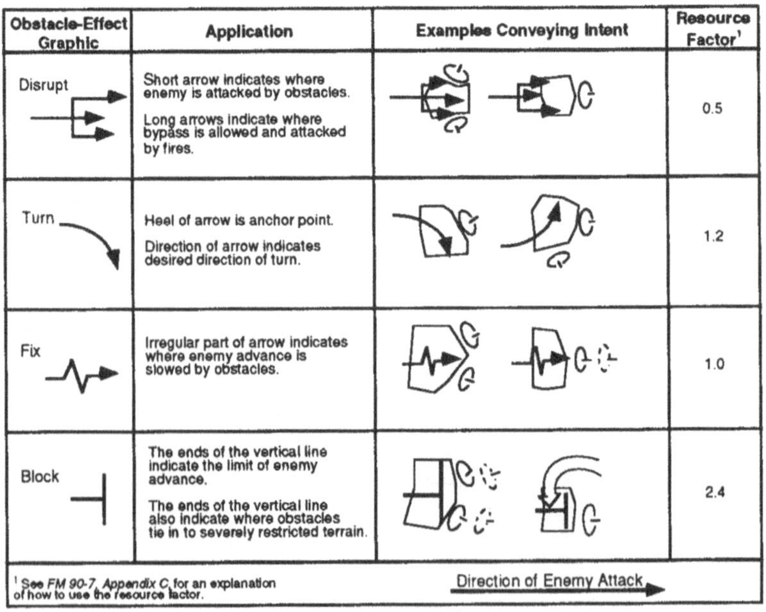

Figure 3-5. Obstacle-effect graphics

- Fix, which—
  - Slows the enemy's advance within a specified area.
  - Causes the enemy to breach repeatedly.
- Turn, which manipulates enemy maneuver in a desired direction.
- Block, which—
  - Stops the enemy along a particular avenue of approach.
  - Defeats the enemy's breaching effort.

Figure 3-6 shows how to employ conventional row minefields to obtain the desired obstacle effects and the planning figures for standard minefields by obstacle effect.

Situational Obstacles

Situational obstacles are "triggered" by an event or situation. Once the general location of the obstacle has been determined, it must be refined to better enhance its effect on the enemy. When planning, preparing, and executing situational obstacles (see *FM 90-7*)—

- Identify the need.
- Plan for appropriate resources.
- Plan the obstacle.
- Integrate the obstacle with friendly fires.
- Identify obstacle execution triggers.
- Withhold execution of the obstacle until it is needed.

Situational obstacles can be used to—

- Attack an enemy's vulnerability.
- Exploit success.
- Separate follow-on enemy forces.
- Provide flank protection.

Tables 3-3 and 3-4, pages 3-14 and 3-15, provide some characteristics of conventional and scatterable mines and are useful in planning minefields. When emplacing standard disrupt, fix, turn, and block minefields, see *Figures 3-7 through 3-12, pages 3-16 through 3-21, and Tables 3-5 through 3-8, pages 3-22 and 3-23,* for information on the employment of SCATMINE systems. FASCAM systems include the air/ground Volcano, Modular Pack Mine System (MOPMS), Flipper, area denial artillery munition (ADAM), and remote antiarmor mine (RAAM).

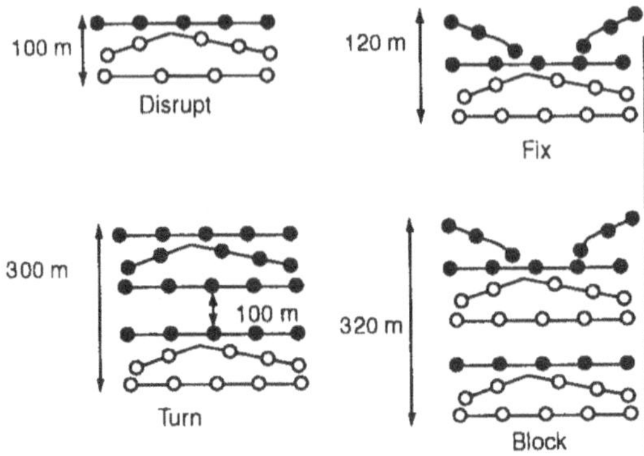

Mines spaced 6 m apart
Rows spaced 50 m apart
Minefield frontage
Disrupt: 250 m
Turn: 500 m
Fix: 250 m
Block: 500 m
o Track-width fuzed mine
● Full-width fuzed mine

AP mine employment

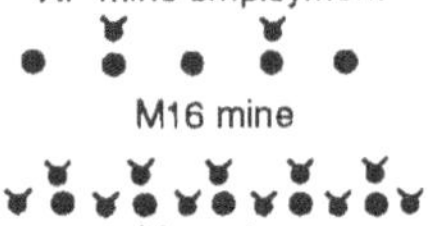

AP mines are—

- Used in block minefields only.
- Used in any two rows of full-width AT mines.
- Patterned based on AP mine type.

	Disrupt	Turn	Fix	Block
Frontage (m)	250	500	250	500
Depth (m)	100	300	120	320
AT full (# rows)	1	4	1	4
AT track (# rows)	2	2	2	2
AP (# rows)	0	0	0	2
IOE (yes or no)	No	No	Yes	Yes
AHD (yes or no)	Yes	No	No	Yes
Plt hrs reqd	1.5	3.5	1.5	5.0
Track-width AT mines	84	168	84	168
Full-width AT mines AP mines M16 AP mines M14	42 0 0	336 0 0	63 0 0	378 84 500
Density AT	0.5	1.0	0.6	1.1
Density AP M16/M14	0	0	0	0.17/1.0

Figure 3-6. Minefield norms

Table 3-3. Conventional mines

Mine	Arming Time	Fuzing	Warhead	AHD	Mine Weight	Mines per 5-Ton Dump	Remarks
M14	NA	Pressure	Blast	No	3.3 oz	6,480	
M16A1	NA	Pressure Trip wire	Bounding fragment	Yes	8 lb	672	
M18A1	NA		Directional fragment	No	3.5 lb	1,782	
M86 (PDM)	50 sec		Bounding fragment	100%	1 lb	—	Self-destruct after 4 hours
M15	NA	Pressure	Blast	Yes	30 lb	90	
M15 w/ M624	NA	Tilt rod	Blast	Yes	30 lb	90	
M21	NA	Tilt rod	Shape charge	Yes	17 lb	192	

OBSTACLE SITING

When determining where to emplace obstacles, the platoon leader should consider the following:

- Supporting the maneuver commander's intent.
- Supporting the CATK plan.
- Supporting the defensive plan.
- Coordinating obstacle locations with the maneuver element, the fire-support officer (FSO), the Operations and Training Officer (US Army) (S3), and the weapons-systems commander.
- Prioritizing the engineer effort.
- Planning the maneuver unit's responsibility.
- Using observation and fields of fire, cover and concealment, obstacles, key terrain, and avenues of approach (OCOKA).

Company/Team Commanders' Coordination

The platoon leader must ensure that key obstacles are covered by indirect fire. These obstacles become priority targets for the artillery. When coordinating with company/team commanders, the platoon leader should consider the following:

- Determining the obstacle intent.
- Determining the company's/team's tactical purpose.

Table 3-4. Scatterable mines

Mine	Delivery System	Arming Time	Fuzing	Warhead	AHD	Self-Destruct Time	Mine Weight	Mines per 5-Ton Dump
M73	155-mm arty (RAAM)	45 sec 2 min	Magnetic	M-S plate	20%	48 hr	3.8 lb	NA
M70	155-mm arty (RAAM)	45 sec 2 min	Magnetic	M-S plate	20%	4 hr	3.8 lb	NA
M75	GEMSS Flipper	45 min	Magnetic	M-S plate	20%	5 days 15 days	3.8 lb	1,600
BLU 91/B	USAF (Gator)	2 min	Magnetic	M-S plate	No	4 hr 48 hr 15 days	3.8 lb	NA
M76	MOPMS	2 min	Magnetic	M-S plate	No	4 hr (recycle 3 times)	3.8 lb	30 modules (510 mines)
Volcano	Ground/air	2 min	Magnetic	M-S plate	No	4 hr 48 hr	3.8 lb	160 canisters (800 mine)
M72	155-mm arty (ADAM)	45 sec 2 min	Trip wire	Bounding fragment	20%	15 days 48 hr	1.2 lb	36 per M692 projectile
M67	155-mm arty (ADAM)	45 sec 2 min	Trip wire	Bounding fragment	20%	4 hr	1.2 lb	38 per M731 projectile
M74	GEMSS Flipper	45 min	Trip wire	Blast fragment	20%	5 days 15 days	3.2 lb	1,600
BLU 92/B	USAF (Gator)	2 min	Trip wire	Blast fragment	100%	4 hr 48 hr 15 days	3.2 lb	NA
M77	MOPMS	2 min	Trip wire	Blast fragment	100%	4 hr (recycle up to 15 days)	3.2 lb	30 modules (120 mines)
Volcano	Ground/air	2 min	Trip wire	Blast fragment	100%	4 hr 48 hr 15 days	3.2 lb	160 canister (160 mines)

- Ensuring that the obstacles the maneuver company/team commanders emplace on the battlefield serve as a combat multiplier and enhance the direct-fire plan.
- Providing forward security during obstacle emplacement.
- Conducting obstacle turnover with the maneuver company/team and ensuring that indirect-fire support is provided after the obstacle is finished.
- Rehearsing lane-closure plans.

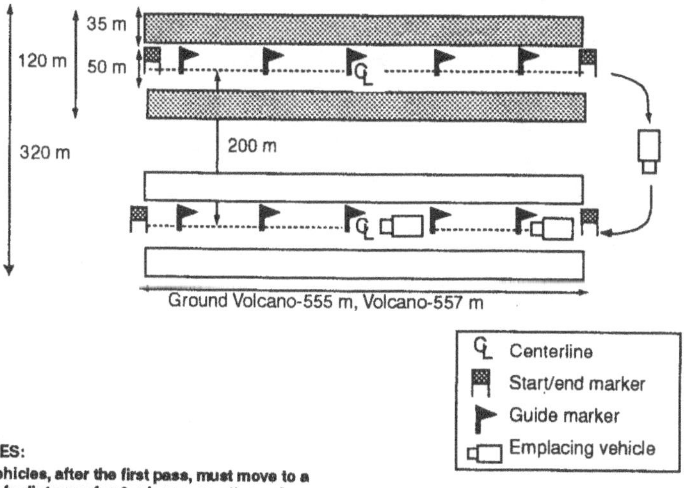

NOTES:

- **Vehicles, after the first pass, must move to a safe distance for 2 minutes to allow mines failing the arming sequence to self-destruct.**
- **Guide markers are offset from centerline.**
- **Start and end markers should be used for air Volcano only.**
- **Turn and block minefields require the dispensing of 80 canisters (40 each side) on each pass, 160 canisters total.**
- **One full load of ground or air Volcano emplaces one turn or block minefield.**

Figure 3-7. Volcano turn and block minefields

Direct-Fire Integration

When integrating obstacles with direct fire, the platoon leader should consider the following:

- Coordinating with the maneuver element overwatching the obstacle.
- Sighting obstacles with weapons.
- Coordinating with the maneuver element for—
  - Forward security.

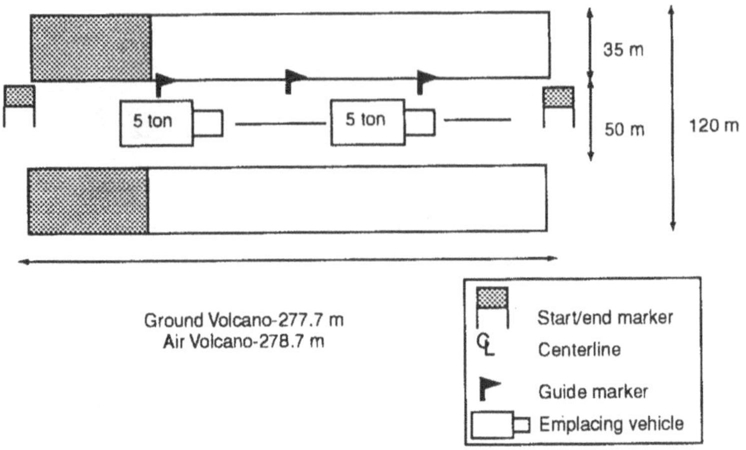

Notes:

- **Guide markers are offset to the left of the centerline to guide the driver and allow the vehicle to remain on the centerline.**
- **Start and end markers should be used for air Volcanos only. Guide markers are not required.**
- **Disrupt and fix minefields require the dispensing of 40 canisters (20 each side).**
- **One full load of air or ground Volcano provides four disrupt or fix minefields.**

Figure 3-8. Volcano disrupt and fix minefields

- Logistics support.
- Obstacle-turnover requirements.

Indirect-Fire Integration

The platoon leader must ensure that key obstacles are covered by indirect fire. These obstacles become priority targets for the artillery. The platoon leader coordinates indirect fire with the company FIST and company/team commander in heavy units. Indirect fire must be coordinated with the FSO and the TF commander if units are light or if the platoon leader is the TF engineer. For more information on

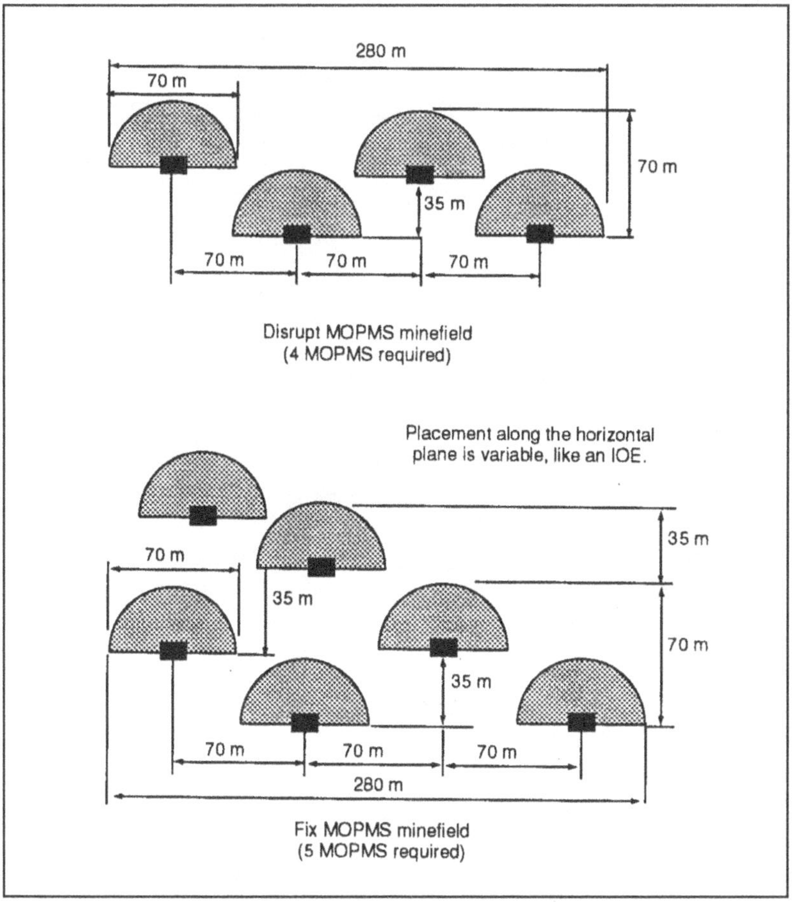

Figure 3-9. MOPMS disrupt and fix minefields

integrating obstacles with indirect fire, see *FM 90-7*. When integrating obstacles with indirect fire, the platoon leader should consider the following:

- Ensuring that there are target reference points (TRPs) on every obstacle.
- Preplanning ADAM and RAAM fires.
- Calling for fire nets (engineers).

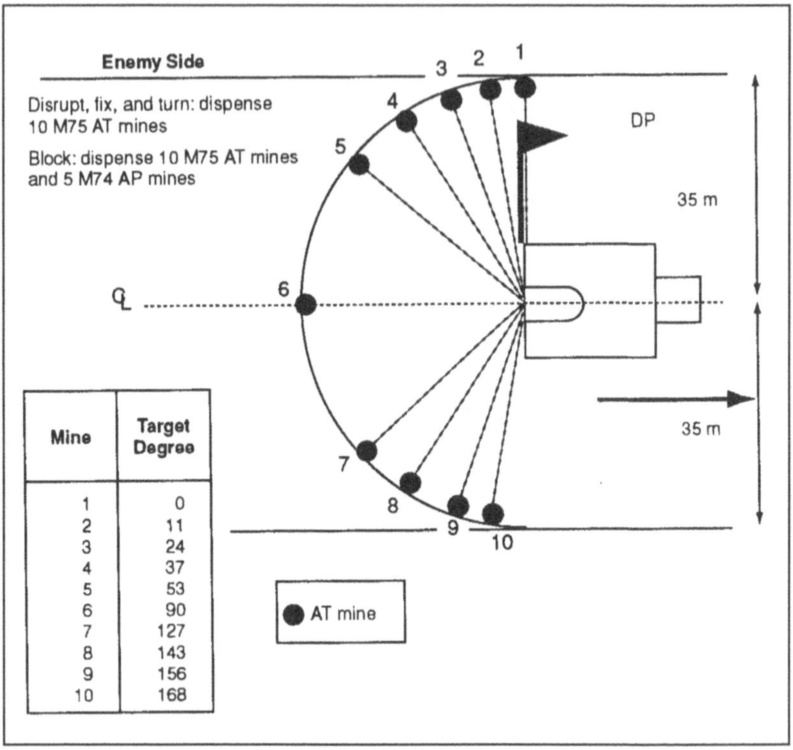

Mine	Target Degree
1	0
2	11
3	24
4	37
5	53
6	90
7	127
8	143
9	156
10	168

Figure 3-10. Flipper TTP

OBSTACLE CAMOUFLAGE (IF REQUIRED)

When camouflaging obstacles, the following should be considered:

- Ensuring that the obstacles emplaced support the intent (sometimes you want the enemy to see the obstacles).
- Using reverse slope siting.
- Burying mines.
- Using night and covert emplacement.
- Minimizing dust signatures from the excavation.

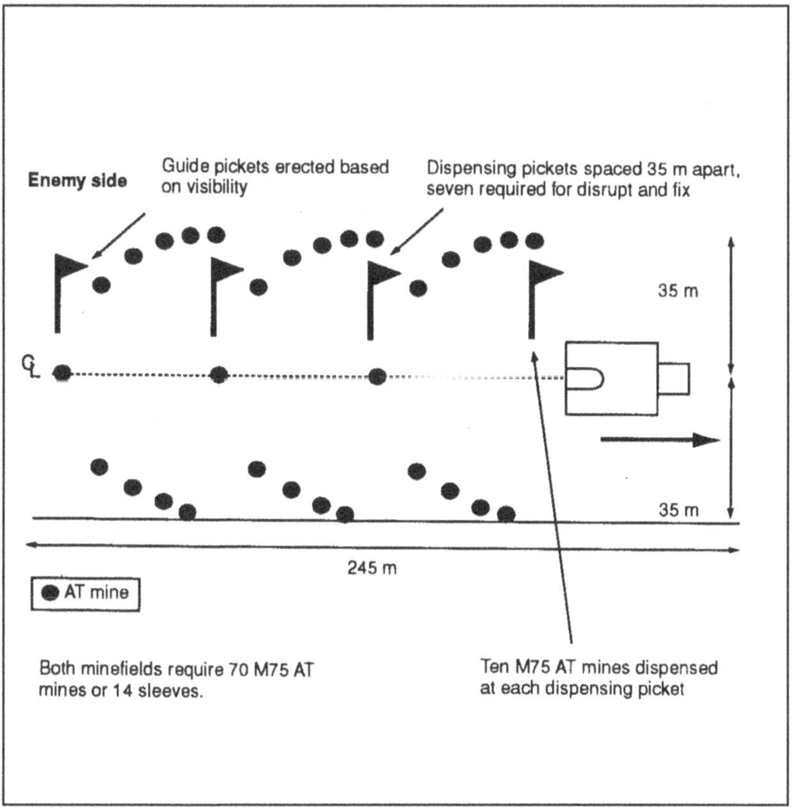

Figure 3-11. Flipper fix and disrupt minefields

OBSTACLE PROTECTION

The following should be considered for the protection of obstacles:

- Denying the enemy covert breaching opportunity.
- Denying the enemy reconnaissance
- Using observation posts (OPs) and counterreconnaissance.
- Displacing forces during low visibility to cover obstacles.

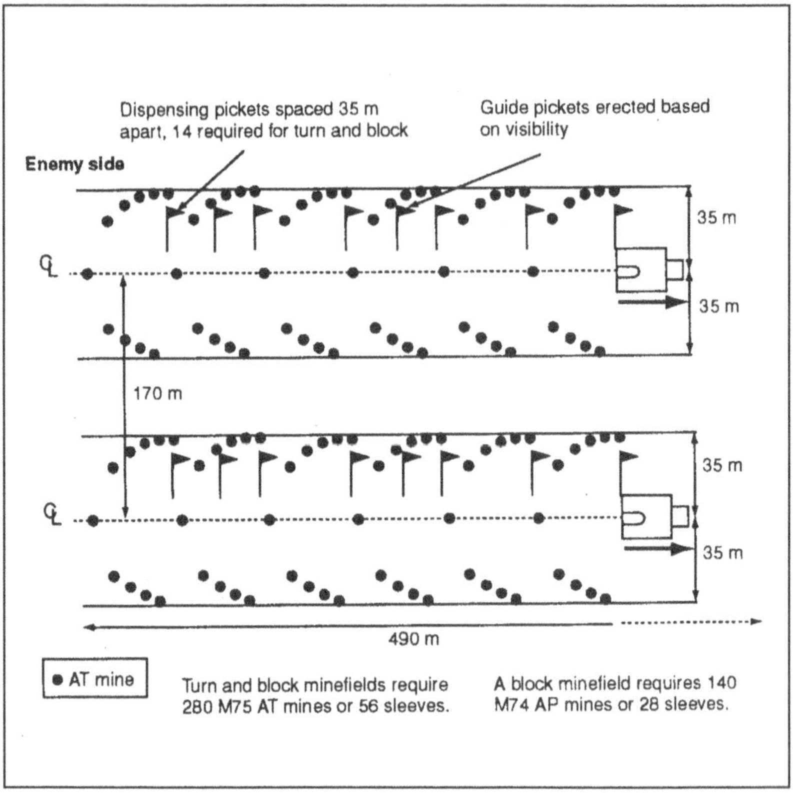

Figure 3-12. Flipper turn and block minefields

- Identifying and destroying the enemy's obstacle breaching equipment and C^2 vehicles.
- Covering the obstacle with direct and indirect fires.
- Patrolling obstacles at night.

SURVIVABILITY

Preparation of fortifications may be difficult. Fortifications in sandy soils often require revetments. It may be impossible to dig in rocky areas, arctic regions, or locations with a high or perched water table. To counter these problems, you may

Table 3-5. Flipper planning factors

Type of Minefield	Depth (Meters)	Front (Meters)	Number of Strips	DPs per Strip	M75 Mines per DP	M74 Mines per DP	Total M75 Mines	Total M74 Mines
Disrupt	70	245	1	7	10	0	70	0
Fix	70	245	1	7	10	0	70	0
Turn	240	490	2	14	10	0	280	0
Block	240	490	2	14	10	5	280	140

Table 3-6. Volcano planning factors

Type of Minefield	Depth (Meters)	Front Ground/Air (Meters)	Number of Strips	Canisters per Strip	Total Canisters	Minefields per Load
Disrupt	120	277/278	1	40 (20 each side)	40	4
Fix	120	277/278	1	40 (20 each side)	40	4
Turn	320	555/557	2	80 (40 each side)	160	1
Block	320	555/557	2	80 (40 each side)	160	1

Table 3-7. RAAM/ADAM planning factors

Type of Minefield	RAAM		ADAM		Width (Meters)	Depth (Meters)
	Area	Linear	Area	Linear		
Disrupt	0.001	0.2	0.0005	0.1	200	200
Turn	0.002	0.9	0.001	0.4	400	400
Fix	0.002	0.4	0.0005	0.1	200	200
Block	0.004	1.6	0.002	0.8	400	400

Table 3-8. Mine dump planning factors

Number of Personnel	Quantity of Mines
2-man team (2 minutes per mine)	25 mines per hour
Squad (8 soldiers)	100 mines per hour
Platoon	300 mines per hour; 3,000 mines per day
Company	6,000 mines per day
NOTE: For planning, soldiers work 10 hours per day.	

have to build up emplacements with existing materials or depressions. *FMs 5-34 and 5-103* provide the standards for dismounted and vehicular field fortifications.

Camouflage is very effective when properly employed. However, you must carefully select patterns and techniques to apply to the local environment. All equipment should have camouflage nets. Other survivability techniques include dispersion and frequent moves.

Direct-fire weapons are more effective from dug-in positions. Logistics areas (brigade support area/division support area (BSA/DSA)) require survivability support. These sites are large, relatively static, and difficult to camouflage. As a result, they are vulnerable to enemy interdiction. Emphasis should be placed on ammunition and fuel dumps.

Digging assets, such as bulldozers, small emplacement excavators (SEEs), or armored combat earthmovers, M9 (ACEs), should be tasked to provide survivability support to these sites. Particular emphasis should be placed on hardening ammunition and fuel-storage locations.

With long-range observation and fields of fire, defensive positions are very vulnerable to offensive fire. This fact, coupled with a lack of natural obstacles, may lead the commander to concentrate the bulk of his engineer effort on survivability positions. Survivability positions enhance the ability of all direct-fire elements to survive indirect-fire preparations and the direct fire from attacking enemy units.

Care should be taken when digging foxholes and tank hide positions in sandy or noncohesive soils since the walls have a tendency to cave in.

PLANNING

Survivability positions are key to surviving any battle on today's modern battlefield. This is especially true where there is a lack of vegetation and natural hiding positions. The TF engineer must know the following information to start planning:

- Number of positions: primary, alternate, and supplementary.
- Type of position: hull-down, turret-down, or hasty.
- Weapons and vehicle priorities.
- Battle-position and EA priorities.
- Unit priorities.

Once the TF engineer determines the number of positions required, then the number of positions that can actually be put in is determined. The engineer unit's capability is based on the following factors:

- Ten-hour days for soldiers.
- Fifteen-hour days for equipment.
- Time available (days).
- Number of blade teams (BTs) available. ACEs, dozers, combat engineer vehicles (CEVs), and bucket loaders can be used to make blade teams. A blade team must include at least one dozer or ACE.
- Work rates.

For work rates, use the planning factor developed for your area of operation. If work rates are unknown, use the following as a guide:

- One hull-down position (HDP) takes 1.5 BT hours.
- One turret-down position (TDP) takes 3.5 BT hours.
- Fifty meters of ATD takes 1 BT hour.

A variety of variables may increase or decrease the time required to dig a vehicle fighting position or ATD. These variables include, but are not limited to, local soils and terrain, operator experience, additional maintenance down time, and the tactical situation. The difference of blade widths may also affect work rates. The M9 ACE has a shorter blade width than the D7 dozer and requires a minimum of two passes to cut the same width.

Survivability assets should be massed on the commander's number one priority. This focuses the survivability effort and allows the TF engineer to monitor construction progress. It also reduces the C^2 burden on the engineer platoon and facilitates rapid transition to changing priorities.

The following equations can be used to determine the number of HDPs and TDPs and the length of ATDs that can be emplaced:

HDP = (# BTs) x (1 HDP/work rate) x (# days) x (15 hrs/day)

TDP = (# BTs) x (1 TDP/work rate) x (# days) x (15 hrs/day)

Meters of ATD = (# BTs) x (ATD m/work rate) x (# days) x (15 hrs/day)

POSITIONS

Coordination between the engineer/equipment operator and the maneuver element is critical to ensure the proper placement of positions. Without effective coordination, much valuable time is wasted redigging positions. The maneuver element, preferably the gunner or tank commander (TC), must tell the engineer/equipment operator the following:

- Type of position (deliberate or hasty).
- Gun target line.
- Sector of fire.
- TRP.

In heavy units, initial vehicle-fighting-position construction starts with the M9 ACE. The ACE has different digging capabilities than the dozer. In many cases, divisional ACEs start with hasty positions, which are then upgraded to improved positions as time allows. Also, the ACEs may start work before corps-level dozers arrive to augment the construction effort. In this situation, the ACEs would dig the hasty positions and the dozers would improve them. The platoon must plan for this augmentation to efficiently integrate corps assets into the construction effort. In heavy units, the company commander must integrate corps assets; however, in light units, the platoon leader must plan for and integrate corps assets.

Figure 3-13, page 3-26, shows the different vehicle fighting positions. Some terrain areas may not accommodate a hull or turret defilade. The use of a modified hasty vehicle-position design or simple marking techniques incorporated with terrain mottling may be required.

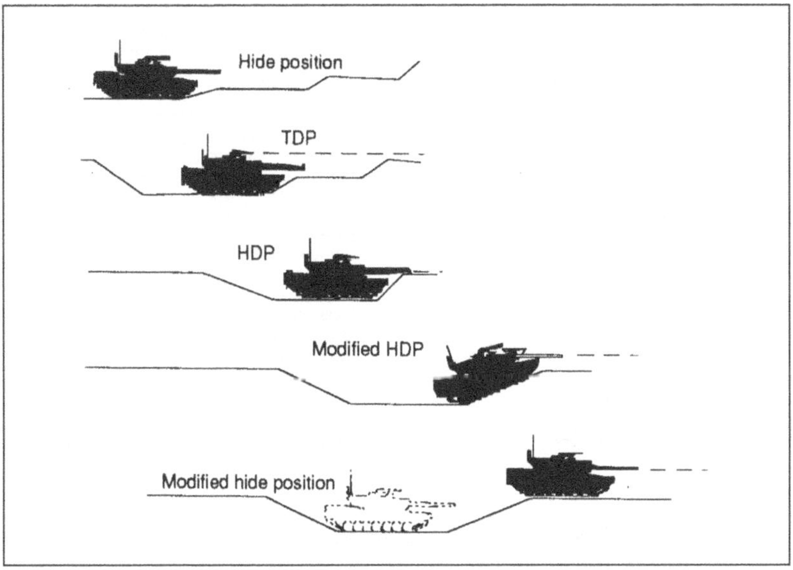

Figure 3-13. Vehicle survivability positions

Each soldier, not just engineers, must also emplace individual and crew fighting positions. Engineers have some limited assets, such as the SEE, that can assist the maneuver commander in accomplishing this mission. Fighting positions are—

- Located to effectively engage the enemy and mutually support each other.
- Designed to allow weapons to be properly fired.
- Constructed to provide adequate cover (see *Table 3-9*).
- Concealed from the enemy.

NUCLEAR, BIOLOGICAL, CHEMICAL (NBC)

Chemical and biological weapons are many countries' preferred weapons of mass destruction. Many of our potential adversaries have an NBC capability. *Figure 3-14 and Tables 3-10 and 3-11, page 3-28,* detail the NBC threat and protective procedures.

Table 3-9. Survivability

Camouflage	Defensive Positions
Soldiers should consider the following when camouflaging: • Breaking up the shape of the object. • Shading the object with natural shadows. • Concealing the shine of the object. • Disrupting the silhouette of the object. • Using folds, gullies, and wadies. • Using reverse-slope positions. • Scaring two to three positions for each vehicle or position. • Using irregular-shaped positions • Building positions a minimum of twice the size of the concealed object. • Exercising noise and light discipline. • Using camouflage for deception.	Soldiers should know and consider the following facts when constructing defensive positions: • Field expedient construction materials, such as corrugated metal, coverts, scrap metal, 40-gallon oil drums, sandbags, and sand-filled ammo boxes, can be used. • Small-arms fire penetrates loose sand 39 inches and sandbag parapets 24 inches. • A minimum of 18 to 24 inches of overhead cover is required to protect against indirect fire. • Revetments are required in sandy soils. • Design details are covered in *FMs 5-34 and 5-103.*

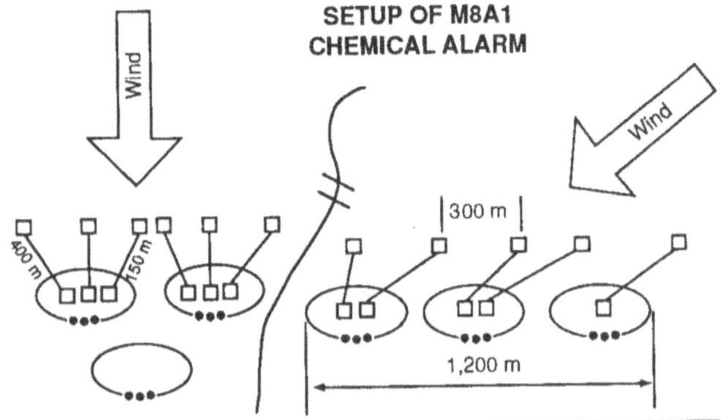

CAM put in operation

1. Remove nozzle's protective cap and place cap onto environmental cap.
2. Place filtered nozzle standoff onto nozzle (do not touch nozzle).
3. Insert nozzle in "H" end of confidence sample for 1 second.
    - Verify that at least 3 bars are displayed.
    - Ensure that the display has no more that 1 bar within 2 minutes.
4. Check in "H" and "G" modes.

Figure 3-14. M8A1 chemical alarm

Table 3-10. Unmasking procedures

Unmasking with M256 Series Kit	Unmasking without M256 Series Kit	MOPP Gear Exchange
This procedure takes 10 minutes. See *FM 3-4* for more information. The leader— • Checks M256 results to ensure that they are negative. • Selects one or two soldiers. The soldiers— • Move to a shady place. • Unmask for 5 minutes. • Clear and reseal their masks. The leader— • Observes the soldiers for 10 minutes. • Gives the all-clear signal if the soldiers show no symptoms. • Watches for delayed symptoms. • Has first aid available.	This procedure takes 25 minutes. See *FM 3-4* for more information. The leader— • Checks the M8 paper for liquid contamination. • Selects one or two soldiers. The soldiers— • Move to a shady place. • Break the seal of their mask, with their eyes open, for 15 seconds. • Clear and reseal their masks. The leader observes the soldiers for 10 minutes. If no symptoms occur, the soldiers • Unmask for 5 minutes. • Clear and reseal their masks. The leader— • Observes the soldiers for 10 minutes. • Gives the all-clear signal if the soldiers show no symptoms. • Watches the soldiers for delayed symptoms. • Has first aid available	Soldiers conduct MOPP gear exchange as a two-man team. See *FM 3-5* for more information. Soldiers— • Decon their individual gear. • Prepare their mask for decon and their overgarment for removal. • Decon their mask hood. • Remove their overgarment and overshoes. • Remove their gloves. • Put on their new overgarment. • Put on their new overshoes and gloves. • Secure their hood. • Secure their gear.

Table 3-11. NBC threat, detection, and protection

Agent Type	Agent Class	Agent Effect	M8A	M256	CAM	M8	M9	Protection Required	Skin Decon	Equip Decon
AC	NP	Blood	N	Y	N	N	N	Mask only	N	N
GA	NP	Nerve	Y	Y	Y	Y	Y	Mask and overgarment	Y	N
GB	NP	Nerve	Y	Y	Y	N	N	Mask and overgarment	Y	N
GD	P	Nerve	Y	Y	Y	Yellow	Y	Mask and overgarment	Y	Y
VX	P	Nerve	N	Y	N	Green	Y	Mask and overgarment	Y	Y
HL	P	Blister	N	Y	Y	Red	Y	Mask and overgarment	Y	Y
HD	P	Blister	N	Y	Y	Red	Y	Mask and overgarment	Y	Y
CG	NP	Choking	N	N	N	N	N	Mask only	N	N

CHAPTER 4
Fire Support

The platoon gets FS in the forms of indirect fires, smoke, and obstacle emplacement. FS protects the platoon while it is executing its mission. The platoon may provide survivability to FS units. The platoon leader must—

- Keep the TOC apprised of all element locations at all times. The FSO gets information on engineer unit locations from the TOC operations section (there are no FISTs with engineers).
- Work closely with the FSO or the FIST to plan and report SCATMINEs, cover friendly obstacles with fire, and provide suppressive fires and smoke on locations of expected enemy obstacles.
- Keep track of the amount of artillery FASCAM allocated to the TF and still available for employment.
- Get the frequencies, call sign, and procedures needed to call for indirect fires for the platoon. Most call for fires will be done by adjusting fires from a target or known point. See *FM 5-34* for call-for-fire procedures. Two examples of call-for-fire procedures follow:
  - From your location, you will adjust fire by distance and direction from the TRP to the target. Give the distance in meters and the direction as you see it. If the target is to the left or right of the TRP as you see it, call LEFT distance in meters or RIGHT distance in meters. If the target is behind the TRP as you see it, call ADD distance in meters and if the target is in front of the TRP, call DROP distance in meters.
  - During breaching operations, you may have to call for smoke. Ensure that the FSO or the FIST knows the location and the duration of your smoked target. The call-for-fire procedures are similar to those discussed above.
- Determine if the maneuver commander plans to provide survivability positions for artillery assets and what the priority is for those positions.
- Ensure that obstacles are synchronized with the indirect-fire plan.

CHAPTER 5
Air Defense

The platoon leader and sergeant should understand how AD weapons can support the platoon and how the platoon can support AD units. The platoon leader should know where the AD umbrella is in relationship to his area of operations and what kind of support he can expect. The platoon can provide its own AD measures, either passive or active. Passive AD measures are used to keep soldiers and equipment undetected by enemy air assets. Active AD measures are used to engage enemy aircraft. The following are active AD measures:

- Using the football-field method for high-performance aircraft (aim 200 meters in front of the aircraft).
- Using the football-field method for helicopters (aim 50 meters in front of the helicopter).
- Aiming weapons slightly above the nose of aircraft that is approaching head on.
- Firing all weapons at the reference point
- Ensuring that the AD-weapons control status is given. Status could be—
  - Hold—weapons are fired only in self-defense or in response to a formal order.
  - Tight—weapons are fired only at aircraft positively identified as hostile.
  - Free—weapons are fired at any aircraft not positively identified as friendly.
- Ensuring that AD warning conditions are given. Conditions could be—
  - White—an air attack is not probable.
  - Yellow—an air attack is probable.
  - Red—an air attack is imminent or in progress.
- Determining if the maneuver commander plans to provide survivability positions for AD assets and what the priority is for those positions.
- Ensuring that the platoon is able to receive AD early warning. When the AD warning condition changes to red, the warning of incoming enemy aircraft and their direction is transmitted over command or operations and intelligence networks. The platoon leader should ensure that the platoon is monitoring the appropriate networks.

CHAPTER 6
Combat Service Support

The platoon's CSS is determined by its supply requirements. The platoon leader and sergeant decide what CSS the platoon needs for an operation and initiate a request. Their request is forwarded to the 1SG of either their parent engineer company or supported maneuver unit, depending on their command and support relationships. For more information on command and support relationships, see *Chapter 2*. The platoon's CSS is normally delivered in a scheduled logistical package (LOGPAC).

RESUPPLY OPERATIONS

A LOGPAC is a resupply element that is formed on the basis of a unit's logistic requirements. Normally, it consists of a petroleum, oils, and lubricants (POL) truck; an ammunition truck (Class IV/Class V supplies); and a supply truck that carries rations, repair parts, and other requested items and pulls a water trailer. LOGPACs are—

- Assembled in the BSA and then moved forward under the control of the support-platoon leader to a logistics release point (LRP).
- Picked up by the 1SG at the LRP and moved forward to a secure area behind the company's position to feed, fuel, and resupply the company.
- Returned by the 1SG to the LRP where the support-platoon leader takes control of it and moves it back to the BSA.

LOGPACs are usually distributed by one of the following methods:

- Service-station method.
- Tailgate method.
- Modification of both.

In the service-station method, the 1SG establishes the LOGPAC in a centrally located and secure site. Successively, each platoon moves to the LOGPAC site. Each squad or crew passes through the various stations before the final inspection and returns to the work sites. Normally, the stations include maintenance support, fuel (POL), food, water, mail, administrative support, medics, and other CSS assets as required by the SOP and the mission, enemy, terrain, troops, and time available (METT-T). *Figure 6-1* shows the service-station method.

The tailgate method differs from the service-station method in two major aspects (see *Figure 6-2)*. In the tailgate method, the LOGPACs—

- Are brought to each platoon's general location by the 1SG.
- Are usually not as comprehensive as the service-station LOGPACs because they are more mobile and spend more time in transit.

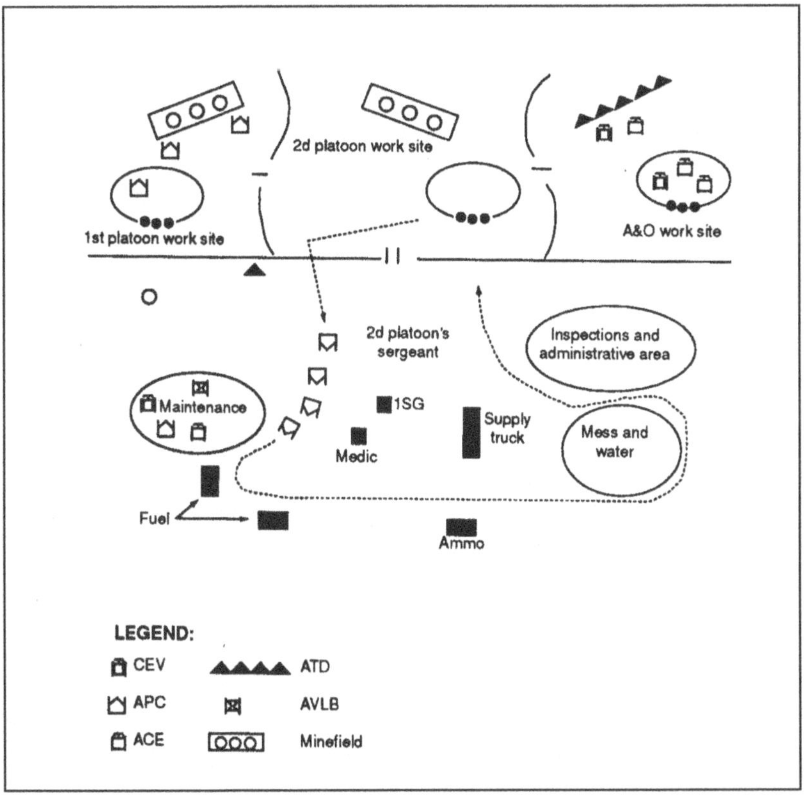

Figure 6-1. Service-station method

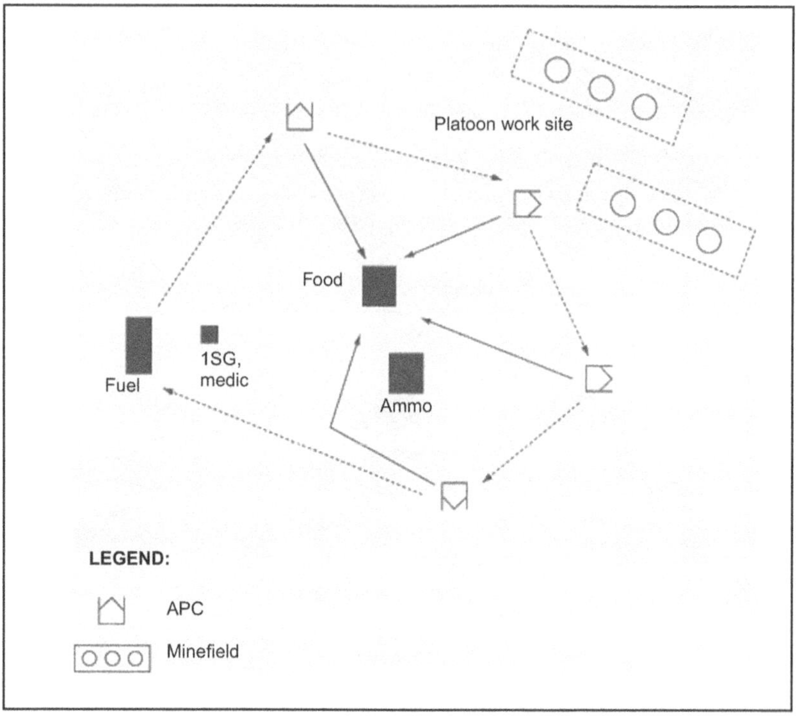

Figure 6-2. Tailgate method

The tailgate method takes more time for the 1SG but is the least interfering of the platoon's work effort. The platoon does not have to spend time moving to the LOGPAC.

Platoons often get resupplied from the maneuver company's/team's LOGPAC using the service-station or tailgate method. For security reasons, maneuver units normally provide LOGPACs to individual vehicles. Therefore, the squad rather than the platoon gets resupplied.

Even when being logistically supported by maneuver elements, platoons sometimes need engineer-specific resupply. Logistics-support responsibilities for command and support relationships are shown in *Table 6-1, page 6-4*.

Table 6-1. Command and support relationships

Relationship	Command and Task Organize	Assign Tasks	Provide Logistics Support	Reorganize as Infantry
Attached	Supported unit	Supported unit	Supported unit	Supported unit
OPCON	Supported unit	Supported unit	Parent unit	Supported unit
DS	Parent unit	Supported unit	Parent unit	Parent unit
GS	Parent unit	Parent unit	Parent unit	Parent unit

SUPPLY CLASSES

Normally, there are 10 classes of supplies; however, during defensive operations, Class IV/Class V supplies require special engineer considerations. Some supply classes require special considerations in resupplying the engineers. Supplies can either be pushed or pulled through the system. Supplies are pushed by predicted requirements and pulled by the needs of the user via a request for resupply.

CLASS I

Class I consists of subsistence and gratuitous health and welfare items. The platoon allocates Class I supplies based on unit strength. Class I supplies are normally pushed. The following are the amounts of water required for a soldier per day for—

- Drinking - 6 gallons.
- Personal hygiene - 2.5 gallons.
- Food preparation - 1.0 gallon.
- Vehicle radiators - 25 percent of the radiator's capacity.

CLASS II

Class II consists of such items as clothing, individual equipment, tentage, hand tools, administrative and housekeeping supplies, chemical-defense and decontamination items, and wet- and cold-weather contingency items. The company allocates Class II supplies based on need. It is normally pulled.

CLASS III

Class III consists of POL, including petroleum fuels, lubricants, hydraulic and insulating oils, preservatives, liquids and gases, bulk chemical products, coolants, deicer and antifreeze compounds, components and additives of petroleum and chemical products, and coal. Supply amounts are based on either offensive or defensive operations. When the TF is preparing to defend, additional Class III supplies are required for all equipment. ACEs require fuel every 8 hours when being heavily used. Additional fuel and package products, such as hydraulic fluid, should be maintained in the combat trains for this purpose. Forward fueling at the work sites should be a standard procedure. Class III supplies are usually pushed.

CLASS IV

Class IV consists of construction materials, including all fortification and barrier materials. These are items for which allowances are not prescribed. Class IV supplies are pulled.

CLASS V

Class V consists of all types of ammunition, including radiological and special weapons, bombs, explosives, mines, fuzes, detonators, pyrotechnics, missiles, rockets, propellants, and other associated items. Engineers need some unique Class V supplies (for example, cratering charges and line charges), which require special coordination with the Supply Officer (US Army) (S4) to get them into the TF's Class V system. Class V supplies are pulled.

CLASS VI

Class VI consists of all personal-demand items, such as candy, cigarettes, soap, and cameras (nonmilitary sales items). Sundry packs are also Class VI items. The platoon allocates Class VI supplies based on need. Class VI supplies are normally pulled.

CLASS VII

Class VII consists of major end items, such as launchers, tanks, mobile machine shops, vehicles, and organizational tool sets. Class VII supplies are demand driven.

CLASS VIII

Class VIII consists of medical material, including repair parts peculiar to medical equipment. Class VIII supplies are normally pulled.

CLASS IX

Class IX consists of repair parts and components, including kits, assemblies, and subassemblies (repairable and unrepairable) that are required for maintenance support of all equipment. The platoon's stock of repair parts is based on vehicle density, the authorized stockage list (ASL), the high-demand prescribed load list (PLL), and specific engineer-equipment requirements. Repair parts for items that the TF does not stock (for example, the M9 ACE or dozer) will have to come from the engineers. The platoon generates requests for repair parts by accurately completing *DA Form 2404*. When the maintenance contact team gets the form, it determines what repair parts are needed and the level of repair. The mechanics may bring the repair parts with them or the crew/operator may get them from the LOGPAC. Class V supplies are pulled.

CLASS X

Class X consists of materials to support nonmilitary programs, such as agriculture and economic development. Class V supplies are pulled.

SPECIAL CLASS IV/CLASS V SUPPLIES

These classes of supplies form the majority of the materials with which engineers construct the obstacles and fortifications required for the defense.

There are two types of Class IV/Class V loads: mission and basic. Mission loads consist of those materials required for a specific mission (for example, a standard fix minefield). Basic loads consist of those materials that the platoon carries to protect itself. The basic load can be used for missions to save time; however, it is to be replenished from the materials in the mission load. Basic loads are pulled and mission loads are pushed.

Mission loads are a TF responsibility regardless of the command and support relationships specified for engineers. The quantity of Class IV/Class V supplies (mines) normally stretches or exceeds the transportation assets of the TF. Palletized standard loads help solve the planning and distribution problem. Class IV/Class V supply points should be selected by their proximity to major obstacles.

Class IV/Class V resupply for the defense is one of the most demanding logistics operations the TF must carry out and requires all the assets that can be made available. A total cooperative effort by the TF, including engineers, is required if the defense is to be adequately resourced.

In the offense, TF planners anticipate Class IV/Class V mission loads for even a hasty defense on the objective. Prepackaging in the BSA speeds movement forward when it is needed. In addition to the Class V supplies that come forward as a part of the LOGPAC, the TF will maintain emergency Class V resupply in the combat trains.

MAINTENANCE PROCEDURES

When the TF is providing logistics support to the engineers, it will also provide organizational maintenance support to them. This causes no particular problem for common equipment. No engineer equipment repairers exist in the maneuver unit's maintenance sections lower than the brigade level except those at the engineer battalion. It may be unrealistic to expect the maneuver unit to provide all logistics support. The platoon may have to coordinate for maintenance assistance from its parent unit. When the engineers are providing logistics support, they will also provide organizational maintenance support.

Doctrinally, recovery is the responsibility of the owning unit. However, it is virtually impossible for engineers working in the TF area to recover their own vehicles. In the defense, the TF will have to provide recovery support at least back to the unit maintenance collection point (UMCP) located near the combat trains. If repairs cannot be made at the UMCP and the time and situation permit, the engineer's maintenance-support team (MST) can come forward and recover the down vehicle. The TF could provide additional recovery back to the BSA for repairs. The engineer battalion maintenance officer (BMO) or technician must ensure that each BSA has direct-support (DS) mechanics capable of repairing engineer equipment.

In the offense, the TF recovers the vehicle to the main supply route (MSR). Depending on the situation, the TF either turns vehicle recovery over to the engineers or recovery assets take the vehicle to the BSA. It must be clear in the TF SOP or order which MST has recovery responsibility for the engineer vehicles. The priority the TF commander places on the engineer assets determines the priority they are recovered in.

Engineer companies should have maintenance contact teams that are familiar with engineer heavy equipment to routinely visit work sites. The communications repairman and armorer can accompany this contact team to help anticipate problems and provide support before the mission is jeopardized.

MEDICAL SUPPORT

Medical support and evacuation must be coordinated for each operation and normally come from the nearest company/team (see *Table 6-2, page 6-8*). The platoon's aidman or combat lifesavers should be required to coordinate with the TF's medics and ensure that the medical requirements of the platoon are planned for and met. In the event of a mass casualty incident, the platoon may have to designate a platoon vehicle as a litter vehicle.

Table 6-2. Command and support relationships and logistics support

Logistics Support	Class III	Class IV/V	Class VIII Medical	Maintenance	Emergency Resupply
Parent unit (OPCON, DS, GS)	Company LOGPAC	Basic loads are resupplied from the LOGPAC. Mission loads are supplied by the supported unit.	Parent unit	Parent unit performs maintenance on all equipment.	Parent unit
Company/team (attached)	Company/team LOGPAC	Basic loads are resupplied from the LOGPAC. Mission loads are supplied by the supported unit.	Company/team	Company/team MST performs maintenance on common equipment. Engineer MST performs maintenance on engineer-specific equipment.	Company/team
TF (attached)	TF support platoon	Basic loads are resupplied from the LOGPAC. Mission loads are supplied by the supported unit.	TF medical platoon	TF maintenance platoon performs maintenance on common equipment. Engineer MST performs maintenance on engineer-specific equipment.	TF support platoon

The platoon leader must—

- Understand the TF's evacuation and medical support plans.
- Understand the engineer company's evacuation and medical support plans.
- Know the locations of the—
  - Battalion aid station.
  - Jump aid station.
  - Personnel transfer points.
  - Casualty collection points.

CHAPTER 7
Command and Control

The platoon leader uses the C^2 processes to ensure that his platoon accomplishes its missions. This chapter provides the TTP needed to command and control the platoon and make sound tactical decisions.

COMMAND AND SUPPORT RELATIONSHIPS

Command relationships can be operational control (OPCON) or attached. Support relationships can be DS or general support (GS). Both supporting and support parties must clearly understand what each of these mean (see *Table 6-2, page 6-8*). The following should be considered when using command and support relationships:

- When OPCON is used, the maneuver element—
  - Commands and controls the platoon.
  - Has the authority to place squads OPCON to companies/teams even though it is not considered advisable to do so with a mechanized platoon. Additionally, the TF cannot attach them to the company.
- When attached is used, the maneuver element—
  - Commands and controls the platoon except for limitations imposed by the attachment order.
  - Has responsibility for the platoon's logistics support except for personnel transfer and promotion.
- When DS is used—
  - The platoon receives its missions from the maneuver element.
  - The engineer company and the engineer battalion support and command the platoon.
  - The platoon works on the tasks that the maneuver element assigns.
  - The TF does not have the authority to split the platoon up and attach squads to companies/teams.
  - Logistics support for the platoon comes from engineer higher HQ.

- When GS is used, subordinate commanders must request support from the engineer commander on a task-by-task basis, and the senior commander sets priorities and assigns tasks to the GS unit. GS is seldom used at TF level. When an engineer commander wants to retain control over the engineers, they are retained in GS of maneuver forces.

The TF always provides the materials necessary to support engineer operations in its sector for offensive and defensive missions. This is a key logistics issue that must be understood at the start of any C^2 discussion. Even though the engineers determine requirements and use the material, supplying it remains a TF responsibility regardless of the command or support relationship (see *FM 5-100*).

COMMUNICATIONS

Engineers must operate on the following frequency modulated networks:

- TF.
- Platoon internal.
- Engineer company.
- Maneuver company/team.

Engineers have a two-net capability to support their communication requirements. In the DS or GS relationship, the engineer platoon should operate the first network on the engineer command network. The second network, the platoon's internal network, should be used to control the platoon's assets. In the attached or OPCON relationship, the engineer platoon should operate the first network on the engineer command network or supported maneuver company/team command network based on the task organization. The platoon leader uses his platoon's internal network to enter other networks for coordination and situation reporting. Before the platoon leader leaves his platoon's internal network, he must first hand over control of the network to the platoon sergeant to control the platoon's assets.

The platoon submits the following reports:

- Routine reports:
  - Commander's situation report (SITREP)—identifies changes to a unit's tactical situation.
  - Sensitive-items report—gives the results of a serial-number check on all sensitive items.
  - Supply-status report—informs the commander and staff of a subordinate unit's supply status.
  - Personnel-status report—is used to maintain the accountability and status of assigned and attached personnel and generate requests for replacements.

  - Maintenance-status report—provides information on the operational status of equipment and is used to generate requests for maintenance and parts.
- Event-driven reports:
  - Spot report (SPOTREP)—identifies any known or suspected enemy activity.
  - Reconnaissance report—provides technical and tactical engineer data on a point/area target.
  - Countermobility report—gives the status of all obstacles.
  - Mobility report—gives the status and conditions of lanes, routes, bypasses, crossing sites, air lines of communication, as required, and ground lines of communication. It also includes status changes and the upgrade/downgrade of previously reported trafficability of locations/sites.
  - Survivability report—provides the detailed status of the unit's survivability level (for example, the number, type, and protective level (hull or turret) of positions).
  - Class IV/Class V special report—is used to keep track of critical engineer classes of supplies. For example, a platoon establishes a mine dump and keeps track of the types and quantities of mines available and the amounts issued.
  - Flash SITREP—is used to notify the next higher HQ when a subordinate unit becomes decisively engaged or has critical shortages that prevent mission accomplishment.
  - NBC report—is used to report NBC events.
  - Shelling report (SHELREP)—provides information on enemy indirect fire.
  - Enemy prisoner of war (EPW) report—is used to account for prisoners captured.
  - SCATMINE report—is used to notify affected units of the emplacement of SCATMINEs.
  - Meaconing, intrusion, jamming, and interference (MIJI) report—is used to notify the Intelligence Officer (US Army) (S2) when the reception of radio signals is hindered, confused, or distorted by any external source or when instructions are received from a station that cannot authenticate.
  - Splash report—is used to report downed friendly aircraft.
  - Casualty report—identifies battle and nonbattle casualties to the parent unit and alerts medical-treatment facilities of incoming casualties.
  - Casualty-feeder report and witness-statement report—provide detailed eyewitness information on a casualty. They are used to account for casualties and provide administrative information for use in notifications and awards.

ENGINEER INTEGRATION

The company commander of a mechanized engineer company normally associated with a battalion/TF is also the staff engineer and advisor to the TF commander. Likewise, in a light TF, the staff engineer advisor to the TF commander is the light engineer platoon leader. The staff engineer integrates engineers into the TF's planning process. He coordinates with other BOS to ensure engineer synchronization. During the preparation of the battle, the staff engineer ensures that soldiers and equipment are in the correct location and on time for linkups and rehearsals. He then executes the engineer mission to support the operation. The platoon is the lowest level engineer unit that can still effectively accomplish independent tasks. The platoon usually operates under the control of an engineer company; however, it can be attached to a maneuver company/team.

The platoon leader must keep both the engineer company and maneuver company/team commanders informed on critical tasks. He must continuously forward and update tactical and technical information upward and laterally by all available means.

In heavy forces, the TF normally receives an engineer company in support, and the company commander is the TF engineer. In light forces and in some special situations in heavy forces, the TF only receives an engineer platoon, and the platoon leader is the TF engineer. When the platoon leader is acting as the TF engineer, he should consider the following information:

- Orders should include when and where the TOC is to be established. If the TOC is to be moved before the platoon arrives, this should also be included.
- Orders should identify an assembly area where the platoon can be consolidated.
- The entire platoon must **not** go to the TOC.

PLATOON LEADER'S RESPONSIBILITIES

Once the platoon leader arrives at the TOC, he should—

- Check in with the S3 or company commander and report the platoon's status and location and any problems.
- Provide the status of equipment and the number of soldiers to the operations noncommissioned officer (NCO).
- Obtain the TF's and the team's signal operation instructions (SOI) and enter the maneuver force's network.

- Ensure that the communications equipment is set for the correct variable for secure operations and is compatible with the supported unit's communications equipment.
- Check with the operations sergeant for any changes to routine or required reports and SOP items.
- Be familiar with the TF SOP. **This is not the time to start studying the SOP.**
- Provide the status of obstacles being constructed.
- Check on format requirements for the periodic report. He should also check to see that TOC personnel are familiar with the reports being sent and know how to post them to their charts or overlays.
- Determine if a specific individual at the TOC has been designated as an expediter or overwatcher for engineers in the sector requiring assistance.
- Provide any terrain, route, or other information acquired from the reconnaissance unit or personnel enroute to the S2 or company/team commander.
- Obtain such information as the current enemy situation, weather and light data, and terrain details, to include topographic products, from the S2 or company/team commander.
- Provide any priority intelligence requirements (PIR) or information requirements (IR) needed for upcoming operations to the S2 or company/team commander.
- Stop at the combat trains and coordinate for classes of supplies and services when joining the maneuver force.
- Obtain frequencies and call signs for the administrative and logistics net control stations (NCSs).
- Provide a platoon roster and identify any critical shortages to the Adjutant (US Army) (S1) or operations sergeant.
- Review the TF SOP for reporting and evacuating casualties.
- Coordinate with the S4 or company/team 1SG and review the platoon's logistics status. The channel used for operational logistics for the platoon varies based on the command and support relationships.

NOTE: The platoon leader should remember that if a problem arises which will prevent the accomplishment of the mission or if normal channels are not working fast enough and logistics become an operational issue, the S3 or commander should be apprised of the problem.

As a minimum, the platoon leader must be present when the—

- Maneuver force receives the WO.
- Commander conducts reconnaissance.
- Commander makes and announces a decision and explains the intent.
- OPORD is received.

MANEUVER FORCE'S CONTROL

Normally, the platoon is kept under the engineer company's control and works for the battalion/TF commander. A major consideration in planning for the use of engineers is to maintain platoon integrity, keeping the entire platoon under the platoon leader's control. When maneuver forces are moving on two axes during mechanized offensive operations, the platoon will not place a part of the platoon on each axis. During the planning process, the engineer planner must weigh the main effort and determine which axis is critical to the mission and requires engineer assets. Placing an engineer NCO with the scout platoon when it is performing reconnaissance can help provide more specific information on the obstacles that the scouts find. Obstacle breaching can thus be preplanned. See *Appendix A* for basic formations, movement techniques, and hand-and-arm signals and *Appendix B* for checklists.

ENGINEER ATTACHMENTS

If the platoon has been augmented with AVLBs, they normally come in pairs. This is for both maintenance and tactical reasons. The AVLBs should be placed with the element that needs a gap-crossing capability. In the deliberate attack, AVLBs can be integrated individually into attack formations. Employment as a pair allows the squad leader to control them. If the AVLBs are not immediately required, they should be returned to the company's control. Curbs on the AVLB must be removed if M1 tanks with rollers are to cross the AVLB, and the roller's dog bone must be raised. If a CEV is attached to the platoon, it normally moves with the platoon.

Heavy equipment and wheeled vehicles should move with the platoon only on tactical road marches over good roads. During a movement to contact or over rougher terrain, METT-T determines where and what heavy equipment moves in the formation.

APPENDIX A

Basic Formations, Movement Techniques, and Hand-and-Arm Signals

The platoon uses a variety of mounted and dismounted formations and movement techniques to maneuver on the battlefield. This appendix gives examples of many of the basic formations, movement techniques, and hand-and-arm signals the platoon leader could use. It is not designed to be all encompassing. For more information on these topics, see *FMs 7-8, 5-34, and 71-1.*

MOUNTED MOVEMENT TECHNIQUES

The mounted engineer platoon must be proficient in moving with its maneuver counterpart. In the following paragraphs, formations, movement techniques, and actions taken during movement for the mounted engineer platoon are discussed.

WEDGE

The engineer platoon almost always maneuvers as part of another larger formation. Normally, this will either be the parent engineer company or a maneuver company/team. *Figure A-1, page A-2,* shows the formation that the platoon is most likely to use. The platoon wedge provides the most defensible formation with the easiest C^2. Ordinarily, the platoon follows a maneuver platoon if task-organized to a maneuver company/team. The platoon might lead if it is part of the engineer company formation. In either case, the wedge is the best formation to use if enemy contact is likely.

The platoon frequently receives assets from the assault and obstacle (A&O) platoon. These could include the AVLB, CEV, or armored vehicle-launched mine-clearing line charge (AVLM). These additional vehicles can strain the C^2 capability of the platoon. *Figure A-2, page A-2,* shows the technique of "tethering". The AVLB, CEV, and AVLM are tethered to individual engineer squads. Each squad leader has the responsibility of commanding and controlling an A&O platoon asset. This improves the C^2 of the platoon and provides each A&O asset with a security element as it moves across the battlefield.

Figure A-3, page A-3, shows a simple wedge formation. Note where the key leaders in the platoon are. The wedge formation uses the wingman concept. The platoon sergeant and leader are both in APCs and each has a wingman. This technique

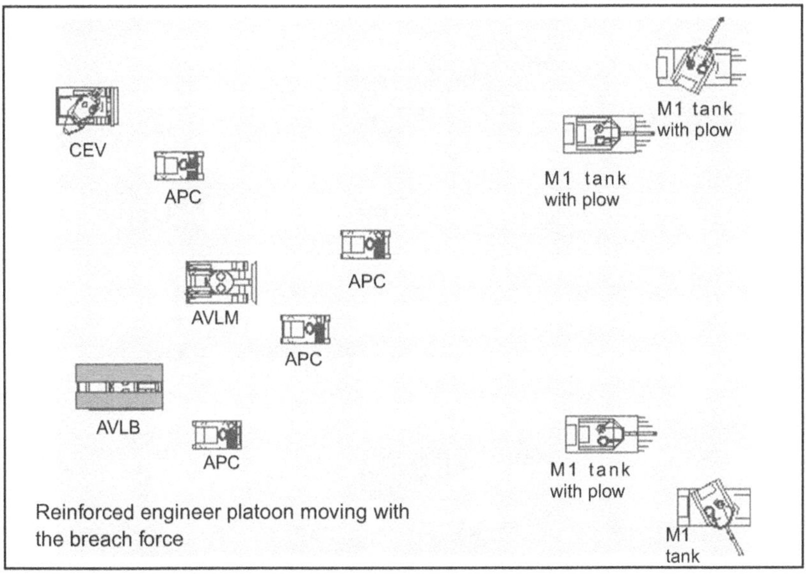

Figure A-1. Engineer platoon with the breach force

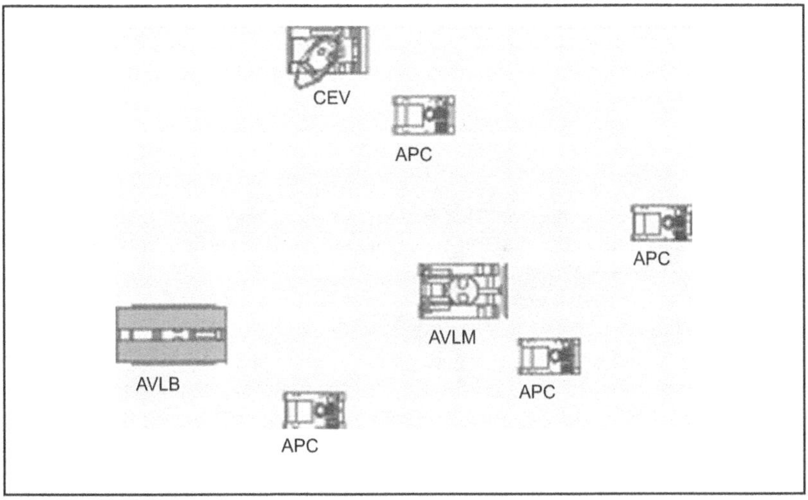

Figure A-2. Modified mounted wedge formation

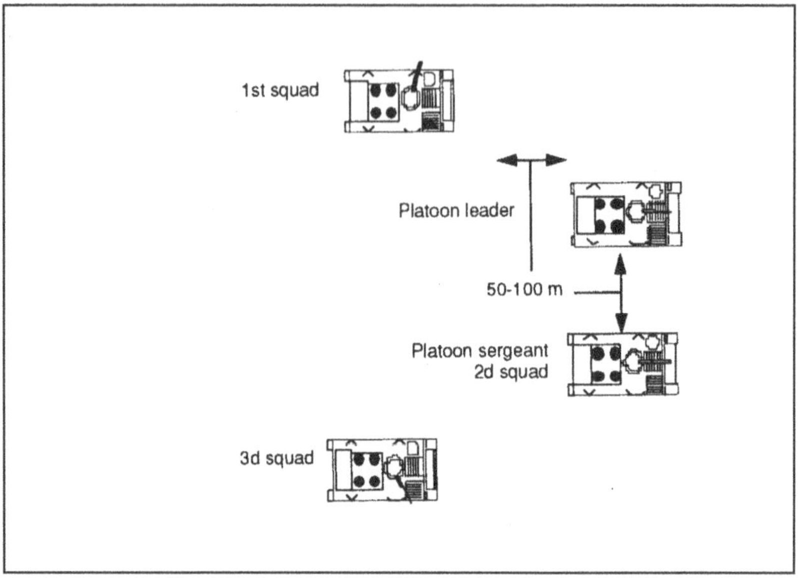

Figure A-3. Mounted wedge formation

also simplifies C^2 and movement. It has the added benefit of placing the most experienced leader in the platoon, the platoon sergeant, forward with the platoon. It also places the platoon sergeant where he can easily take over in the absence of the platoon leader.

COLUMN

Figure A-4, page A-4, depicts a column formation. This formation is used when enemy contact in not expected. This formation maximizes C^2 and the speed of the formation. Normally, the platoon transitions from the column to the wedge as enemy contact becomes more likely.

LINE

Figure A-5, page A-4, shows the platoon in a line formation. This formation is designed to maximize the forward firepower of the platoon. The platoon transitions from the wedge to the line as enemy contact becomes eminent. Engineer platoons do not generally use this formation. However, if the platoon is tasked to suppress a dismounted threat while another platoon maneuvers, this formation works well.

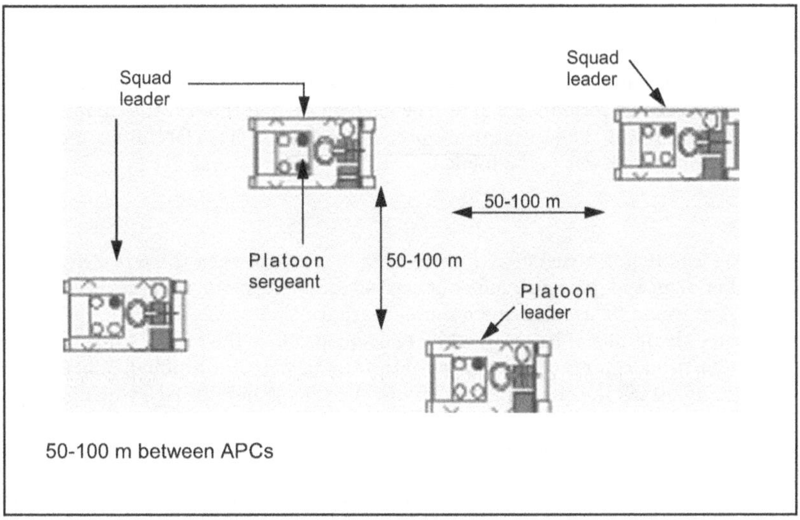

Figure A-4. Mounted column formation

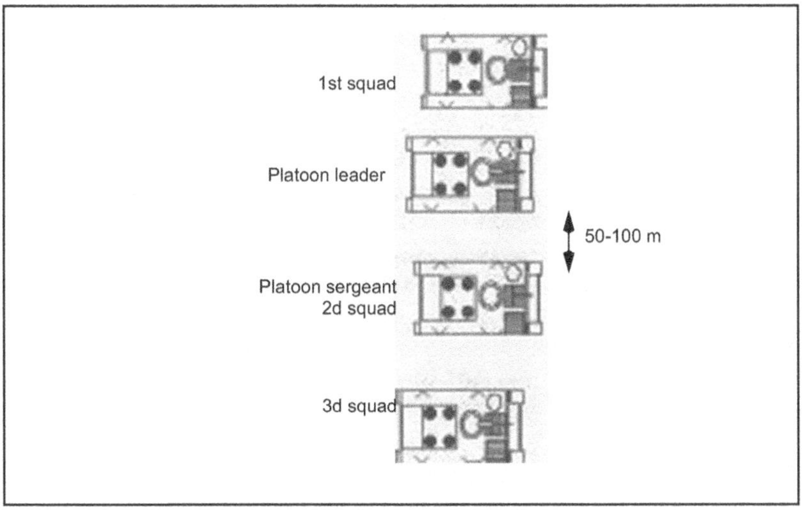

Figure A-5. Mounted line formation

ECHELON

The echelon formation is used when the platoon is on the flank of the company and the enemy threat is also from the flank. *Figure A-6* shows an echelon right formation. The echelon formation can also be used on the left flank of the company and, in that case, it would be the mirror image of *Figure A-6*. This formation maximizes the platoon's firepower to the flank.

V

The V-formation is a variation of the wedge. It is used when there is a significant threat of command-detonated mines or explosives. The formation shown in *Figure A-7, page A-6,* would be used during a route-clearance operation. It allows the platoon to secure the flanks of the road while two squads clear the route. This formation also identifies command-detonated-mine firing wires or ambushes before the enemy can attack the element on the route. It is not generally used when there is a significant enemy threat.

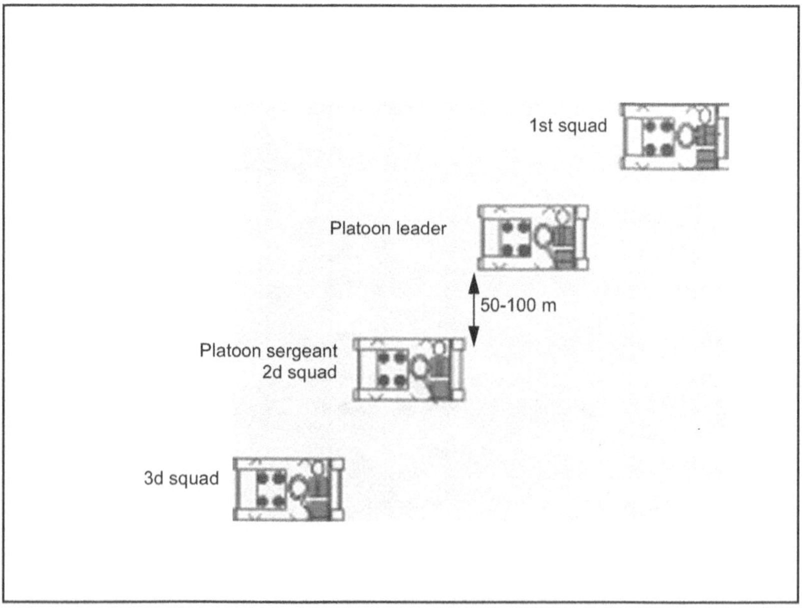

Figure A-6. Mounted echelon formation

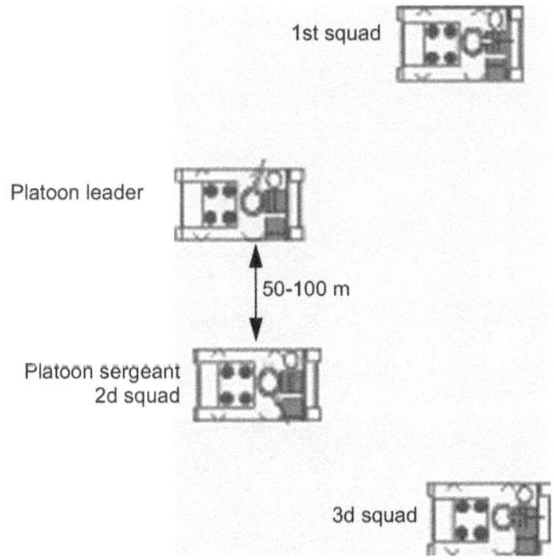

Figure A-7. Mounted V-formation

DISMOUNTED MOVEMENT TECHNIQUES

The dismounted engineer platoon must be proficient in moving with its maneuver counterpart. In the following paragraphs, formations, movement techniques, and actions taken during movement for the dismounted engineer squad and platoon are discussed.

SQUAD FORMATIONS

The squad formation is built from the fire-team wedge (see *Figure A-8).* The interval between soldiers in the wedge formation is normally 10 meters. The wedge expands and contracts depending on the terrain. When rough terrain, poor visibility, or other factors make control of the formation difficult, the squad uses the modified wedge. In this formation, the normal interval between soldiers is reduced so that all team members can still see their team leader and each team leader can see their squad leader. The sides of the wedge can contract to the point where the wedge resembles a single file. When moving in less rugged terrain, where control is easier, soldiers expand the formation or resume their original positions.

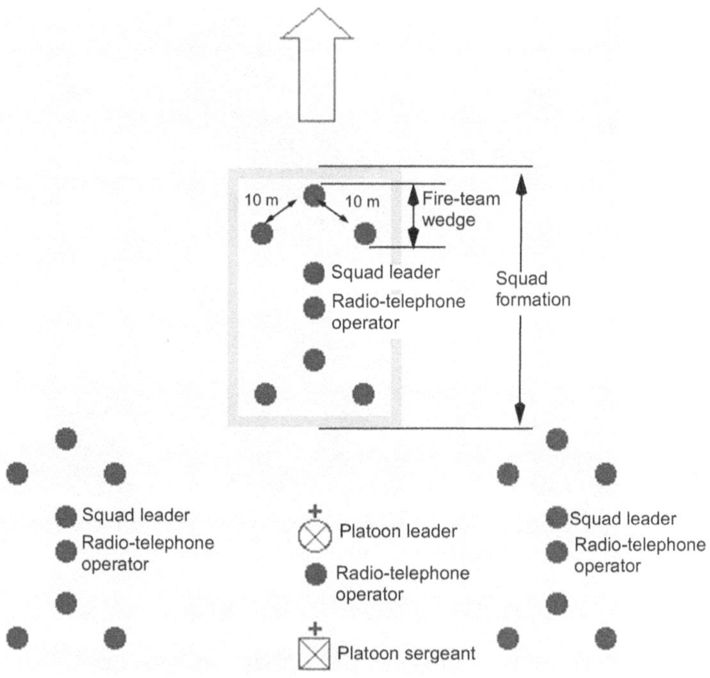

NOTE: Unlabeled positions are dismounted soldiers.

Figure A-8. Dismounted-platoon wedge formation

PLATOON FORMATIONS

Platoon formations are built from multiple squad formations. These formations include the platoon column, line, V, and wedge. The platoon leader selects the best formation based on his METT-T analysis. *Table A-1, page A-8,* shows a comparison of each formation and its use.

Table A-1. Comparison of platoon formations

Dismounted Movement Formations	Movement Use	Movement Characteristics			
		Control	Flexibility	Fire Capability/ Restriction	March Rate
Column	When the platoon performs primary movement formations	Good for maneuver (fire and movement)	Provides good dispersion laterally and in depth	Allows limited firepower to the front and rear, high volume to the flank	Good
Line	When the enemy situation is unknown and the leader wants all soldiers forward for maximum firepower to the front	Difficult	Is minimal	Allows maximum firepower to the front, little to the flanks and rear	Slow
V	When the enemy situation is vague, but contact is expected from the front	Difficult	Provides two squads up front for immediate firepower and one squad to the rear for movement upon contact from the flank	Allows an immediate heavy volume of firepower to the front or flanks	Slow
Wedge	When the enemy situation is vague, but contact is not expected	Difficult but better than the platoon V and platoon line	Enables the leader to make contact with the smallest element and still have two squads to maneuver	Provides a heavy volume of firepower to the front or flanks	Slow but faster than the platoon V
File	When visibility is poor due to terrain or light	Easiest	Is the most difficult formation from which to maneuver	Allows immediate fires to the flanks; prevents focused fires to the front and rear	Fastest

Wedge

The dismounted wedge is used when the enemy situation is vague and enemy contact is likely (see *Figure A-8, page A-7*). This formation allows a large volume of fire around the formation. Generally, at least one squad or element will be free to maneuver from the wedge after contact.

Column

The column formation is the platoon's primary movement formation (see *Figure A-9, page A-10*). It provides good dispersion both laterally and in depth and simplifies control. The lead squad is the base squad for fire control.

Line

The line formation allows the delivery of maximum fire to the front but little fire to the flanks (see *Figure A-10, page A-11*). This formation is hard to control and does not lend itself well to rapid movement. It is the basic platoon assault formation during an attack.

V

The V-formation has two squads up front to provide a heavy volume of fire on contact (see *Figure A-11, page A-12*). It also has one squad in the rear that can either overwatch or trail the other squads. This formation is hard to control, and movement is slow.

File

The file formation gives maximum control to leaders and is used for speed during movement (see *Figure A-12, page A-13*).

Traveling

The dismounted engineer platoon supporting a maneuver TF normally travels as part of one of the larger maneuver elements. When breaching assets are needed forward with the lead company, the engineer platoon travels behind the lead platoon for security (see *Figure A-13, page A-14*). If each maneuver company needs a breaching capability, the dismounted engineer platoon can be broken into squad-sized elements, where each squad travels with a maneuver company. In this case, the platoon HQ travels with either the battalion tactical operations center or the main effort for C^2.

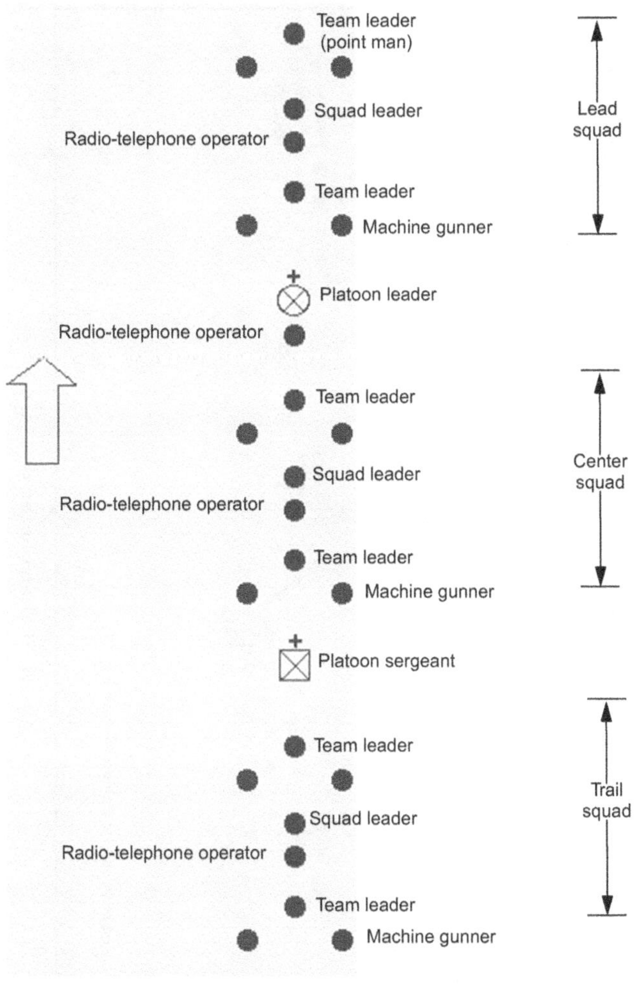

NOTE: Unlabeled positions are dismounted soldiers.

Figure A-9. Dismounted column formation

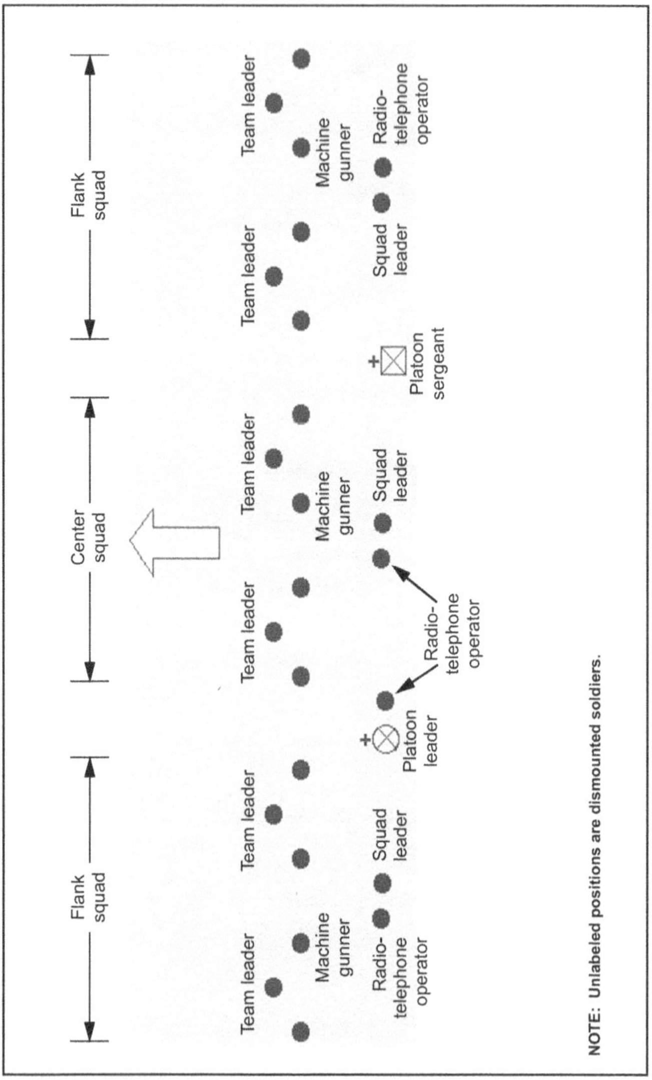

Figure A-10. Dismounted line formation

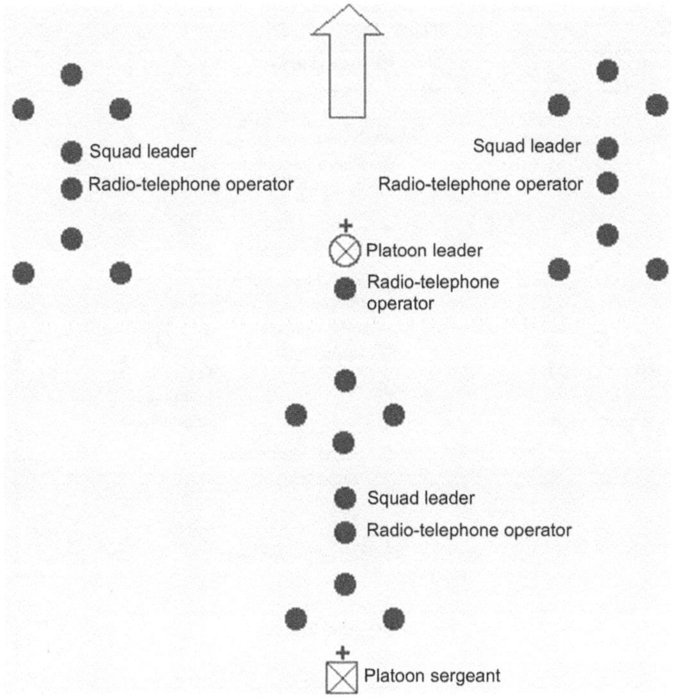

NOTE: Unlabeled positions are dismounted soldiers.

Figure A-11. Dismounted V-formation

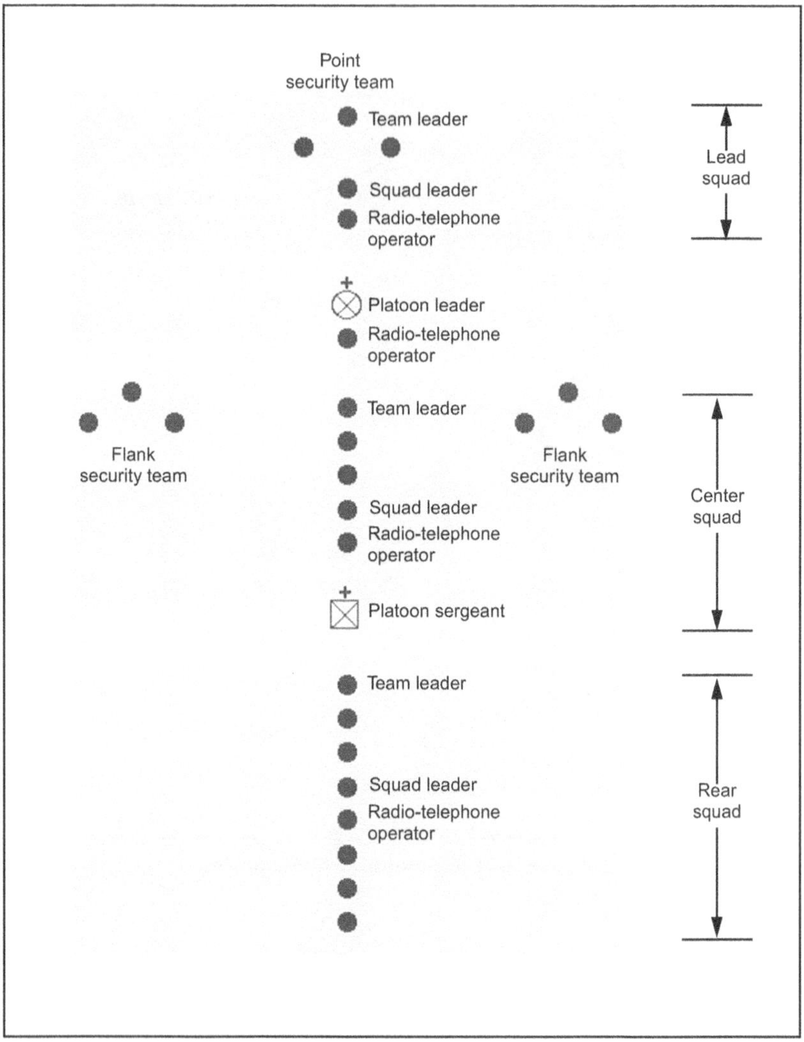

Figure A-12. Dismounted file formation

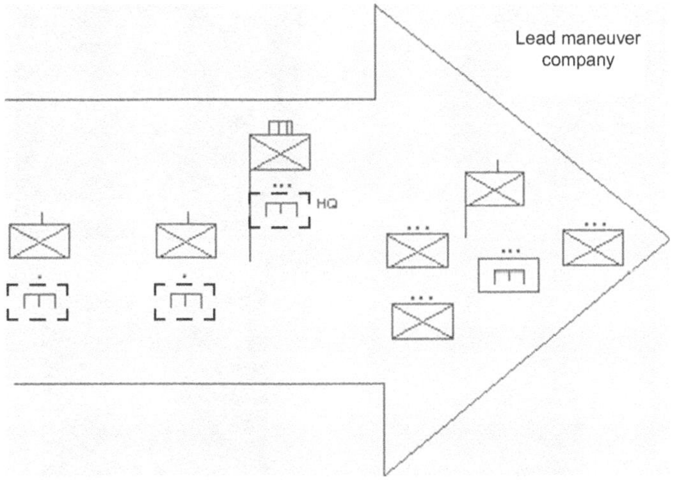

NOTE: The engineer platoon traveling with a TF could travel as part of the lead infantry company to provide immediate breaching capability, or the platoon could be broken down into squad-sized elements to travel with each maneuver company.

Figure A-13. Dismounted traveling formation

HAND-AND-ARM SIGNALS

During many operations, the platoon has to use hand-and-arm signals for C^2. *Figures A-14 and A-15, pages A-15 through A-20,* show many of the command hand-and-arm signals used when the platoon is mounted or dismounted. These signals are normally used when either radio listening silence is in effect or stealth is needed.

DAY | NIGHT

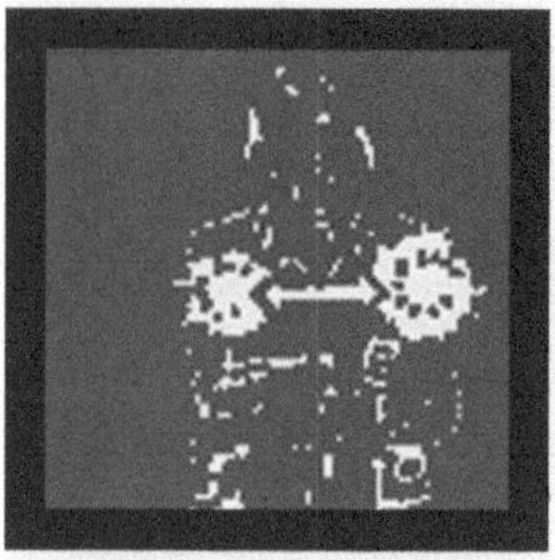

Stop or stop engines

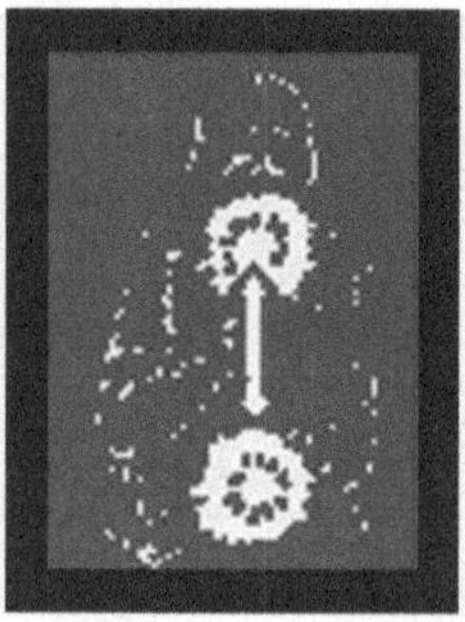

Move forward, move out, increase speed, or double-time

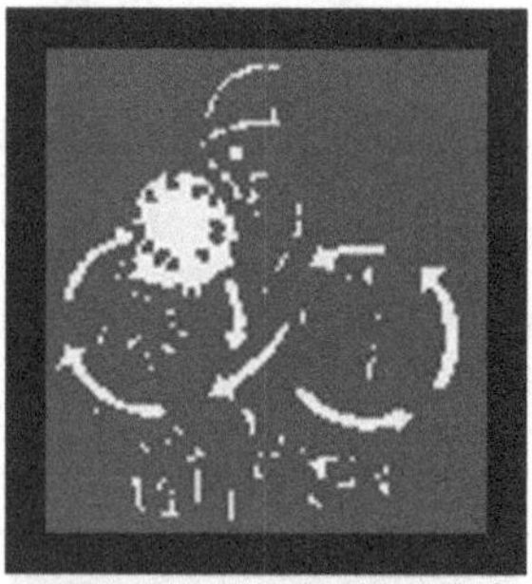

Start engine or prepare to move

Figure A-14. Hand-and-arm signals

DAY

NIGHT

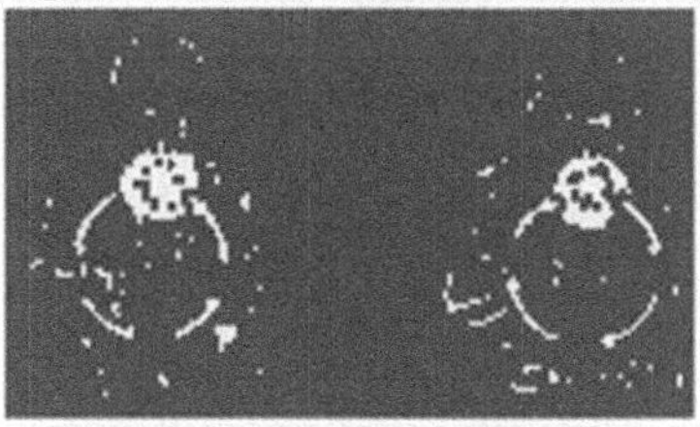

Left

Right

Turn left or right

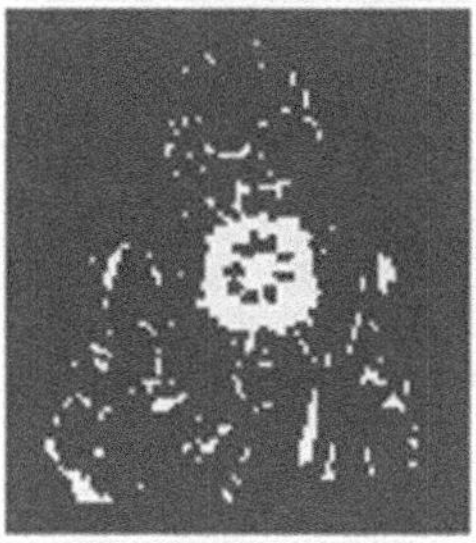

Slow down

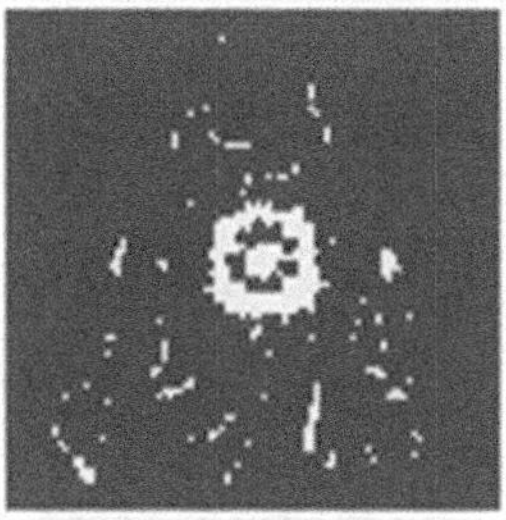

Blink

Move in reverse

Figure A-14. Hand-and-arm signals (continued)

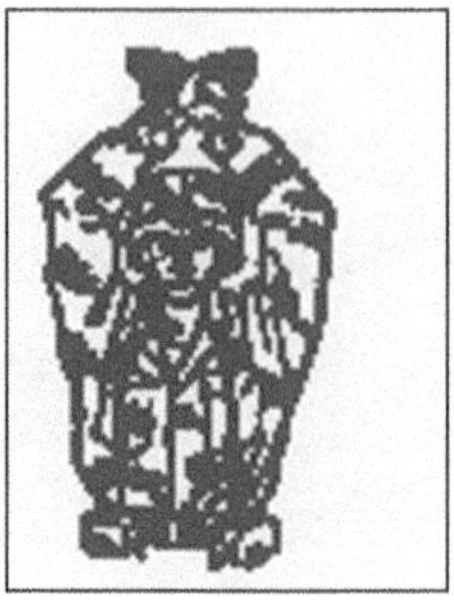

Disregard previous command

Attention

I do not understand

I am ready

Close the distance between vehicles and stop

Figure A-14. Hand-and-arm signals (continued)

Figure A-14. Hand-and-arm signals (continued)

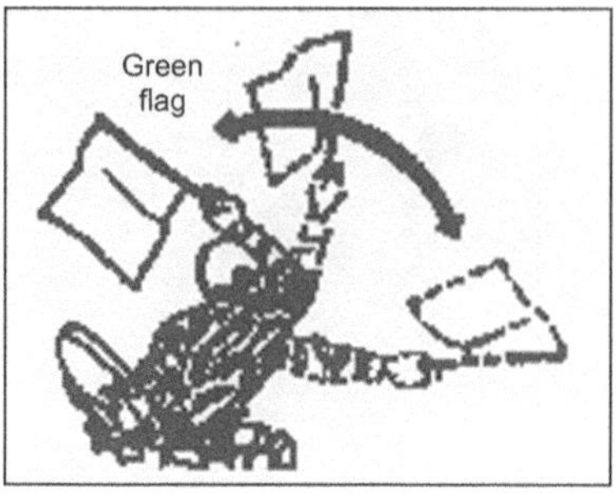

Move out

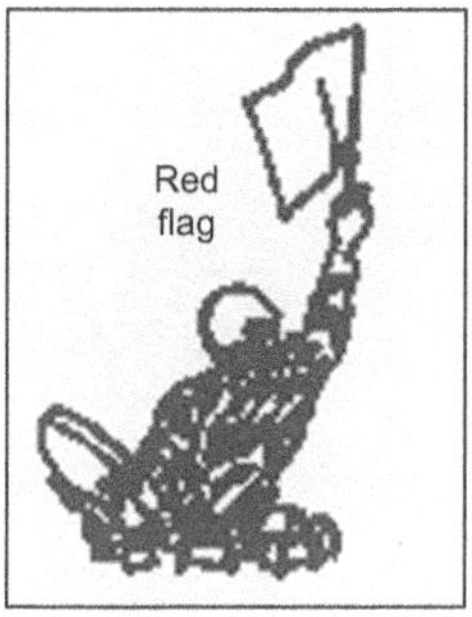

Enemy is in sight

Form a line

Figure A-14. Hand-and-arm signals (continued)

Mounted Movement Formations	Hand-and-Arm Signals	Formation Examples
V		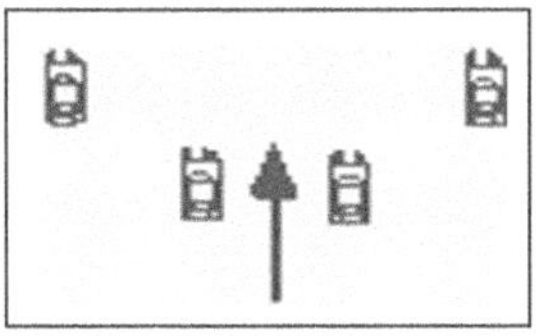
Echelon right (left)		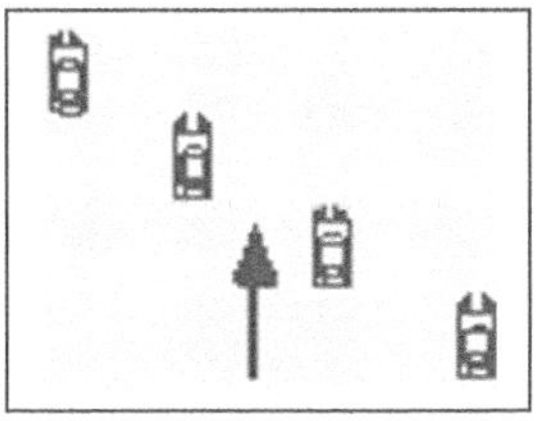
Wedge		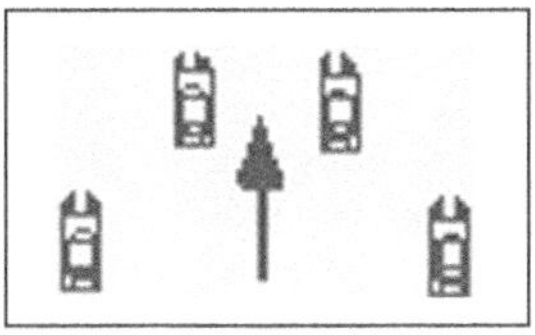
Line	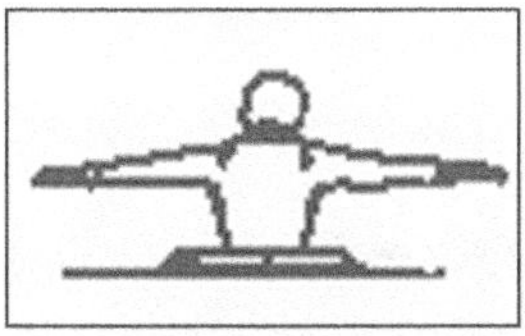	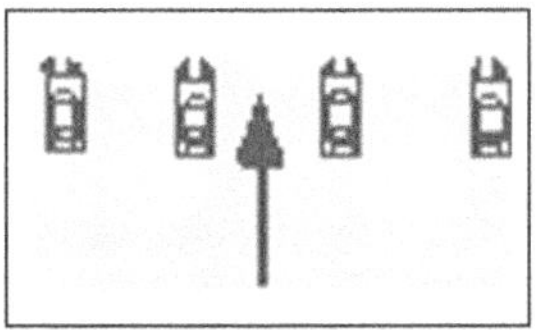
Column	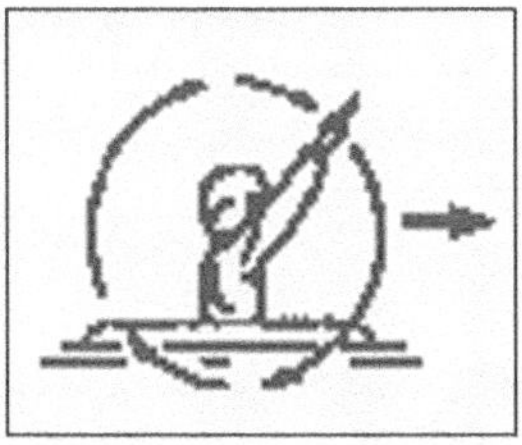	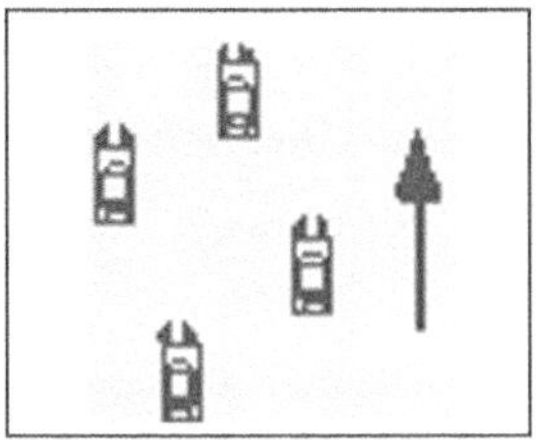

Figure A-15. Mounted movement formations and hand-and-arm signals

APPENDIX B

Checklists

This appendix gives several checklists that the platoon and its leaders will find useful. These checklists are not all encompassing and should be used to augment local SOPs, not replace them. Where units do not have standardized checklists, these may be helpful in developing local unit SOPs.

LOCAL SECURITY AND JOB-SITE SECURITY CHECKLISTS

When the platoon halts, the platoon leader establishes and maintains local security. He assigns each squad a sector of the perimeter to ensure mutual support and to cover all gaps by observation and fire. The platoon leader designates OPs and squad leaders designate OP personnel. OPs communicate with the platoon's command post and warn the platoon of the approaching enemy before it is attacked. Frequently, the platoon will be in a tactical halt while waiting for an operation to start.

When the platoon is halted, the platoon leader establishes a priority of work, to include—

- Positioning crew-served weapons, protective obstacles, and chemical alarms.
- Constructing individual and crew-served fighting positions, if the halt is for an extended period.
- Setting up wire communications between the squads and the platoon's command post. Radio silence should be maintained at the halts for OPSEC.
- Coordinating with adjacent units.
- Preparing range cards.
- Clearing fields of fire.
- Distributing ammunition, rations, water, supplies, and special equipment.
- Conducting PMCS on weapons, vehicles, and equipment.
- Inspecting the platoon's members and equipment.

- Rehearsing critical aspects for the upcoming mission.
- Test firing small-arms weapons, if the tactical situation permits.
- Conducting personal hygiene and field sanitation.
- Instituting a rest plan.
- Completing the work priorities as time permits.

Job-site-security actions are similar to those actions that are taken at a halt. However, job-site security is specifically designed to ensure that the platoon is able to work without being harassed.

Typically, the platoon will use this checklist when constructing obstacles and field fortifications. Before moving to the job site, the platoon leader informs everyone of the warning signals, code words, and pyrotechnics signals to be used. Upon arrival at the job site, the platoon—

- Occupies a job-site overwatch position.
- Dispatches reconnaissance/minesweeping/NBC teams to clear the job site.
- Establishes local security.
- Establishes escape routes; identifies possible enemy avenues of approach; and places OPs, chemical alarms, and crew-served- and antitank-weapons positions to cover enemy avenues of approach.
- Divides the job site into defensible sectors and assigns them to the squads.
- Establishes communications with the engineer company.
- Starts work.

PRECOMBAT CHECKLISTS

Each individual should have—

- Load-bearing equipment.
- A chemical protective mask with an M256 chemical-detection kit, an M13 decontamination kit, a waterproof bag for the mask, M8 chemical-detection paper, and three chemical-antidote injectors.
- A kevlar helmet with a cover and a band.
- An individual weapon with magazines and the zero posted in the pistol grip and a cleaning kit.
- A ruck sack and duffle bags packed with military clothing and table of allowance (TA)-50 equipment.

- Wire gauntlets.
- A fragmentation vest.

Each leader should have—

- A map and a compass.
- Signal operating instructions.
- His unit's SOP.
- *FMs 5-34 and 5-10* in their uniform cargo pockets.
- *FMs 20-32, 90-13-1, and 90-7* in their ruck sack.
- Other FMs, as required or needed.
- A watch and a solar-powered calculator.
- A notebook, pens, and pencils.
- Binoculars.
- Night-vision goggles.
- Casualty cards.
- A laminated range card.
- *DA Forms 1355 and 1355-1-R.*

All vehicles should have—

- Material, tied down and loaded according to the load plan.
- Fuel tanks, full.
- Preventive maintenance checks and services (PMCS), completed and dispatched.
- Packaged POL products, stowed.
- Water and extra fuel cans, full.
- Meals, ready-to-eat (MREs), stored.
- Basic-issue items (BIIs) and additional authorized list (ALL) items, inventoried, cleaned, and stowed.
- Spare parts and tires or track blocks, stowed.
- First-aid kits, inspected and stowed.
- A vehicle log book, the vehicle's TM, and *DA Forms 2404,* stowed.

- A camouflage screen system, stowed.
- Chemical-detection equipment, mounted.
- A tow cable with locking pins, stowed.
- VS17 marking panels for emergency marking.
- Ammunition evenly distributed and demolitions properly segregated and prepared for use.

UNIFORM-, DUFFLE-BAG-, AND RUCK-SACK-PACKING CHECKLISTS

A field uniform should include—

- A kevlar helmet with a cover and a band.
- A fragmentation vest.
- A pistol belt with ammunition pouches.
- A canteen with a cover and a cup, on the right hip.
- Ear plugs.
- A flashlight, on the right suspender strap (optional).
- A holster with a lanyard (if armed with a pistol), on the right side of the pistol belt.
- A bayonet, on the left side of the pistol belt.
- A compass, on the left side of the pistol belt.
- An M17A1 protective mask, on the left hip, and an M25 mask under the left arm.
- A battle dress uniform (BDU) with black boots.
- A field jacket.
- A chemical-decontamination kit connected to the protective mask.
- A nerve-agent antidote.
- A first-aid pouch, on the left suspender strap.

One duffle bag should contain—

- Three sets of BDUs.
- Three pairs of underwear.

- Three pairs of wool socks.
- Three T-shirts.
- One towel.
- Two laundry bags.
- One waterproof bag.
- One pair of boots with shoe polish, a brush, and a sewing kit.

The additional duffle bag should contain—

- Five tent pegs (10 for officers).
- Three tent poles (6 for officers).
- One tent rope (2 for officers).
- One sleeping bag.
- One pair of trouser suspenders.
- Two pairs of coveralls.
- One field jacket with a liner.
- One shelter half (2 for officers).
- One wool blanket.
- One pair of field pants.
- One pair of field pants with a liner.
- One cold-weather mask.
- One pair of extreme cold-weather boots.

The ruck sack should contain—

- Wet-weather pants and a parka.
- Overshoes.
- A weapons cleaning kit.
- Rations.
- A waterproof bag.
- A mission-oriented protective posture (MOPP) suit with gloves.
- Cold-weather mittens, winter use only.
- A parka with a liner and a hood, winter use only.

- A wool sweater, winter use only.
- A pile cap, winter use only.
- A scarf, winter use only.
- A towel and toilet articles.
- A pair of underwear.
- A T-shirt.
- Two pairs of wool socks.
- A BDU hat.
- An entrenching tool.

APPENDIX C

Breaching Case Studies

FM 90-13-1 and *Chapter 3* provide doctrinal overviews for breaching, breaching tenets, and breaching theory. The following case studies are designed to provide the platoon with instructive examples of dismounted and mounted breaching. Either case study could be applied to many different field applications. For example, the dismounted case study could be used to examine covert breaching, assault breaching, or deliberate breaching by light engineers. Likewise, the mounted case study could apply to both deliberate and in-stride breaching. These case studies are not designed to be prescriptive. They are designed to illustrate the thought process that the platoon leader must go through to successfully breach.

All breaching operations follow a similar pattern. They differ in technique. During breaching, the—

- Platoon leader determines the obstacle location and its composition.
- Platoon leader selects the breach site and lane locations.
- Platoon leader orders the breach to start, if the platoon is the primary reduction force.
- Squad prepares to breach, if the platoon is the backup reduction force.
- Squad breaches the obstacle.
- Squad marks the lane.
- Platoon leader reports the lane location to the company commander.

DISMOUNTED CASE STUDY

The company commander orders the dismounted platoon leader that is attached to him to breach an obstacle. The company is assaulting an enemy objective. The platoon leader moves to a location where he can see the obstacle and determines that it is a concertina fence and antitank minefield. He orders a squad forward to conduct reconnaissance and ascertain the best lane locations. The squad leader reports the locations and marks them with orange tape. The squad has determined that there are no antipersonnel mines in the minefield. The squad leader reports this to the platoon leader (see *Figure C-1, page C-2*).

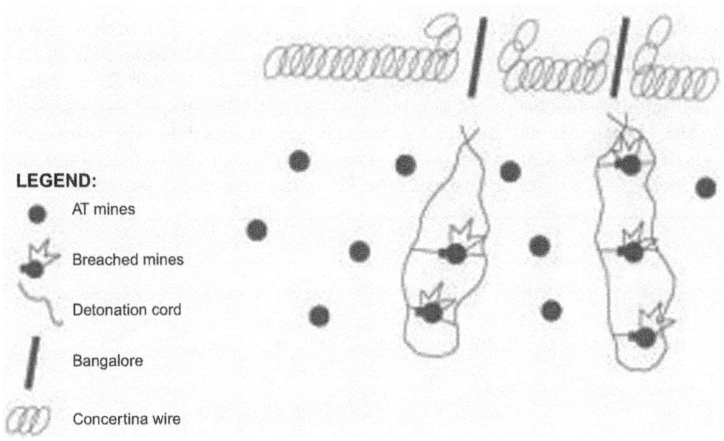

1 – The platoon leader determines the obstacle location and its composition.
2 – The platoon leader selects the breach site and lane locations.
3 – The squad breaches the concertina fence using cutters or Bangalores.
4 – The squad breaches the minefield using hand-emplaced explosives.
5 – The squad marks the lanes for the assault force.
6 – The platoon leader reports to the company commander that the lane is open.

Figure C-1. Dismounted case study

The platoon leader then orders the forward squad to remain at the obstacle as guides. He orders his second squad to come forward with two Bangalore sets. The first squad guides the second squad to the lane locations. Both squads breach the lanes in the concertina using the Bangalores and then move through the minefield placing charges to destroy antitank mines in the proposed lanes. The squads move to safety and initiate the explosives destroying the mines. Both squads provide security while the third squad comes forward to mark the lanes and guide the infantry company through the breach. The platoon leader keeps the third squad in reserve in case one of the squads fail or require additional help. However, in this scenario that was not necessary. The platoon leader observes the lanes and reports their completion to the infantry company commander who then starts his assault. The platoon leader then switches to the engineer company frequency and gives his company commander a situation report in reference to the lanes.

MOUNTED CASE STUDY

The engineer platoon is fighting as part of an engineer company/team. The company/team, which includes the engineer line platoon, the A&O platoon, and a tank platoon equipped with track-width mine plows, is the TF breach force. The company commander decides to use the tank plows, initially for speed and security, to create the breach site and secure the far side of the obstacle. Then the engineer platoon widens and marks the lanes. The A&O platoon leader places the far-lane recognition signals and initially orients the TF on the breached lanes (see *Figure C-2).*

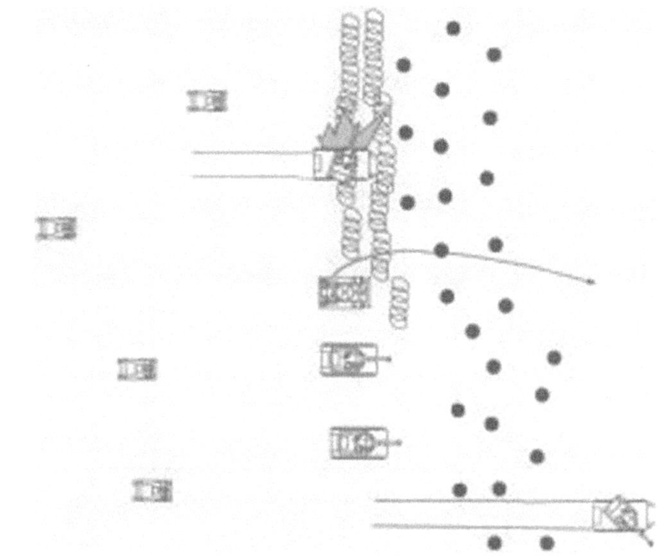

1 – The platoon leader determines the obstacle location and its composition.

2 – The platoon leader selects the breach site and lane locations.

3 – The tank crew breaches the obstacle with the tank plows.

4 – The platoon leader orders the AVLM forward to breach with the MICLIC when the enemy destroys a tank plow.

5 – The squad marks the lanes for the assault force.

6 – The platoon leader reports to the company commander that the lane is open.

Figure C-2. Mounted case study

In this scenario, the engineer company commander orders the engineer platoon forward to identify the obstacle location and its composition and potential lane locations. The platoon leader orders a dismounted element forward to conduct his reconnaissance. He reports his reconnaissance to the company commander. He states that the obstacle is a mixed minefield with antitank and antipersonnel mines, and a partial, multiple row of concertina fence is forward of the minefield. The platoon leader also reports two potential breach sites with an alternate site between them.

The engineer company commander orders the tank platoon to breach two lanes, based on the engineer platoon leaders' reconnaissance. He orders the engineer platoon to be prepared to back up the tank platoon. The northern tank equipped with a mine plow is destroyed by direct fire as it starts to breach. The engineer platoon leader observes this and orders his AVLM forward to breach with the MICLIC. The AVLM and the other tank plow successfully breach the obstacle.

The platoon leader orders two squads forward to mark each lane. These squads widen the lanes, per the initial plan, with hand-emplaced explosives after the assault force passes through. The platoon leader keeps one squad in reserve, with the AVLM, to counter any enemy attempts to close the lanes with artillery-delivered mines. He reports to the engineer company commander that the lanes are marked.

Glossary

&	and
+	plus
#	number
%	percent
1SG	first sergeant
1st	first
2d	second
3d	third
A&O	assault and obstacle
AA	axis of advance
AAR	after-action review
ABF	attack by fire
AC	hydrogen cyanide
ACE	armored combat earthmover, M9
AD	air defense
ADA	air-defense artillery
ADAM	area denial artillery munition
AHD	antihandling device
ALL	additional authorized list
ammo	ammunition

AP	antipersonnel
APC	armored personnel carrier
app	appendix
arty	artillery
ASL	authorized stockage list
aslt	assault
AT	antitank
AT-3, AT-5, AT-8	antitank missiles
ATD	antitank ditch
attn	attention
AVLB	armored vehicle-launched bridge
AVLM	armored vehicle-launched MICLIC
BDU	battle dress uniform
BII	basic-issue items
BMO	battalion maintenance officer
BMP	Soviet tracked amphibious infantry combat vehicle
BMP-1, BMP-2	Soviet infantry fighting vehicles
BOS	Battlefield Operating Systems
BSA	brigade support area
BT	blade team
C^2	command and control
CA	combined arms
CAM	chemical agent monitor
CATK	counterattack
CDR	commander

CEV	combat engineer vehicle
CG	phosgene
℄	centerline
COA	course of action
CP	checkpoint
CSS	combat service support
cu	cubic
DA	Department of the Army
decon	decontaminate/decontamination
DP	dispersing point
DS	direct support
DSA	division support area
EA	engagement area
EPW	enemy prisoner of war
FASCAM	family of scatterable mines
FEBA	forward edge of the battle area
FIST	fire-support team
FLOT	forward line of own troops
FM	field manual
FMFM	fleet Marine force manual
FS	fire support
FSO	fire-support officer
ft	foot (feet)

GA	tabun
GB	sarin
GD	soman
GEMSS	Ground-Emplaced Mine-Scattering System
GS	general support
HD	distilled mustard
HDP	hull-down position
HEMTT	heavy expanded mobility tactical truck
HL	mustard lewisite
HMMWV	high mobility multipurpose wheeled vehicle
HQ	headquarters
hr	hour(s)
IOE	irregular outer edge
IPB	intelligence preparation of the battlefield
IR	intelligence requirements
km	kilometer(s)
lb	pound(s)
LD	line of departure
LOGPAC	logistical package
LRA	local reproduction authorized
LRP	logistics release point
m	meter(s)

M-S	Miznay-Schardin
METT-T	mission, enemy, terrain, troops, and time available
MICLIC	mine-clearing line charge
MIJI	meaconing, intrusion, jamming, and interference
min	minute(s)
mm	millimeter(s)
MOPMS	Modular Pack Mine System
MOPP	mission-oriented protective posture
MP	military police
MRB	motorized rifle battalion
MRC	motorized rifle company
MRE	meal, ready-to-eat
MRP	motorized rifle platoon
MSR	main supply route
MST	maintenance support team
N	no
NA	not applicable
NAI	named area of interest
NBC	nuclear, biological, chemical
NCO	noncommissioned officer
NCS	net control station
No	number
NP	nonpersistent
OBSTINTEL	obstacle intelligence
OCOKA	observation and fields of fire, cover and concealment, obstacles, key terrain, and avenues of approach

OP	observation post
OPCON	operational control
OPLAN	operation plan
OPORD	operation order
OPSEC	operations security
ORP	objective rally point
oz	ounce(s)
P	persistent
PCC	precombat check
PCI	precombat inspection
PDM	pursuit-deterrent munition
PIR	priority intelligence requirements
PL	phase line
PLD	probable line of deployment
PLL	prescribed load list
plt	platoon
PMCS	preventive maintenance checks and services
POL	petroleum, oils, and lubricants
psn	position
R&S	reconnaissance and surveillance
RAAM	remote antiarmor mine
recon	reconnaissance
reqd	required
S&P	stake and platform
S1	Adjutant (US Army)

S2	Intelligence Officer (US Army)
S3	Operations and Training Officer (US Army)
S4	Supply Officer (US Army)
SCATMINE	scatterable mine
sec	second(s)
SEE	small emplacement excavator
SHELREP	shelling report
SITREP	situation report
SOI	signal operation instructions
SOP	standing operating procedure
SOSR	suppress, obscure, secure, reduce
SPOTREP	spot report
T/C/S	tasks, conditions, and standards
T72/80	Soviet tank
TA	table of allowance
TC	tank commander
TCP	traffic-control post
TDP	turret-down position
TF	task force
TLP	troop-leading procedures
TOC	tactical operations center
TRP	target reference point
TTP	tactics, techniques, and procedures
UMCP	unit maintenance collection point
US	United States

USAF	United States Air Force
VX	VX
w/	with
WO	warning order
WP	white phosphorous
wt	weight
XO	executive officer
Y	yes

Index

References

SOURCES USED

These are the sources quoted or paraphrased in this publication.

Army Publications

FM 3-4. *NBC Protection* (FMFM 11-9). 29 May 1992.

FM 3-5. *NBC Decontamination* (FMFM 11-10). 17 November 1993.

FM 5-34. *Engineer Field Data*. 14 September 1987.

FM 5-100. *Engineer Combat Operations*. 22 November 1988.

FM 5-103. *Survivability*. 10 June 1985.

FM 7-8. *Infantry Rifle Platoon and Squad*. 22 April 1992.

FM 20-32. *Mine / Countermine Operations*. 30 September 1992.

FM 71-100. *Division Operations*. 16 June 1990.

FM 90-7. *Combined Arms Obstacle Integration*. 29 September 1994.

FM 90-13-1. *Combined Arms Breaching Operations*. 28 February 1991.

FM 21-60. *Visual Signals*. 30 September 1987.

DOCUMENTS NEEDED

These documents must be available to the intended users of this publication.

Department of the Army Forms

DA Form 1155. *Witness Statement on Individual*. 1 June 1966.

DA Form 1156. *Casualty Feeder Report*. 1 June 1966.

DA Form 1355. *Minefield Record*. March 1987.

DA Form 1355-1-R. *Hasty Protective Minefield Record (LRA).* July 1975.

DA Form 2028. *Recommended Changes to Publications and Blank Forms.* 1 February 1974.

DA Form 2404. *Equipment Inspection and Maintenance Worksheet.* 1 April 1979.

***FM 3-34**(FM 5-100)

Field Manual
No. 3-34

HEADQUARTERS
DEPARTMENT OF THE ARMY
Washington, DC, 2 January 2004

Engineer Operations

Contents

---

*This publication supersedes FM 5-100, 27 February 1996, and FM 5-114, 13 July 1992.

Page

Page

Page

Page

Preface

Field Manual (FM) 3-34 is the Engineer Regiment's capstone manual for operating in today's operational environment within the framework of the Army transformation and, although focused at the operational level, is applicable for all levels of war. Engineers are a combat arm (a branch comprised of combat, combat support [CS] and combat service support [CSS] components) that enables joint and maneuver commanders to achieve their objectives through strategic movement and tactical maneuver by providing unique combat, geospatial, and general engineering capabilities. It has been the engineer creed to support the maneuver commander since June 16, 1775, when the Continental Congress organized an Army with a chief engineer and two assistants. Engineers contributed to the hardest fought battles in the Revolutionary War, including Bunker Hill, Saratoga, and the final victory at Yorktown. At the end of the Revolutionary War, the engineers were mustered out of service. However, their unique skills were realized and they were called back to active duty in 1794 when Congress organized a Corps of Artillerists and Engineers and later in 1802 as a separate Corps of Engineers. Today's *FM 3-34* includes engineer doctrine that has evolved for over 200 years.

FM 3-34 is the capstone doctrinal manual for engineer operations and is linked to joint and Army doctrine to ensure its usefulness for all joint and Army level commanders and staff. All other engineer FMs *(see Appendix A)* are based on the principles and tenets found in this manual and are synchronized with their respective joint publications. These principles and tenets are founded on the successful employment of engineers, past and present. In today's complex operational environment, the engineers' warfighting focus produces a full-spectrum force that meets the needs of the land component commander (LCC) in war, conflict, and peace.

The primary audience for *FM 3-34* is engineer commanders and staffs down to and including engineer companies, maneuver force commanders, and battalion and task force (TF) organizations. The focus includes Army Service Component Command responsibilities for conducting operations as part of a multinational force. Information contained in this manual will assist multinational forces and other services and branches of the Army to plan and integrate engineer capabilities. This doctrine also will assist Army branch schools in teaching the integration of engineer capabilities into Army operations. Engineer involvement is a given for nearly every military operation. The degree of involvement will include one or more of the roles associated with engineers performing combat, CS, or CSS missions.

While the nature of war remains constant throughout history, the conduct of war is continually changing in response to new concepts, technologies, and requirements. The contemporary threat is continually evolving and adapting to friendly engineer capabilities. No matter how many engineer capabilities are embedded into Army systems, it is the engineer soldier that must recognize shortfalls and develop new concepts and methods to overcome any doctrine, organization, training, materiel, leadership and education, personnel, and facilities (DOTMLPF) shortfalls. Failure to adapt to these changes may result in the engineer branch not being ready to confront the challenges of future threats. Therefore, we must recognize that it is the adaptable and professional engineer soldiers of the Regiment that are most important to our future. *FM 3-34* furnishes the authoritative foundation for subordinate engineer doctrine and terminology, force design, materiel acquisition, professional education, and individual and unit training. This manual introduces several new terms, including assured mobility, geospatial engineering, maneuver support (MANSPT) *(see Appendix B)*, and field force engineering (FFE) *(see Appendix C)*.

FM 3-34 is built directly on the concepts of *FMs 3-0, 3-90, and 3-07* blending key points of *Joint Publications (JPs) 2-03, 3-0 3-15, 3-34, and 4-04* into its approach to ensure that Army elements of a joint force use all engineer assets to their fullest extent. Given the magnitude of doctrinal changes in recent years, you will need to be familiar with these documents to effectively use *FM 3-34*. This manual addresses engineer roles and functions within a multinational operation, under potentially multinational or interagency leadership and within diverse command relationships. Finally, this manual focuses on the key functional planning considerations for employing engineers at the strategic, operational, and tactical levels of war.

The proponent for this publication is HQ TRADOC. Send comments and recommended changes on Department of the Army (DA) Form 2028 directly to Commandant, United States Army Engineer School (USAES), ATTN: ATSE-DOT-DD, Directorate of Training, 320 MANSCEN Loop, Suite 336, Fort Leonard Wood, Missouri 65473-8929.

NOTE: The bibliography lists FMs by the new number, followed by the old number in parentheses.

Unless this publication states otherwise, masculine nouns or pronouns do not refer exclusively to men.

Chapter 1

The Army and the Role of Engineers

Essayons (Let us try!)
--Motto of the Corps of Engineers

The Army organizes, trains, and equips its forces to fight and win the nation's wars and achieve directed national objectives. The Army also protects the nation's sovereign borders and national interests against aggressors. The engineer's role in the Army's mission of national defense is critical. This chapter outlines the role of the engineer and the mission essential tasks doctrinally performed by engineers, defining the engineer battlespace functions across the spectrum of conflict. This chapter also discusses how engineer leaders interact not only with other Army forces but also with joint, interagency, and multinational organizations to perform the engineer battlespace functions.

SECTION I - THE ENGINEER REGIMENT

1-1. Army forces are normally the decisive component of land warfare in joint and multinational operations. The engineers are Army enablers for success in these operations. They organize and fight with the Army's maneuver forces to win the nation's wars and achieve its national objectives. They also support the Army's ability to deter war by maintaining their deployability and war-fighting skills, while simultaneously leading the world in advances in technology, such as standoff detection systems, intelligent munitions, and doctrinal concepts such as assured mobility. The engineer force is always ready to respond as a combat-ready force prepared to deal with the full spectrum of potential operations. Engineer forces can be tailored to support operations in austere environments, with little or no infrastructure, and provide mobility and enhance force protection through countermobility and survivability. They also provide geospatial (formerly topographic engineering) and general engineering support across the spectrum of potential operations. Engineers support light, heavy, and special operations forces (SOF) in all types of terrain and operational environments (OEs). This provides the joint forces' land, sea, and air component commander the greatest flexibility to package a force that can rapidly deploy, assist in deterring adversaries, and preclude our enemies from gaining an operational advantage in an area of operations (AO). Engineers are trained to operate with and support other service requirements, integrate with other service engineers and contractors, and assume command and control (C2) of other services or branches.

1-2. As Army forces fight and win the nation's wars, they also deter them. The Army's war-fighting focus enables a diverse (full-spectrum) force to meet the needs of the joint force commander (JFC) in war, conflict, and peace. In war, Army forces form the nucleus of the land component—imposing their will on enemies and causing them to collapse. In conflict, Army forces deploy quickly into an AO to deter adversaries and potential enemies from establishing their forces and to preclude them from gaining an operational advantage. If deterrence fails, Army forces defeat the enemy, terminate conflict to achieve national objectives, and establish self-sustaining postconflict stability. Early movement of Army forces retains initiative and freedom of action by providing the JFC complementary means to conduct decisive offensive operations at a time and place of the commander's choosing. If theater circumstances require it, Army forces provide the means to block the enemy's offensive and deliver the counteroffensive blow necessary to win as rapidly as possible. In peace, Army forces train for war and provide military support to civil authorities when necessary. They also help shape the international security environment through engagement activities and nationally, they provide support to civilian authorities both at home and abroad in response to homeland security (HLS) for natural or man-made disasters. Regardless of the type of commitment of Army forces, the degree of engineer participation is likely to be high.

1-3. The Engineer Regiment contributes to the Army's war-fighting abilities and focus. It consists of all Active Component (AC) and Reserve Component (RC) engineer organizations (as well as the Department of Defense [DOD] civilians and affiliated contractors and agencies within the civilian community) with a diverse range of capabilities. The Chief of Engineers leads the Engineer Regiment and is triple-hatted as the chief of the engineer branch, the staff officer advising the Chief of Staff of the Army (CSA) on engineering matters, and the Commander of theUnited States Army Corps of Engineers (USACE). The AC of the Engineer Regiment consists of USACE and AC military engineer units within the combatant commands (COCOMs) and major Army commands (MACOMs). The RC consists of the Reserve and National Guard and provides support to the theater engineer commands (ENCOMs). The RC engineer force consists of more than three fourths of all military engineer forces and includes a wide range of specialized capabilities. Additionally, certain types of units are found only in these two components. For example, facility engineer detachments (FEDs) are only found in the RC. The Regiment is joint in its integration capabilities and supports the planning, preparing, and executing of joint operations. The Regiment is experienced at interagency support and leveraging nonmilitary and nongovernmental engineer assets to support mission accomplishment. At the operational/strategic level, the Regiment is represented as shown in *Figure 1-1*. The Regiment is represented by the various engineer organizations and capabilities reflected in *Figure 1-2, page 1-4* at the tactical/operational level. *Appendix D* provides a more in-depth view of the organizations depicted in *Figure 1-2*.

THE THREEFOLD BRANCH

1-4. The main component of the Engineer Regiment is the Engineer Branch. Engineer officers and engineer enlisted soldiers with combat engineer military occupational specialties (MOSs) are combat arms soldiers. The Engineer Branch reinforces and complements the effects of the other branches

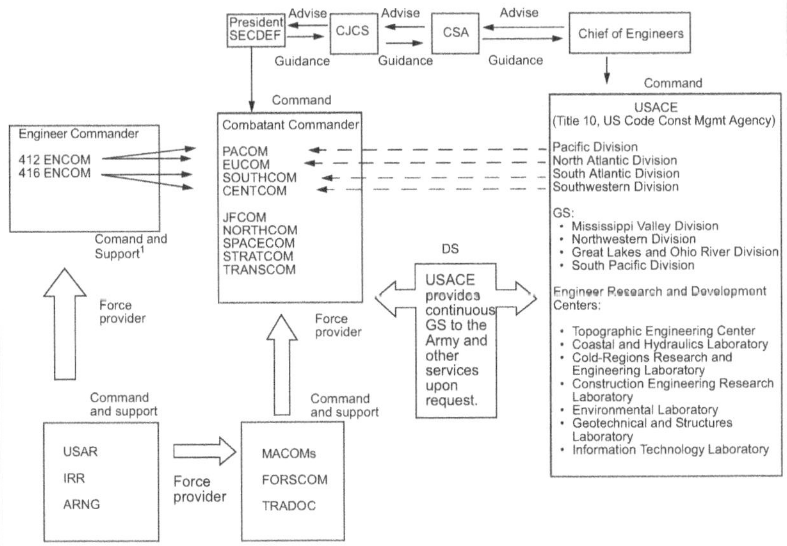

[1]Command and support to the land component commander for the combatant commander.

Figure 1-1. The Engineer Regiment at the Strategic/Operational Level

in the Army and the other services. While most branches are identified as being either combat, CS, or CSS, engineers are identified in all of these categories and have significant overlap within the branch and the roles, missions, and functions that they perform. The only other branch that shares this distinction and challenge is the Aviation Branch.

COMBAT ARMS

1-5. Combat arms are those units and soldiers who close with and destroy enemy forces or provide firepower and destructive capabilities on the battlefield. Many engineer units perform combat arms roles. The commander task-organizes combat engineer units with maneuver units and integrates them into a combined arms formation. The engineer units provide demolition and reduction capabilities to the combined arms team. Additionally, engineer units can fire and maneuver to employ direct-fire weapons systems to aid in employing obstacles and breaching obstacles. Regardless of the mission, armored engineer vehicles are combat vehicles and provide a significant contribution to the combat power of the entire formation.

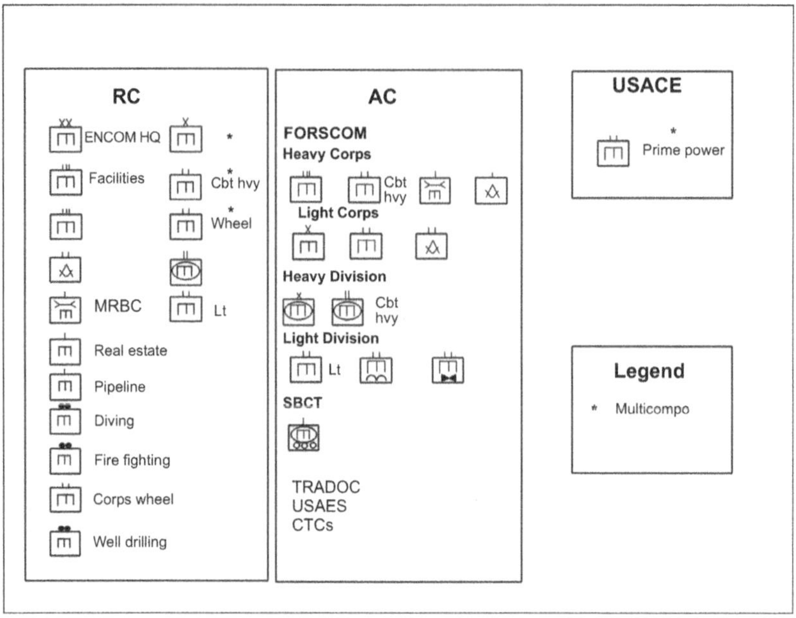

Figure 1-2. The Engineer Regiment (by type) at the Tactical/Operational Level

1-6. Combat engineer units have a secondary mission to fight as infantry. Reorganizing and employing engineers as infantry requires serious consideration. Engineers provide far more combat power in their primary mission than when reconfigured as infantry. Stopping engineer work may reduce the combat power of the commander's entire force. Because of the long-term impact, reorganizing engineers to fight as infantry is an operational issue and careful consideration should be given to what level of command is authorized to order reorganization (normally corps commander or higher). Reorganization implies withdrawing engineer units to a rear area to issue infantry equipment (such as antitank [AT] systems) and providing them with other combined arms maneuver battalion assets (such as a fire support team and enhanced combat medical support). It also implies a period of training with these newly acquired assets.

COMBAT SUPPORT

1-7. CS engineer elements provide the maneuver commander with significant amounts of additional combat power and capabilities not typically organic to the maneuver division. These elements support the maneuver commander using established command and support (C&S) relationships. CS engineers include combat heavy battalions, corps combat battalions, and bridge companies. CS encompasses critical combat functions provided by units and soldiers, in conjunction with combat arms units and soldiers, to secure victory. Engineers operate as integral members of the combined arms team to provide full combat, geospatial, and general engineering capabilities. CS engineers may also advise the maneuver commander on the effective use of terrain; construction efforts; improvement and maintenance of routes, bridges, and airfields; and reorganization to fight as infantry, when required.

1-8. In offensive operations CS engineers concentrate their efforts to support the maneuver by reducing and crossing obstacles and gaps, assisting in the assault of fortified positions, and emplacing obstacles to protect the flanks of friendly attacking forces. In defensive operations, CS engineers reinforce the terrain to anchor the defense in critical areas, maximize the effects of defenders' fires, provide maximum protection to friendly fighting positions, and facilitate the movement of counterattack forces. They may also upgrade routes and construct life support areas.

1-9. Geospatial capabilities are typically a CS element represented by topographic battalions, companies, and terrain teams that manage the flow of geospatial information from the unit level to the level of national service organizations, such as the National Imagery and Mapping Agency (NIMA). Topographic units develop geospatial products (to include detailed terrain analysis products, maps, and digital terrain data) that assist commanders from company to theater levels. Geospatial information and services assist the commander with building a common operational picture (COP) by identifying avenues and routes, covered and concealed areas, obstacle locations, key terrain features, engagement areas (EAs), unit positions, and possible target areas.

COMBAT SERVICE SUPPORT

1-10. CSS engineers sustain the momentum of the force they support. The primary role of tactical CSS units is to sustain Army forces. The CSS effort is successful only when it concentrates and supports forces by focusing on sustaining and reconstituting tactical units. General engineering is the primary CSS engineer function. It involves a variety of roles—civil engineering, fire fighting, maintaining lines of communications (LOCs), employing force protection, assessing area damage control (ADC), managing inland waterways, generating power, and providing environmental support. Much, but not all, of the Regiment's reach capabilities tend to be in the category of CSS.

1-11. CSS engineers at the operational level include theater construction engineering; real estate support; contract services; facilities development; well drilling; base camp development and management; and pipeline, rail, and port construction. USACE elements actively participate in support of engineer CSS functions.

MISSION-ESSENTIAL TASKS

1-12. The Army's mission-essential tasks derive from statutory requirements, operational experience, strategies for the employment of the nation's military forces, and operational requirements of the combatant commanders. They are the operational expression of the Army's core competencies contained in *FM 1 (100-1)*. Army forces develop their battle-focused mission-essential task lists (METLs) in concert with *FM 7-15*. To perform these tasks, the Army continuously integrates DOTMLPF in accordance with Chairman of the Joint Chiefs of Staff Instruction (CJCSI) (S) 3010.02A.

1-13. The six Army mission-essential tasks are—

- Shape the security environment.
- Respond promptly to crisis.
- Mobilize the Army.
- Conduct forcible entry operations.
- Dominate land operations.
- Provide support to civil authorities.

1-14. Engineers are critical to meeting these Army mission-essential tasks. These tasks use engineers in unified action with the other services and in conjunction with multinational and interagency organizations. Engineers are primary players as the Army responds to missions across the spectrum of operations and in line with their mission-essential tasks. Army forces expand the JFC's range of military options in full-spectrum operations, the range of operations Army forces conduct in war and military operations other than war (MOOTW).

1-15. The Engineer Regiment's mission-essential tasks derived from the Army's METL are—

- Shape the security environment.
- Respond promptly to crisis.
- Mobilize engineer forces.
- Support forcible entry operations.
- Support assured mobility to dominate land operations.
- Provide support to civil authorities.
- Provide quality, responsive engineering services to the nation.

SHAPE THE SECURITY ENVIRONMENT

1-16. Engineers shape the security environment by supporting the combatant commander's vision. They engage other nations by working with allies and potential coalition partners in engineer institutions, by participating in international exercises, and by supporting stability operations in foreign countries. Engineers can often lead combatant commander engagement initiatives through military-to-military contact, instruction, or construction projects.

RESPOND PROMPTLY TO CRISIS

1-17. Engineers support the LCC to organize operations in time and space to present the enemy with simultaneous and multidimensional threats via land, air, sea, and space. Some of these forces will come from elements stationed in Germany, Korea, and other outside the continental United States (OCONUS) locations. Engineer forces are ready to deploy by land, sea, air, and rail quickly and with all their capabilities. As combatant commanders develop plans to enter an AO rapidly, engineers provide a myriad of capabilities, such as joint rapid airfield construction and repair, port construction, diving, bridge and road construction/repair, and prime power support. These capabilities are provided by modular packages that meet the specific needs of the mission. Engineers also contribute to strategic responsiveness by providing geospatial terrain products for the commander to visualize the land dimension early as a foundation for base planning.

MOBILIZE ENGINEER FORCES

1-18. Engineers maintain the preponderance of their military force in the United States Army Reserve (USAR) and Army National Guard (ARNG). These reserve forces provide unique skills that in some cases are only, or primarily, in the RC (well drilling, fire fighting). RC engineers are ready to deploy and fulfill the combatant commander's needs or requirements for war or national emergencies. Engineer RC forces provide the Secretary of Defense (SECDEF) a means to expand the Army to confront unforeseen threats to national security. RC forces can also provide peacetime mobilization of engineer units to support HLS and other combatant commander engagement policies.

SUPPORT FORCIBLE ENTRY OPERATIONS

1-19. Engineers provide the Army combat (mobility, countermobility and survivability [M/CM/S]), geospatial, and general engineering capabilities to strike contested areas from the air, land, and sea. Engineer forces support establishing lodgments to project follow-on forces and sustain the maneuver force. During the Cold War, forces fell in on predeployed equipment and supplies in mature theaters with host nation support (HNS). Since then, deployments have become increasingly expeditionary, which requires a greater engineering effort to develop secure ports and forward operating bases (FOBs) in austere or underdeveloped regions with largely unskilled indigenous labor. Engineer forces with airborne and air assault forces support assured mobility by restoring seized airfields and developing storage sites. Engineers can also support Army forces involved with amphibious operations with general engineering support for logistics-over-the-shore (LOTS) operations.

SUPPORT ASSURED MOBILITY TO DOMINATE LAND OPERATIONS

1-20. Army forces are the dominant land force in the world. Engineers support decisive, shaping, and sustaining operations to facilitate success. They support assured mobility to maneuver forces to close with and destroy the enemy to achieve campaign objectives in support of national objectives. Geospatial engineering provides the combatant commander with detailed terrain and other geospatial products that facilitate visualization of the AO. Combat engineering provides freedom of movement to maneuver forces throughout the AO by providing bridging, clearing, detecting, and neutralizing mines; and removing natural and man-made obstacles. Engineers provide unimpeded movement through terrain and interdict threat mobility from interfering with Army maneuvers. They enhance force protection for vulnerable assets by providing survivability support. They provide general engineering support from the joint level down to the most forward tactical units.

PROVIDE SUPPORT TO CIVIL AUTHORITIES

1-21. The Army can adapt and tailor its war-fighting capabilities to complement and support civil authorities and agencies at home (under the guidance of the President and SECDEF) and abroad (according to the Status of Forces Agreement [SOFA] and other formal agreements between the United States [US] and other host nations [HNs]). Engineers provide a full range of engineering capabilities and expertise to reinforce or fill critical requirements beyond the immediate capabilities of civil authorities and agencies. Army engineers are uniquely postured to provide this support by accessing the analytical and construction expertise of USACE, ENCOMs, USAES, and other engineer organizations through TeleEngineering and other consultation. This reach consultation, in combination with unique engineer unit capabilities, can enable commanders to restore basic support infrastructure (transportation networks, utility systems, and facilities), alleviate refugee suffering, and reinstate local governmental authority. Engineers in the AC and RC forces have robust capabilities to support disaster relief, crisis resolution, and HLS. Capabilities include, but are not limited to, prime power, bridging, fire fighting, well drilling, and the broad spectrum of FFE. Engineers continue sustained support until civil authorities no longer require the Army's assistance. USACE supports this requirement with specialized personnel. Engineers, while supporting full-spectrum operations, may have to support civil authorities while simultaneously conducting combat operations. Engineers at the joint level are linked to interagency elements and facilitate a seamless transition between operations.

PROVIDE QUALITY, RESPONSIVE ENGINEERING SERVICES TO THE NATION

1-22. Engineers support the nation by planning, designing, building, and operating water resources and other civil-works projects (navigation, flood control, environmental protection, and disaster response) to facilitate national requirements and prevent disasters. USACE, as the Title 10 Construction Management Agent, also designs and manages the construction of military facilities for the Army and the Air Force. Engineers also provide design and construction management support for other DOD and federal agencies. This is a critical responsibility for interagency and international services.

1-23. Engineers support the Regiment's METL through operations characterized by the engineer battlespace functions at the operational level addressed in *JP 3-34*. Combat engineering enables the commander to maneuver forces freely (mobility), attack the enemy's ability to maneuver (countermobility), and support force protection (survivability). Geospatial engineering provides key geospatial information and services, including analysis products of the terrain within the operational area to enhance the picture within and throughout the theater. General engineering provides for the facilities and infrastructure that play a critical role in shaping the AO and are essential to force projection for decisive operations. Army engineer units maintain critical capabilities within each of these functions that independently may support, or be combined or leveraged with, other service engineers to support joint operations.

1-24. For a closer look at how the engineers perform these functions across the battlefield operating systems (BOSs), *see FM 7-15*. While geospatial engineering is not specifically identified in *FM 7-15* under a particular BOS, it is generally aligned with and supports the intelligence BOS. All of these relationships are further discussed in *Chapter 4* of this manual. Remember that the Army Universal Task List (AUTL) is focused on the tactical level. For linkage to joint operations, refer to the *Universal Joint Task List (UJTL), CJCSM 3500.04C*. The relationship of geospatial engineering to intelligence is better documented in this manual, *CJCSM 3500.04C*.

FULL-SPECTRUM OPERATIONS

1-25. Army doctrine includes the range of full-spectrum operations (*see Figure 1-3, page 1-10*). The 2001 Quadrennial Defense Review (QDR) describes a capabilities-based approach to defense planning that provides broader military options across the operational spectrum, from pre- to postconflict operations. The force-sizing construct–1-4-2-1–takes into account the number, scope, and simultaneity of tasks assigned to the military. It sizes the force for the defense of the US homeland by (1) providing forward deterrence in four critical regions, (4) conducting simultaneous war-fighting missions in two regions, (2) preserving the President's option to call for decisive victory in one of those conflicts, (1) and participating in multiple, smaller contingency operations. Army commanders at all echelons may combine different types of operations simultaneously and/or sequentially to accomplish missions in war and MOOTW. For each mission, the JFC and Army component commanders determine the emphasis Army forces place on each type of operation. Offensive and defensive operations normally dominate military operations in war and some smaller-scale contingencies (SSCs). Stability and support operations will tend to dominate in operations that include certain SSCs and peacetime military engagements (PMEs) and in MOOTW.

1-26. Full-spectrum operations include offensive, defensive, stability and support operations (*see Figure 1-4, page 1-11*). Missions in any contemporary environment require Army forces to conduct or be prepared to conduct any combination of the following primary operations.

OFFENSIVE OPERATIONS

1-27. Offensive operations aim at destroying or defeating an enemy. Their purpose is to impose US will on the enemy for decisive victory. The primary focus for engineers enabling operational and tactical maneuver through combat engineering support to lines of communication and combined arms and breaching river crossing operations.

DEFENSIVE OPERATIONS

1-28. Defensive operations defeat enemy attacks, buy time, economize forces, and develop conditions favorable for offensive operations. Defensive operations alone normally cannot achieve a decisive victory. Their purpose is

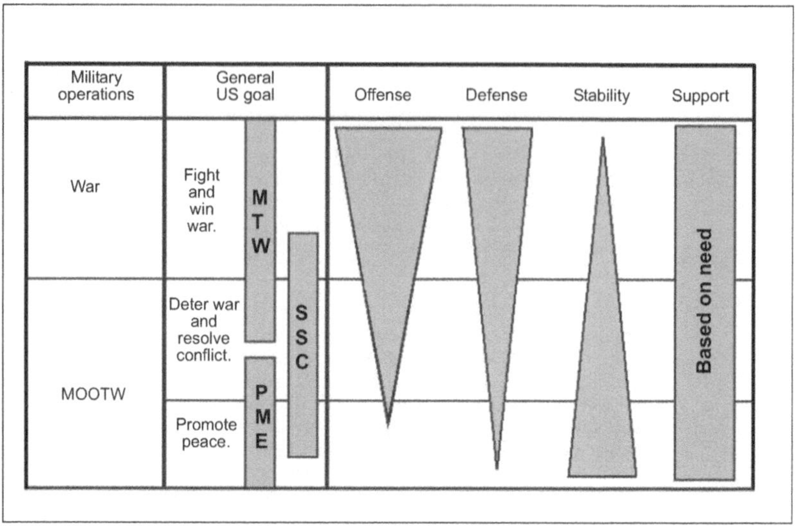

Figure 1-3. Range of Army Operations

to create conditions for a counteroffensive that will regain the initiative. The primary focus for engineers is on combat engineering to enable combined arms obstacle integration and ensure mobility to friendly repositioning or counterattacking forces.

STABILITY OPERATIONS

1-29. Stability operations promote and protect US national interests by influencing diplomatic, civil, and military environments. Engineers conducting missions provide stability, security, and resources to assist in disaster or theater response. Rapid and effectively emplaced sustainment operations reduce human injuries/fatalities and harden infrastructure. Regional security is supported by a balanced approach that simultaneously enhances regional stability and economic prosperity. Army force presence promotes a stable environment. Engineers are focused on assisting in stabilizing a region by improving the infrastructure and integrating with and supporting maneuver forces. While engineers may be focused on one or more of the engineer functions, all functions operate simultaneously in support of stability operations.

SUPPORT OPERATIONS

1-30. Support operations employ Army forces to assist civil authorities, foreign or domestic, as they prepare for or respond to crises and relieve suffering. Domestically, Army forces respond only when directed by the SECDEF. ARNG forces may be activated by order of the state governor.

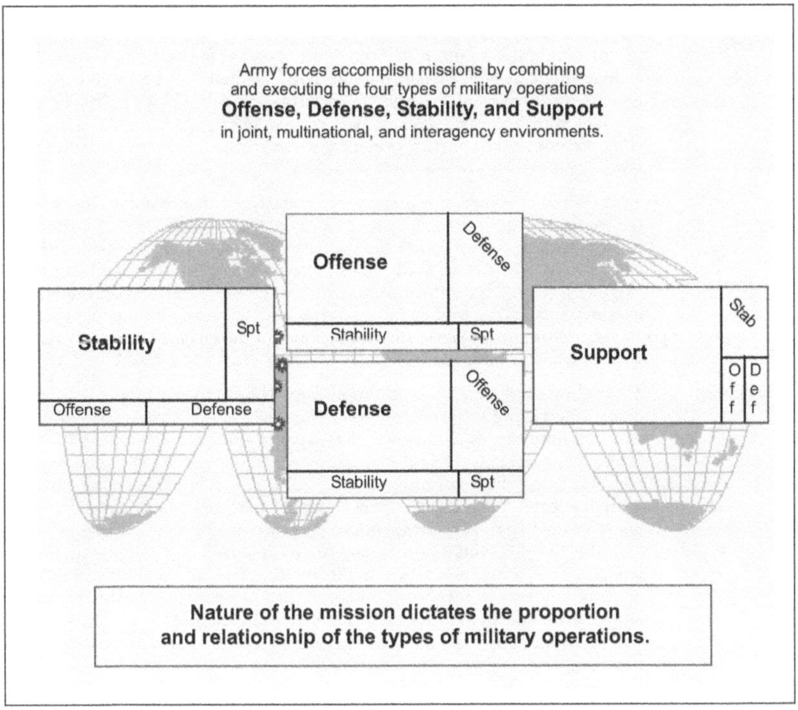

Figure 1-4. Full-Spectrum Operations

Engineers conducting missions provide stability, security, and resources to assist in disaster or theater response. Rapid and effectively emplaced support operations reduce human injuries and fatalities and harden infrastructure. Army forces operate under the lead federal agency (LFA) and comply with provisions of US law, to include the Posse Comitatus Act and the Stafford Act. Typically, the focus for engineers will be on general and geospatial engineering, although all of the engineer functions may be required.

CONDUCTING FULL-SPECTRUM OPERATIONS

1-31. When conducting full-spectrum operations, commanders combine and sequence the four types of operations, as appropriate, to accomplish the mission. As missions change from promoting peace to deterring war, and from resolving conflict to engaging in war, the combinations of and transitions between these operations require skillful planning, preparation, and execution. Engineers may also be required to support civil authorities conducting HLS. Missions can range from providing disaster relief support and humanitarian assistance to urban search and rescue. Operations designed to accomplish more than one strategic purpose may be executed simultaneously, sequentially, or both. For example, within a combatant commander's area of responsibility (AOR), combat and CS engineers may be executing large-scale offensive operations while CS and CSS engineers conduct stability operations. It is very possible that within the combat zone (CZ), various engineer units may also be simultaneously participating in stability operations or support operations.

1-32. When conducting full-spectrum operations, commanders must balance environmental protection and mission requirements. Military environmental-protection principles do not necessarily override other operational factors; rather they are a standard part of the military decision-making process (MDMP). The mission parameters for the operational area identify and quantify the time and resources devoted to environmental protection. Commanders must analyze environmental considerations and impacts in concert with mission requirements and force protection *(see FM 3-100.4).*

1-33. Commanders allocate different proportions of their force to each type of operation during different phases of the mission. Large units are likely to conduct simultaneous offensive, defensive, stability, and support operations. Units at progressively lower echelons receive missions that require fewer combinations. For example, an ENCOM supporting the joint force land component may allocate two corps engineer battalions to the attack (offense), while a third engineer battalion assists in securing and upgrading a port and airfield complex that sustains the force (defense). Around the airfield and port, engineer units may assist in distributing food and providing facilities for medical support to refugees (support). Still other corps engineer units may support the equipping and training of HN forces (stability). Engineer units will be supporting each of these operations with one or more of the engineer functions.

1-34. In the same manner that commanders combine and sequence the four types of military operations, engineers will contribute varying portions of each of the three engineer battlespace functions (combat, general, and geospatial) to each of the four types of military operations. The type, variety, and relative percentages will vary from mission to mission across the spectrum of conflict. *Figure 1-5* highlights the generic relationships between and among the engineer functions and the four types of military operations.

1-35. Everyday, the Army trains soldiers and units while developing leaders. Battle-focused training (BFT) for combat, CS, and CSS tasks prepares soldiers, units, and leaders to fight and win the nation's wars. Training is the cornerstone of success. It is a full-time job for commanders in peacetime, and it continues while deployed in full-spectrum operations as well. Training to high standards is essential for a versatile force. Every commander, soldier, and unit in a strategically responsive Army must be trained and ready to deploy, fight, and win. Upon alert, engineer forces deploy immediately, conduct operations, and complete any mission-specific training in country, if necessary. Commanders conduct abbreviated mission rehearsal exercises (MREs)... if necessary, based on available time and resources. The expectation is that engineer units train, alert, deploy, and sustain.

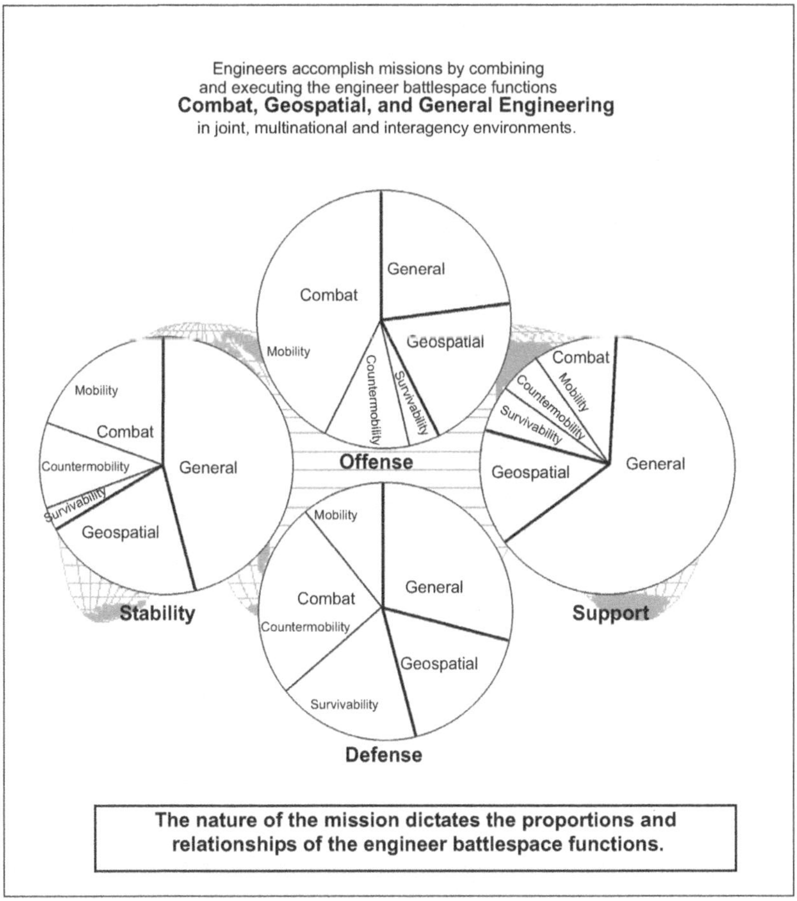

Figure 1-5. Engineer Contributions Across the Spectrum of Conflict

1-36. The METL development process outlined in *FM 7-0 (25-100)* and *FM 7-1 (25-101)* describes the link between mission and training. Commanders focus their METL, training time, and resources on combat tasks, unless directed otherwise. Because Army forces face diverse threats and mission requirements for full-spectrum operations, commanders may need to adjust their METL temporarily from battle-focused tasks and manage their training focus and resources to attain a readiness posture for anticipated missions. MACOM and numbered Army and corps commanders determine the battle focus, resources, and METL that maintain the required readiness posture for anticipated offense, defense, stability, and support (ODSS) operations in war or MOOTW. MACOM commanders decide for operational-level units, and corps commanders decide for corps units. Commanders at lower levels conduct BFT unless otherwise directed. Geospatial and general engineering tasks performed tend to be very similar across the spectrum of conflict for offensive, defensive, stability, or support missions. The primary variable is the condition, with some adjustment to the standard.

1-37. The full-spectrum operations training process is all a part of the cycle—train, alert, (train), deploy, and sustain. After alert, additional specialized training may be required. An example is the training to meet the unique or specialized skills associated with a mission to support another government agency in performing a firefighting mission.

BATTLEFIELD ORGANIZATION

1-38. Within the discussion of the operational framework *(see FM 3-0)* is the concept of the battlefield organization. The battlefield organization *(see FMs 3-0 and 3-07)* is allocating forces in the AO by purpose. Battlefield operations consist of three all-encompassing categories of operations: decisive, shaping and sustaining. The purpose unifies all elements of the battlefield organization by providing common focus for all actions. Commanders organize forces according to purpose by determining whether each unit's operation will be decisive, shaping, or sustaining. This decision is the basis of the concept of operations. When circumstances require a spatial reference, commanders describe the AO in terms of deep, close, and rear areas. These spatial categories are especially useful in operations that are generally contiguous and linear and feature a clearly defined enemy force. Combat engineering primarily tends to support the categories of decisive and shaping operations. Geospatial engineering supports all of the categories with a primary focus on the categories of decisive and shaping operations. General engineering may support all three, and while it is not a separate engineer function, sustainment engineering often describes those engineer missions that support the category of sustaining operations.

DECISIVE OPERATIONS

1-39. Decisive operations are those that directly accomplish the task assigned by the higher headquarters (HQ). They conclusively determine the outcome of major operations, battles, and engagements. In stability operations, decisive operations are usually those that achieve and maintain stability, protect lives and property, or promote peace. Unlike decisive operations in offense or defense, decisive operations in stability operations do not always have immediate impacts; sometimes it takes years to achieve results. In support operations, decisive operations normally prevent or mitigate the effects of natural or man-made disasters. They relieve or reduce conditions such as disease, hunger, or privation. Decisive operations could include stabilizing areas by providing security for personnel, facilities, or capabilities; rendering certain services to populations; or reestablishing critical infrastructure. In certain missions, the collective engineer missions may constitute the decisive operation. In all decisive operations, commanders must analyze environmental considerations and impacts in concert with mission requirements and force protection. As with other considerations, the importance of environmental considerations should be clearly articulated in the higher commander's guidance.

SHAPING OPERATIONS

1-40. Shaping operations, at any echelon, create and preserve conditions for the success of decisive operations. Shaping operations include lethal and nonlethal activities conducted throughout the AO. They support the decisive operation by affecting enemy capabilities and forces or by influencing enemy decisions. Shaping operations use all elements of combat power to neutralize or reduce enemy capabilities. They may occur before, concurrently with, or after the start of the decisive operation. They may involve any combination of forces and occur throughout the AO. In stability operations, shaping operations often convert temporary gains into long-term political successes. The capabilities required to exploit stability often differ from those needed to achieve stability. This may require such measures as rotating different types of units in and out of the AO as the operation progresses through different stages. In support operations, shaping operations may include influencing perceptions, ideas, and information, as well as maintaining legitimacy. Shaping operations will entail transferring tasks to civilian agencies or the local government. For military engineers, this may also include transferring responsibilities to USACE, civilian contract engineers, or HN organizations.

SUSTAINING OPERATIONS

1-41. Sustaining operations are operations at any echelon that enable shaping and decisive operations by providing CSS, rear area and base security, movement control, terrain management, and infrastructure development. Engineering performed in support of sustainment operations is often referred to as sustainment engineering. The vast majority of this support will be general engineering and include at least some geospatial engineering.

SECTION II - ENGINEER BATTLESPACE FUNCTIONS

1-42. The five traditional engineer functions are organized at the operational level into three engineer battlespace functions: combat (M/CM/S), geospatial, and general engineering. They are critical in understanding how the Regiment supports the Army and the foundation of engineer doctrine. The current trend of military operations is an ever increasing US involvement in global responses. Operations Restore Hope and Continued Hope in Somalia, Operation Restore Democracy in Haiti, and Operation Joint Endeavor in Bosnia are all examples of the expanding role of the Engineer Regiment in support of the Army and joint operations. Couple this with the global trend toward quantitatively smaller but technologically and qualitatively better military forces and the need for relevant doctrine becomes apparent. Engineer integration into staff planning requires increased emphasis, since synchronizing the operation or battle is increasingly complex. Engineer C2 must occur rapidly to be responsive to the dynamics of an ongoing operation. Integrating environmental considerations must begin early in the MDMP to ensure that appropriate emphasis is given to these factors. Terrain analysis and its products assist in faster planning and provide a unique perspective of the AO. Requirements for fortifications and protective shelters and the Army's standards for the quality of life (QOL) for its soldiers have gained importance. Obstacle systems retain their importance, while aggressive doctrine within *FM 3-0* places greater emphasis on operational mobility, obstacle breaching, and rapid gap crossing. It is the engineer battlespace functions that will allow the Regiment to address these needs.

COMBAT ENGINEERING

1-43. Combat engineering (M/CM/S) is focused on the support of combined arms maneuver. Combat engineering enhances operational movement, maneuver, and force protection by facilitating M/CM/S operations. At the tactical level, these three subordinate functions are the basis for the engineer tasks included in the M/CM/S BOS, as articulated in *FM 7-15*. Combat engineering, as well as general engineering tasks, provide operational mobility tasks and countermobility to the force at large. The distinction is made by what level of effects are gained from the engineer tasks accomplished. At the operational level, engineer mission analysis must be closely linked with higher-level logistics and movement control planning to ensure that the force sustainment is not encumbered from a lack of operational mobility. The principles of engineer operations still apply, regardless of the level at which they are conducted.

MOBILITY

1-44. Mobility is the first component of combat engineering and includes operations ranging from engineer reconnaissance to countermine/ counterobstacle operations and from gap crossing to the construction of combat roads and trails. It enables the force commander to maneuver tactical units into advantageous positions over the enemy. In the attack, engineers aggressively execute drills to reduce enemy obstacles and assault and destroy enemy fortifications. The commander designates routes for ground forces well in advance of their intended use so that engineer units can upgrade the routes, as necessary, and keep them open or repaired. Engineers may be involved with port construction and improvement of the reception and staging areas for deploying forces. Engineers also prepare field sites that the Army, Air Force, and Marine Corps aviation assets use to support operations. *(See FMs 3-34.2 [90-13-1]* and *90-13* for more information on techniques and procedures for mobility.) Mobility applies at strategic, operational, and tactical levels of war.

1-45. Mobility enables the force commander to maneuver units into advantageous positions. The Army commander relies on mobility to achieve surprise and mass at the critical time and to maintain momentum. Operation Desert Storm provides an excellent example of operational mobility (typically included under general engineering). As it became apparent that Iraqi forces were content to occupy Kuwait and brace for the coming attack, COCOM planners formulated the offensive plans for the sweep north that included two Army corps. The turning movement around Kuwait required displacing the XVIII Airborne Corps from its defensive positions in central Saudi Arabia to the Northwest on the Iraqi border. Marshaling areas to upload tracks and road enhancements allowed for the rapid displacement of the corps into tactical assembly areas (AAs) without providing the Iraqi force with a hint of the coalition's intentions. Army nondivisional engineers aided in breaching the elaborate Iraqi defensive system, thereby allowing divisional engineers to remain integrated with the maneuver force. In a similar vein, the bridge across the Sava River into Bosnia (and the accompanying crossing site support areas) displayed the criticality of operational mobility during Operation Joint Endeavor.

1-46. Mobility includes countermine/counterobstacle, gap crossing, combat roads/trails, forward aviation combat engineering (FACE), and engineer reconnaissance. Obstacle reduction and lane marking equipment are used to increase synergy and survivability on the future battlefield.

COUNTERMOBILITY

1-47. Countermobility is the second component of combat engineering. It augments natural terrain with obstacle systems according to the commander's concept, adding depth to the battle in space and time by attacking the enemy's ability to maneuver its forces. With its movement disrupted, turned, fixed, or blocked, the enemy is vulnerable to friendly forces. Countermobility includes mine systems, obstacle development, and emplaced obstacle planning and control, which will be enhanced through digital communications links in the future. Intelligent minefields with turn on/turn off and sensor capabilities will provide real time intelligence along with increased situational awareness to the combined arms team in the future as well. Engineers advise the force commander on the best means to reinforce the terrain and emplace obstacles that support the force commander's plan. *FM 90-7* is the primary reference for countermobility planning; however, for more information on tactics and techniques for countermobility, *see FM 20-32* and *JP 3-15.*

1-48. Engineers ensure obstacle integration through the proper exercise of obstacle C2, focusing on obstacle emplacement authority, the authority that a unit commander has to emplace reinforcing obstacles, and obstacle control. The LCC or ARFOR commander usually has the authority to emplace obstacles. Generally, the ARFOR commander delegates the authority to corps commanders who may further delegate it to division commanders. Obstacle control ensures that obstacles support current and future operations. The ARFOR commander's control mechanisms to ensure that subordinate commanders do not emplace obstacles which will interfere with future operations are known as obstacle zones and obstacle-free restrictive areas. The nature of obstacle integration from the ARFOR level to company and/or team level leads to echelonment in obstacle planning. At each lower level, engineers conduct more detailed planning. Operational planning consists of developing obstacle restrictions and granting obstacle emplacement authority to subordinate elements. Obstacle planning requires engineers at each level to provide subordinate units with the right combination of positive control and flexibility. The engineer is also an important advisor/partner in deep-targeting discussions and the coordination focal point concerning obstacle barriers and mines for joint managers and coalition forces. Timely, accurate reporting of obstacles from the emplacement unit all the way to the ARFOR HQ—

- Enhances fratricide avoidance—the fourth component of protection *(see FM 3-0).*
- Allows for dissemination as boundaries change or units pass through areas occupied by others.
- Provides critical information in planning the forward passage of lines (FPOL) and rearward passage of lines (RPOL).
- Enhances demining operations at the conclusion of contingency operations.

SURVIVABILITY

1-49. Survivability is the third component of combat engineering. It provides cover and mitigates the effects of enemy weapons on personnel, equipment, and supplies while simultaneously deceiving the enemy regarding the intentions of the force. Survivability operations range from employing camouflage, concealment and deception to the hardening of facilities, C2 nodes, and critical infrastructure. Engineers may be called on to mass their skills and equipment to augment combat units in developing defensive positions into fortifications or strong points and improve defensive positions. More often, however, engineers participate in and provide staff advice on camouflage, concealment, and deception (CCD) measures and the hardening of facilities to resist the destruction of C2 facilities (as part of integrated plans), air-defense weapons systems, and support structures within the communications zone (COMMZ). Within a missile threat environment, engineers provide field fortification support to harden key assets against missile attacks. ADC and force protection includes survivability engineering applications to HN facilities and US-operated facilities as protective measures against terrorist or extremist groups that threaten US forces or national interests. *See FM 3-34.112 (5-103)* for more information on techniques and procedures for survivability and force protection. Survivability also includes providing concealment and protective shelter from the effects of enemy weapons. Engineers—

- Have the technical knowledge, skills, and equipment to assist other units in developing defensive positions into fortifications and improve defensive positions.
- Provide technical advice on camouflage.
- Prepare fighting positions beyond the combat units' organic capabilities.
- Harden facilities to resist destruction by the enemy.
- Provide the equipment support necessary to establish nuclear, biological, and chemical (NBC) decontamination points and assist in route and area decontamination.
- Use digital position navigation systems to mark survivability positions positively on the battlefield.

GEOSPATIAL ENGINEERING

1-50. Geospatial engineering support is provided based on requirements that are determined and validated by the geospatial information and services (GI&S) officer at the joint level and by the engineer filling that role at tactical levels. That support encompasses those tasks that provide GI&S to commanders and staffs across the full range of military operations. Geospatial engineering provides terrain analysis, digitized terrain products, nonstandard tailored map (NTM) products, map production, precision survey, terrain data management, and baseline survey data—all of which significantly contribute to providing a COP within the joint force. This provides products that assist in identifying key and decisive terrain as well as feasible courses of action (COAs) from a terrain perspective. NIMA produces digital terrain and feature data. The Defense Logistics Agency (DLA) distributes this data to units. The geospatial engineer can request imagery (both NTM and commercial) sources through the Assistant Chief of Staff, G2, (intelligence) (G2) or directly, which can be used for spatial and temporal reasoning or multispectral analysis (CCD or inframetric) products that are customized to meet particular operational requirements. Imagery enhances three-dimensional (3-D) and fly-through perspectives. This support is developed and provided during all phases of an operation throughout the theater of operation (TO), in accordance with *JP 2-03*.

1-51. All engineer officers are charged to be terrain experts and advise commanders on how to conceptualize the battlespace effectively. They must be supported by terrain analysts to fully assist others to use the terrain more effectively. As the proponent for mobility, engineers identify and/or recommend avenues and routes, restricted terrain, danger areas, and potential obstacle locations, environmental hazards, EAs, and unit battle positions. During times of crisis or operations, source data may come directly from NIMA or coalition partners. Terrain teams use digital terrain data to develop a detailed terrain analysis. *See FM 3-34.230 (5-105)* for more information on geospatial techniques and procedures.

1-52. Geospatial engineering provides commanders with information about the terrain and assists them in conceptualizing the battlespace more accurately to make knowledgeable decisions. At the ARFOR or LCC level, the topographic battalion and the planning and control detachment integrate echelons above division (EAD) support to Army forces as well as joint and multinational forces. The staff of the Stryker brigade combat team (SBCT) contains an organic terrain team that provides products to support the brigade commander. The heavy division staff's terrain teams are distributed down to brigade level and provide products for division- and brigade-level planning. Creative and productive use of digital terrain data aids the ARFOR or LCC's visualization of the battlefield, by determining factors such as, but not limited to—

- Avenues and routes for friendly/enemy forces.
- Terrain limitations to enemy capabilities.
- Obstacle zone locations.
- Environmentally significant areas (water resources, hazards).
- Major EAs.

- Unit positions.
- Deep operation targets and their impact on future operations.
- Rescue operation parameters.
- Flood prediction models.
- Mission planning and rehearsal data.

1-53. *See FM 3-34.230* and *JP 2-03* for more information on geospatial support to Army and joint systems. GI&S support requirements, products, and capabilities are included in the GI&S annex (joint) or appendix (Army) to each contingency plan (CONPLAN), operation plan (OPLAN) or operation order (OPORD).

GENERAL ENGINEERING

1-54. General engineering encompasses the construction and repair of LOCs, main supply routes (MSRs), airfields, utility systems, and logistic facilities to support joint and Army military operations. It may be performed in direct support (DS) of combat operations such as battle damage repair (BDR). Facilities are fundamental to the success of force projection and the conduct of military operations and play a critical role in shaping the AO and infrastructure to support the force. When facilities and real estate can be obtained through HNS and commercial leases, or through international agreements (SOFAs), facilities acquisition and real estate management become an important component of general engineering. For further information, *see FM 5-104* and *JP 4-04.*

1-55. Typically, general engineering helps establish and maintain the infrastructure necessary for sustaining military operations in theater. General engineering tasks—

- May include construction or repair of existing logistics-support facilities, supply and line of communication (LOC) routes, airfields, ports, water wells, power plants, pipelines, and base camps/force bed down.
- May be performed by a combination of joint engineer units, civilian contractors and HN forces.
- Usually require large amounts of construction materials, which must be planned and provided for in a timely manner.
- May include the production of construction materials.

1-56. *See FM 5-104, FM 3-100.4;* and *FM 7-15* for additional information on general engineering. General engineering tasks include—

- Constructing and/or repairing the following:
  - Existing logistics support facilities.
  - Supply routes, airfields, and heliports.
  - Railroads.
  - Ports.
  - Water wells.
  - Utilities (electric, heat, and water) and sanitation (sewage, hazardous waste, and solid waste).

  - Power plants.
  - Pipelines.
  - Base camps/force bed down.
- Providing electrical distribution expertise.
- Providing expertise on environmental considerations and protection.
- Supporting ADC.

1-57. General engineering support begins with the supporting requirements for the initial reception of the force projection Army (receiving equipment and soldiers).

- Is maintained throughout the operation, providing the infrastructure for the logisticians to sustain the force.
- Provides the support structure to redeploy the force.
- Ends with environmental restoration and the return of the facilities that were used by the deployed forces to HN control.

1-58. As the force advances forward on the battlefield, the corps's rear boundary will be drawn forward continually. Therefore, general engineering also invokes force protection through operations such as demining after the tactical breach and assisting explosive ordnance teams in clearing unexploded ordnances (UXOs) within the expanded COMMZ to the extent necessary to conduct military operations safely.

1-59. At times, the military strategy may be to extend general engineering support to restore facilities, power, and life-support systems within the infrastructure of the combatant countries. This effort aids in the recovery and the transition to preconflict conditions. Central to planning and executing these tasks is construction standards. The challenge is in establishing measures of success and conditions for transitioning to civil support. Within the modern framework of operations, all these efforts are likely to be performed by a combination of joint engineer units, civilian contractors, and HN forces. Once again, these efforts require that large amounts of construction materials and specialized resources be planned and provided for in a timely manner.

1-60. Army leadership can achieve and enforce environmental protection standards more easily when environmental considerations are integrated into the Army's decision making process and activities. Planning for all Army operations and strategies should include efforts to minimize the release of hazardous substances into the environment, protect cultural and natural resources, and prevent pollution. By maintaining environmental stewardship, relations with the HNs are enhanced and cleanup efforts in postconflict periods can be minimized. Guidelines on requirements for overseas operations are outlined in the SOFA with HNs or the DOD overseas environmental baseline guidance document (OEBGD). *See FM 3-100.4* for more information.

SECTION III - UNIFIED ACTION: JOINT/INTERAGENCY/MULTINATIONAL ENVIRONMENT

DESCRIPTION

1-61. In full-spectrum operations, Army engineers operate as part of a joint force and often within a multinational and interagency environment. The term *Unified action* describes the wide scope of actions (including the synchronizing of activities with governmental and nongovernmental agencies) taking place within unified commands, subordinate unified commands, or joint task forces (JTFs) under the overall direction of the commanders of those commands. Public law charges combatant commanders with employ military forces through unified action. Under unified action, commanders integrate joint, single-service, special, and supporting operations with interagency, nongovernmental, and multinational forces (to include United Nations [UN]) operations *(see JP 0-2).*

1-62. Unified action transfers subordinates to the combatant commander under COCOM authority. Multinational, interagency, and nonmilitary forces work with the combatant commander through cooperation and coordination. Regardless of the task or the nature of the threat, combatant commanders employ air, land, sea, space, and SOFs and coordinate with multinational and interagency partners to achieve strategic and operational objectives. They formulate theater strategies and campaigns, organize joint forces, designate operational areas, and provide strategic guidance and operational focus to subordinate commanders. The aim is to achieve unity of effort among many diverse agencies in a complex environment. Subordinate JFCs synchronize joint operations in time and space, direct the action of other military forces (multinational operations), and coordinate with governmental and nongovernmental organizations (NGOs) (interagency coordination) to achieve the same goal. The conduct of operations by ENCOMs is almost always in a joint environment.

THE LEVELS OF WAR

1-63. The three levels of war—strategic, operational, and tactical—are doctrinal perspectives that clarify the links between strategic objectives and tactical actions, although there are no finite limits or boundaries between them. Understanding the interdependent relationship of all three levels helps commanders visualize a logical flow of operations, allocate resources, and assign tasks. Actions within the three levels are not associated with a particular command level, unit size, equipment type or force, or component type. Instead, actions can be defined based on their effect on or contribution to achieving strategic, operational, or tactical objectives *(see Figure 1-6).* Engineers support all three levels of war. The senior engineer at a given echelon typically supports the commander at that level, but in some cases, the engineer may be the commander. We will use the framework of the levels of war in our discussion of engineer operations.

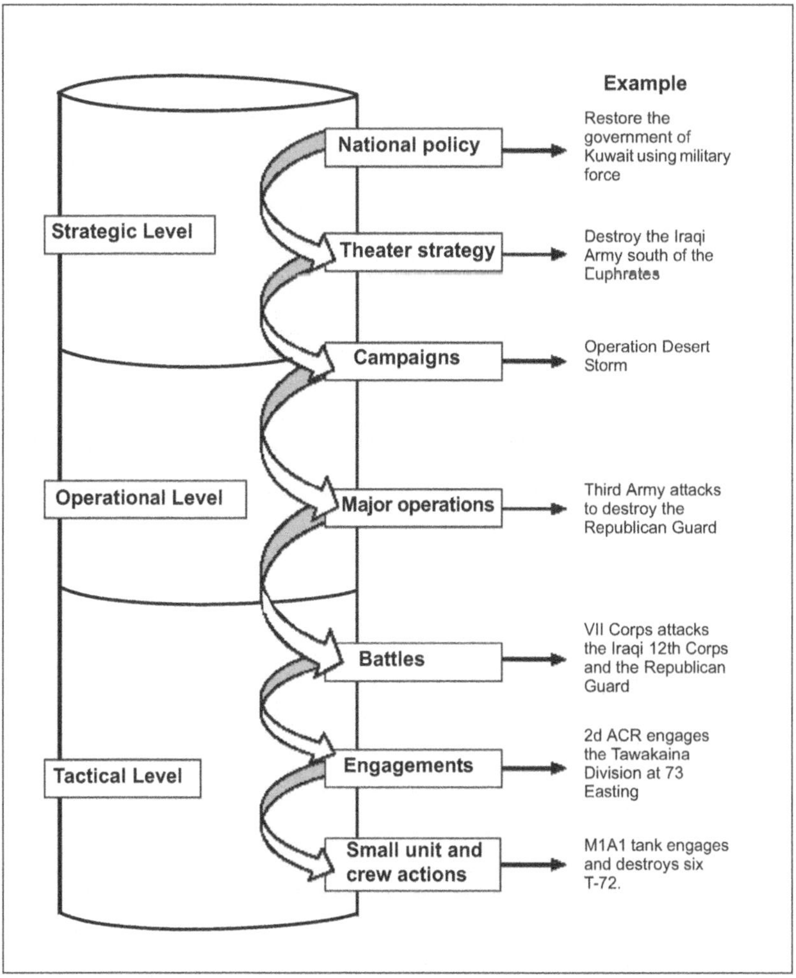

Figure 1-6. Levels of War

THE STRATEGIC LEVEL

1-64. The strategic level is the level of war in which a nation, often as a member of a group of nations, determines national and multinational security objectives and guidance and develops and uses national resources to accomplish these objectives. The term *strategy* is the art and science of developing and employing armed forces and other instruments of national power in a synchronized fashion to secure national or multinational objectives. The Office of the Secretary of Defense translates policy into national strategic military objectives. These military objectives facilitate theater strategic planning. Military strategy, derived from policy, is the basis for all operations *(see FM 3-0* and *JP 3-0).* The engineer organizations that operate at this level are typically the ENCOMs and USACE.

1-65. Engineer activities include force planning, engineer policy and doctrine, and planning and execution of operations and exercises. During operations, strategic engineer activity focuses primarily on the means and capabilities to generate, mount, sustain, and recover forces. Additionally, infrastructure development is a critical aspect of enabling and sustaining force deployments and places a heavy demand on engineer requirements. Engineers advise on terrain, infrastructure, force generation, priorities of engineer support, engineer considerations affecting joint targeting, humanitarian assistance, environmental protection, engineer interoperability, input to the rules of engagement (ROE), and force protection.

THE OPERATIONAL LEVEL

1-66. The operational level of war is the level at which campaigns and major operations are planned, conducted, and sustained to accomplish strategic objectives within theaters or AOs. It links the tactical employment of forces to strategic objectives. The focus at this level is on *operational art*—the use of military forces to achieve strategic goals through the design, organization, integration, and conduct of theater strategies, campaigns, major operations, and battles. Operational art determines when, where, and for what purpose major forces will be employed to influence the enemy disposition before combat. It governs the deployment of those forces, their commitment to or withdrawal from battle, and the arrangement of battles and major operations to achieve operational and strategic objectives. The engineer organization operating at this level and functioning as the senior engineer is typically the ENCOM or the corps engineer brigade.

1-67. Operational art is translated into an OPLAN through operational design. A well-designed plan and successfully executed operation shape the situation for tactical actions. Executed skillfully, a good plan increases the chances of tactical success by creating advantages for friendly forces and disadvantages for the enemy. A flexible plan gives tactical commanders freedom to seize opportunities or react effectively to unforeseen enemy actions and capabilities. Flexible execution maintains the operational initiative and maximizes tactical opportunities *(see FM 3-0).*

1-68. Joint operational engineers assist in preparing and shaping the joint operational area. They develop geospatial products for the commander and staff and develop recommendations to the commander on joint fires, preserving and protecting the force, and sustaining the infrastructure.

THE TACTICAL LEVEL

1-69. Tactics is the employment of units in combat. It includes the ordered arrangement and maneuver of units in relation to each other, the terrain and the enemy, to translate potential combat power into victorious engagements and battles. An engagement is a small tactical conflict between opposing maneuver forces, usually conducted at brigade level and below. Engagements are usually short, lasting minutes, hours, or a day. A battle consists of a set of related engagements that last longer and involve larger forces than an engagement. Battles can affect the course of a campaign or major operation *(see FM 3-90* for more information). The tactical echelon is where many engineer commanders and units, especially combat and CS engineers, will operate.

1-70. Tactical engineers focus on providing geospatial products for the commander to conceptualize the battlespace and make decisions. They provide combat engineering support to tactical maneuver forces. Tactical-level engineers are not solely an activity in the land component. They may support amphibious operations by developing beachheads and air operations by constructing and repairing airfields.

JOINT/INTERAGENCY/MULTINATIONAL OPERATIONS

1-71. In unified action, Army forces synchronize their actions with other participants to achieve unity of effort and accomplish the combatant commander's objectives. The capabilities of joint, multinational, and interagency partners expand strengths, compensate for limitations, and provide operational and tactical depth to Army forces. Given their unique capabilities and the flexible nature of the engineer branch to perform combat, CS, and CSS functions, Army engineers are frequently the unit of choice committed to support these operations. The Engineer Regiment's METL supports these operations as well.

JOINT OPERATIONS

1-72. An operation that includes forces of two or more services under a single commander is a joint operation. Land operations and joint operations are mutually enabling—land operations are inherently joint operations. Joint integration allows the JFC to attack an opponent throughout the depth of a JFC's AO to seize the initiative, maintain momentum, and exploit success. Effective joint integration does not require joint commands at all echelons but does require an understanding of joint synergy at all levels of command. Joint synergy extends the concept of combined arms synergy familiar to land forces. The strengths of the different services combine to overcome each service's limitations and reinforce each other's effects. The combination of multiple and diverse joint force capabilities creates military power more potent than the sum of its parts. The focus of joint operational engineers is to provide the combatant commander operational maneuver options through repair maintenance and construction of aerial pods of debarkation (APODs) and sea port of debarkation (SPODs), enabling river crossing operations and LOC repair or construction. Army engineers provide unique capabilities and strengths to the joint arena. The primary focus of joint engineer operations is to achieve the commander's intent by coordinating engineer support throughout the joint AO.

THE OTHER ARMED FORCES

1-73. US armed forces defend US national interests worldwide. The Constitution commits the federal government and, by extension, the armed forces to protecting the US and US territories. Through Title 10, United States Code (USC) and DOD Directive (DODD) 5100.1, Congress organized the national defense and defined the function of each branch of service. All US armed forces—Army, Air Force, Navy, Marine Corps, Coast Guard, and SOFs —are required to provide globally responsive assets to support combatant commanders' theater strategies and the national security strategy. The capabilities of the other armed forces complement those of Army forces. During joint operations, they provide support consistent with the JFC's directed missions.

EMPLOYING ARMY ENGINEERS IN JOINT OPERATIONS

1-74. Joint doctrine describes the employment of all US military forces in joint operations. Army force commanders are always subordinate to a JFC or are, themselves, designated as the JFC with elements from the other armed forces subordinate to them. The C2 relationships between those forces are the key to effective joint military operations.

Army Engineers in Unified Commands

1-75. Except for forces exempted by the SECDEF, military departments assign all forces—to include nonfederalized ARNG and unmobilized USAR forces—under COCOM of combatant commanders. The Joint Strategic Capabilities Plan apportions major Army forces, by type, to combatant commanders for deliberate planning. In addition to forces assigned in peacetime, Army forces are allocated to combatant commanders in response to crises. The SECDEF, through the Chairman, Joint Chiefs of Staff (CJCS), directs other combatant commanders to reinforce the supported combatant commander with augmentation forces.

Chain of Command

1-76. The SECDEF exercises authority and control of the armed forces through a single chain of command with two branches. One branch goes from the SECDEF to combatant commanders to the various service component commands and subordinate joint commands for the conduct of operations and support. The other branch goes from the SECDEF to the military departments to their respective major service commands. An administrative control (ADCON) relationship exists between the secretary of the military department to the respective service component commands to carry out their Title 10 responsibilities of recruiting, manning, equipping, training, and providing service forces to the combatant commanders. Although the service branch of the chain of command is separate and distinct from the operating branch, the Army service component commander (ASCC) and the ARFOR operate within the combatant commander's chain of command in the theater.

Command Relationships

1-77. At the theater level, when Army forces operate outside the US, they are assigned under a JFC *(see JP 0-2; JP 3-0;* and *FM 100-7)*. A JFC is a combatant commander, subunified commander, or JTF commander. At the theater level, the combatant commander provides strategic direction and operational focus to forces by developing strategy, planning the theater campaign, organizing the theater, and establishing command relationships for effective unified action. The JFC plans, conducts, and supports the campaign in the theater of war, subordinate theater campaigns, major operations, and battles. The four joint command relationships are COCOM, OPCON, tactical control (TACON), and support with only a combatant commander authorized to exercise COCOM over a joint force. *Chapter 7* contains a more in-depth discussion of C&S relationships and their application at all echelons for engineers.

The Army Service Component Commander

1-78. An ASCC is responsible for Army Title 10 requirements in support of a combatant commander. This includes recruiting, organizing, supplying, equipping, training, servicing, mobilizing, demobilizing, administering, and maintaining the construction, outfitting, and repair of military equipment; the construction, maintenance, and repair of buildings, structures, and utilities; and the acquisition of real property. The ASCC may also be responsible for significant DOD and combatant commander-designated common-user logistics (CUL) functions. The ASCC provides administrative and logistic services to assigned Army forces and the ARFORs of subordinate JFCs. When appropriate, the ASCC may delegate authority for support tasks to a single theater support command (TSC) or another subordinate Army HQ, such as the ENCOM or the medical command (MEDCOM), when the focus of support suggests this as the best solution. USACE is also often involved with supporting the ASCC, and will generally operate through the ENCOM, if one is present.

The Army Force

1-79. Designated ASCCs, other numbered armies, and corps HQ can serve as the base for ARFORs, joint force land components, and JTFs. Division HQ with augmentation may serve as the ARFOR to a JTF in SSCs. Army force HQ can perform in a variety of capacities and rapidly reorganize to meet diverse mission requirements. This versatility is a prerequisite for full-spectrum operations. However, all Army organizations require augmentation to function as a joint HQ. The optimum ARFOR organization and C2 structure is based on the mission, the OE, and the JFC's C2 requirements. It is entirely possible that in some specialized support missions, the senior engineer may also be the ARFOR commander. The ARFOR staff engineer at echelons above corps (EAC) is not dual-hatted as the staff engineer and commander. There is no organic engineer commander at these echelons. Therefore, the senior engineer commander task-organized at that level will be responsible for the

C2 of all subordinate engineer units, and the staff engineer will continue to synchronize and plan engineer operations.

INTERAGENCY COORDINATION

1-80. The intricate linkages among the instruments of national power demand that commanders consider all capabilities and agencies to help achieve the common end state. Interagency coordination forges a vital link between military operations and activities conducted by such organizations as US government agencies; agencies of partner nations; NGOs; and regional, international, and UN organizations, as well as agencies of an HN. Theater strategies routinely employ the capabilities of the entire US interagency network. The National Security Act of 1947 establishes an interagency process for national security-related issues. The National Security Council provides national oversight of the interagency process *(see JP 3-08)*. Because of their unique capabilities, combat, CS, and CSS engineers will frequently be involved with interagency coordination involving all of the engineer functions.

MULTINATIONAL OPERATIONS

1-81. Although the US sometimes acts unilaterally, it pursues its national interests through alliances and coalitions when possible. In Operation Desert Shield and Operation Desert Storm, more than 800,000 military personnel from 36 nations combined their wills, forces, and resources to oppose the Iraqi armed forces. The coalition increased the size of the overall force, shared the cost of waging the war, and enhanced the legitimacy of the strategic aims. These multinational operations demonstrated the advantage of successful multinational warfare over the unilateral efforts of a single state. It is the responsibility of the senior engineer commander to coordinate engineer operations and ensure interoperability issues. Multinational operations are conducted within the structure of an alliance or a coalition *(see JP 3-16* and *FM 100-8)*.

Chapter 2

The Operational Environment

A general in all his projects should not think so much about what he wishes to do as much as what his enemy will do; that he should never underestimate this enemy, but he should put himself in his place to appreciate difficulties and hindrance the enemy could interpose; that his plans will be deranged at the slightest event if he has not foreseen everything and if he has not devised means with which to surmount the obstacles.

Frederick the Great: Instructions to his Generals, iii, 1747

To win the next war, we have to think differently.

President George W. Bush 2001

An OE, as defined in *JP 1-02,* is a composite of all conditions, circumstances, and influences which affect the employment of military forces and bear on the decisions of the unit commander. An OE consists of both conventional and unconventional threats. These threat strategies dictate the key variables affecting operations, define the characteristics of the AO, and link to their strategic aims and will use a broad range of ways and means to enable their strategies. Operational and tactical designs are the manifestation of the application of those ways and means. *JP 1-02* also states that the contemporary operational environment (COE) is the overall operational environment that exists today and in the near future (out to the year 2020). It is imperative that engineers understand the OE to best inform the commander at all levels how to effectively employ engineer capabilities to support the Engineer Regimental METL.

GENERAL DESCRIPTION

2-1. The general current OE that exists in today's world is expected to exist until a peer competitor (similar to the Soviet Union during the Cold War) arises. There are multiple threats to US interests, ranging from regional powers to transnational groups interested in terrorism, illicit drug trading, or other illegal activities. They are a mix of radicalism and advanced technological capabilities. This broad range of activities threatens US interests and citizens at home and abroad.

2-2. Our current force of corps and divisions is resourced, designed, organized, and equipped to deploy to a mature theater, with HNS. Before the publication of *FM 3-0*, the Army was trained to meet an echeloned, doctrinally based enemy who shaped a linear battlefield with complex obstacles. Now we are challenged to deploy to austere, underdeveloped theaters to face either conventional forces or nonlinear, asymmetric terrorist adversaries across the spectrum of conflict.

CRITICAL VARIABLES

2-3. The Army and its engineers will encounter a variety of conflicts in a number of different OEs. The OE is a function of a number of variables. There are eleven critical factors or variables that define the nature of the OEs in which those conflicts or other US military activities may occur. These factors are variables, because the exact nature of the conditions, circumstances, and influences that make up the OE will vary according to the particular situation. The variables include—

- Physical environment.
- Nature and stability of the state.
- Military capabilities.
- Technology.
- Information.
- External organizations.
- Sociological demographics.
- Regional and global relationships.
- National will.
- Time.
- Economics.

2-4. These variables are interrelated and sometimes overlap. Different variables will be more or less important in different situations, but they are all common to any OE. Nevertheless, the collective content of these variables will define any OE the Army and its engineers could face, whether they are involved in PME, SSCs, or major theater war (MTW). Each OE is different because the content of the variables is different. The most difficult aspect of analyzing the OE is that the variables will evolve over time. They will not remain fixed but will morph and change. Therefore we can expect the physical environment we are operating in to change over time.

ASYMETRIC, ADAPTIVE APPROACHES

2-5. The concept of asymmetric warfare is critical to understanding the COE. Asymmetry is a condition of ideological, cultural, technological, or military imbalance that exists when there is a disparity in comparative strengths and weaknesses. In the context of the COE, asymmetry means an adaptive approach to avoid or counter US strengths without attempting to oppose them directly, while seeking to exploit weaknesses. The asymmetric approach is not a new phenomenon, but given the position and capabilities of the US as opposed to its potential enemies, it is more likely to be used against the US and its allies by other nations and nonstate adversaries. Potential opponents will seek to avoid US strengths while exploiting perceived US weaknesses.

2-6. Various countries and nonstate entities have studied how the US fights and have begun to devise ways to fight a technologically superior force, if necessary, and win. Most nations of the world and other actors of consequence have devoted considerable effort in studying the US and how we fight. Since it is difficult to predict whom we might have to fight, we don't always have the luxury of having studied these nations or other actors in detail. We must be prepared to refocus quickly, learn fast, and apply what we have learned rapidly. Flexibility and initiative are key to being able to adapt our DOTMLPF solutions to meet the challenges of a given adversary and the asymmetric approaches that he may apply against us.

THREATS AND OTHER INFLUENCES

2-7. Engineers must be prepared to go into any OE and perform its full range of missions while dealing with a wide range of threats and other influences. Some threats come in the form of nation-states; this may be a country or a coalition of countries. Threats can also come from entities that are not states; these can include insurgent, terrorist, drug trafficking, and other criminal organizations. The nonstate actors may use force of arms to further their own interests and threaten the interests of the US or other nation-states. Nonstate threats may exist in isolation or in combination with other nonstate or nation-state threats.

2-8. No single nation-state actor is expected to present a peer or near-peer threat to the US until 2020 or beyond. However, this does not mean that the US and its armed forces will not face serious challenges in the next two decades. The sum total of all the possible conflicts and the level of difficulty of those conflicts could present a challenge equivalent to that of having a near-peer competitor. The net effect of several OEs when considered in its totality from a global, strategic perspective stretches our combat power in ways we have never before encountered. Remember that combat power is composed of maneuver, firepower, leadership, protection, and information. *See FM 3-0* for an in-depth discussion of combat power and its subordinate elements.

2-9. We must be ready to counter all possible threats and, at the same time, be prepared to deal with various third-party actors, such as international humanitarian relief agencies, news media, refugees, and civilians on the battlefield. These groups may be working in cooperation with us or allied efforts or may be operating independently. It is essential that commanders define our relationships with these organizations and include them in the ROE, especially regarding their status within our force protection structure. The engineer effort should be attentive to Title X limitations and how they affect potential cooperation between our organizations. When not properly integrated, their presence can change or constrain the nature of the conflict and can influence the outcome.

2-10. As depicted in *Figure 2-1, page 2-4*, recurring consideration needs to be given as to where a particular geopolitical actor is on the continuum of relative interests and what the affect of that is on our forces. Adversaries share a common characteristic—the capacity and intent to conduct violence against Army forces to negatively influence mission accomplishment. All potential adversaries need to be evaluated in this light. Using the continuum of relative interests, we may find a neutral population becomes an obstacle or adversary over time. An aide such as an NGO could also become an obstacle over time. Therefore, it is imperative that we maintain a current knowledge of all parties in an AO and discern where they are in relation to the continuum and how that might affect engineer operations.

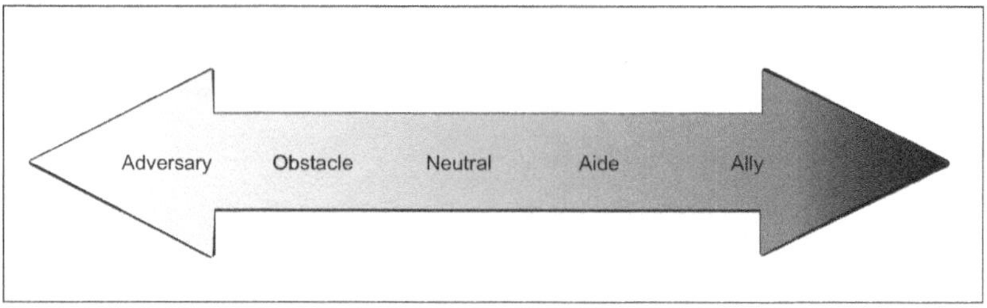

Figure 2-1. Continuum of Relative Interests

WARFIGHTING IN THE CONTEMPORARY OPERATIONAL ENVIRONMENT

2-11. Most participants in conflicts around the world will have no intent to fight the US—they look for ways to keep us out of conflicts or from staying involved. They will not fight us in the same manner as their peers or lesser forces in their region. We can expect potential adversaries to adapt their methods of fighting, using a combination of principles, to include—

- Controlling access into the region.
- Changing the nature of the conflict.
- Employing operational shielding.
- Controlling the tempo.
- Neutralizing technological overmatch.
- Causing politically unacceptable casualties.
- Allowing no sanctuary.
- Entering into urban areas.

2-12. These principles attempt to exploit perceived weaknesses of vulnerabilities in the US force's activities, force structure, or ROE.

IMPLICATIONS FOR EACH OF THE LEVELS OF WAR

STRATEGIC IMPLICATIONS

2-13. Based on threats and the National Security Strategy, the combatant commander should continue to promote their engagement strategies. Engineers will support the commanders with planning support, including geospatial studies; advice on the integrating environmental considerations; infrastructure construction; training; and assistance to international disasters. In times of war, engineers provide a rapid strategic response. Engineers from the AC and RC are prepared to deploy on short notice with their units or just provide tailored capabilities. RC personnel mobilize for specific periods of time to provide their unique skills. Engineers improve interoperability by standardizing with allies in the North Atlantic Treaty Organization (NATO); American, British, Canadian, and Australian (ABCA) coalition; and other international organizations to simplify and facilitate interoperability between international and service engineers and maneuver units. By potentially deploying to austere environments, engineers with limited haul capabilities or manpower can access theater engineers or USACE to contract for the Logistics Contract Augmentation Program (LOGCAP), HN, and local support (and contractors). They can use FFE and the associated reach capabilities to access analyses or expertise on other general-engineering-related challenges. They can also reach engineer centers of expertise for unique solutions to unforeseen and one-of-a-kind challenges or reach laterally to other engineer organizations and assets for expertise and support.

OPERATIONAL IMPLICATIONS

2-14. Operational implications. Many threats or threat forces (rogue organizations) can only fight the US and its allies using asymmetric techniques to nullify technological advantages. However, they may also possess or gain access to weapons of mass destruction (WMD) that cause significant implications for operations, to include tremendous environmental considerations. Threat forces may initially fight force on force and then integrate their forces into urban areas and use civilians to complicate operations. Engineers must conduct their battlefield assessment and work closely with intelligence officers to analyze the threat during the intelligence preparation of the battlefield (IPB). In today's OE, engineers have difficulty predicting an enemy course of action (COA) based on battlefield geometry alone. Engineers may describe an enemy force more in terms of functions (fixing, assaulting, exploiting, shielding) of its subelements and not in terms of where it might be found on a linear battlefield (2d echelon, main defense belt). US engineers will have to develop methods to discern threat engineer patterns of behavior. Engineers have to match training and material capabilities to specific requirements. Operational engineer commanders have proponency for training soldiers on common soldier skills such as mine awareness, detection, avoidance, and extraction and environmental considerations before, during, and after combat operations.

TACTICAL IMPLICATIONS

2-15. Tactical implications. Tactical missions are more complex than ever when preparing to fight both symmetric and asymmetric threats over a noncontiguous AO. Terrain analysis and an understanding of threat capabilities are more important than ever as engineers template potential asymmetric capabilities. Since engineers do not have one tool to accomplish all tasks or counter every threat impediment, they need to gather very specific threat information. They must be able to discern and identify patterns. Engineers work to develop specific detection strategies based on the threat reinforcing obstacle type. The obstacle intelligence (OBSTINTEL) collection plan can then be integrated into the intelligence, surveillance, and reconnaissance (ISR) collection plan and resourced appropriately. The proliferation of mines around the world requires engineers to continuously develop new DOTLMPF procedures to address evolving countermine requirements. Training and Doctrine Command (TRADOC) engineers work closely with United States Army Forces Command (FORSCOM) engineers and other service engineers to address lessons learned and new DOTMLPF procedures. Engineers will have to operate mine information coordination cells (MICCs) and establish databases to track existing mines and other obstacles such as the tactical minefields. Engineers will establish close relationships with explosive ordnance disposal (EOD) units to facilitate a COP of mine/UXO/improvised explosive device (IED) threats.

IMPACT OF THREAT ENGINEERS ON OUR DOCTRINE

2-16. As stated, threat forces may oppose the Army symmetrically or asymmetrically, as seen in Bosnia and Afghanistan. Engineers can anticipate dealing with increased mines, booby traps, IEDs and UXOs throughout linear or nonlinear AOs. The threat may no longer come from mines on the ground but may now include side and top attack (offensive) mines that are command-detonated, remotely or via hard wire. The threat's countermobility will be dispersed throughout the AO to impede friendly maneuver ability and attack our will to fight. The possibility of an opposing threat using the urban environment to diminish US capabilities is increasing. *See FM 3-06(90-10)* and *FM 3-06.11(90-10-1)* for more information. Recent urban combat experiences in Northern Ireland, Grozny, Jenin, Afghanistan, and Iraq reveal an accelerated adaptation to countermeasures developed by friendly forces in urban areas. Experience shows the threat adapting and using IEDs to complicate our detection and neutralization efforts.

2-17. Examples of recent IEDs found around the world include remote detonating devices using electronic, radio control, or cell phones to initiate the systems. These methods are all shared between terrorist organizations via the Internet. What British Forces encounter in Northern Ireland may soon confront our soldiers in another urban AO. Some IED techniques include—

- Coupling. Coupling links one mine or explosive to another, usually with detonating cord. When the first device is detonated, it detonates the linked explosive. This technique is often used to defeat countermine equipment such as mine rollers. When the linked devices are directional fragmentation mines, they can create a large, lethal EA.
- Boosting. Boosting involves stacking buried low-metal mines on top of one another and fuzing the farthest from the surface. This reduces the probability of detection and increases the force of the blast.
- Sensitizing. Sensitizing uses AT mines. On some nonmetallic AT mines, the pressure plate can be cracked and the spring removed to reduce the pressure required to initiate the mine. Similarly, the pressure plate can be removed from metallic AT mines to obtain the same effect. Alternatively, a pressure-fuzed antipersonnel (AP) mine can be placed on top of an AT mine, creating a very large AP mine.
- Daisy-chaining. Daisy-chaining uses command-detonated AP mines. Enemy forces link the mines with trip wires or detonating cord. When the initial mine is detonated, the other mines will detonate. This creates a large, lethal EA.

2-18. Threat forces may also use environmental damage as a weapon. Enemies, using our sense of environmental stewardship and our desire to avoid/prevent environmental disaster, may deliberately release environmental hazards or hide them in structures that, when attacked, may release unacceptable environmental damage. A good example is the deliberate burning of oil wells in Iraq during Desert Storm. Threat forces could use industrial chemicals not banned by international law but which could cause widespread contamination if released. Rebels used these chemicals against the Russians in Grozny.

2-19. Another threat operational capability includes area inundation. This is the idea of releasing large bodies of water from dam systems to temporarily flood areas as operational countermobility. Threat regions with a large number of dam and dike systems are a planning consideration for potential engineer bridging requirements.

2-20. The threat commander understands the importance of Army engineers and will focus assets to eliminate their capabilities, especially when they operate independently and are most vulnerable. He will develop new techniques to counter friendly mobility and countermobility operations. Army engineers must coevolve with their adversaries and adapt new procedures concurrently in the AO. Using internal training or mobile training teams (MTTs) in the field will facilitate this necessary use of new techniques and equipment.

2-21. Engineers will be challenged to understand the OE they face and apply their knowledge and capabilities to support the force. They should understand the variables that constitute the OE to best understand how to leverage the engineer battlespace functions to support the Regimental METL. Future adversaries have developed a wide range of asymmetric capabilities that will challenge the engineer functions. It is imperative that engineers understand how to analyze the threat differently than in the past and develop solutions that consider all the variables associated with the OE.

Chapter 3

Engineer Battlespace Functions and Regimental Capabilities

The European operations were frequently referred to as an Engineer war. The fact that Engineers so ably performed all tasks which fell to their lot indicates the basic soundness of Engineer doctrine, Engineer training, and European troop organization.

Final Report of the Chief of European Theater of Operations 1942-1945

The three primary engineer battlespace functions are combat (M/CM/S), geospatial, and general engineering. These functions, when integrated within the operational framework in support of the elements of combat power, assist the commander in achieving successful decisive, shaping, and sustaining operations. At the strategic level, *JP 3-34* combines the combat engineering functions of M/CM/S, while *JP 4-04* explains civil engineering (general engineering in the Army). *JP 2-03* describes engineer functions and the contributions of geospatial engineering. As these functions are used at the operational and tactical levels, they tend to blur and overlap. Therefore, they are addressed as five separate functions to differentiate the capabilities within each function. For instance, a MSR upgrade for tactical units in their zone or sector is a mobility function but a LOC upgrade at the operational and strategic levels is a general engineering function. The engineer commander has the responsibility to integrate safety and environmental considerations when implementing these functions. The staff engineer is also the primary integrator of environmental considerations the planning process and writing of the environmental appendix to the OPLAN/OPORD. Regimental capabilities exist to support each of the echelons of command and across the spectrum of conflict. Which of those capabilities support an echelon of command depends upon the requirements and challenges associated with mission, enemy, terrain and weather, troops and support available, time available, and civil considerations (METT-TC).

SECTION I - ENGINEER BATTLESPACE FUNCTIONS

3-1. The engineer battlespace functions are the baseline functions that contribute to supporting the METL of the Engineer Regiment. The functions are each generally aligned with a specific BOS, although they have impact in and across the other BOSs. Combat engineering is aligned primarily with the M/CM/S BOS, geospatial engineering with the intelligence BOS, and general engineering with the CSS BOS *(see Figure 3-1)*. Each BOS is also related to and a part of assured mobility.

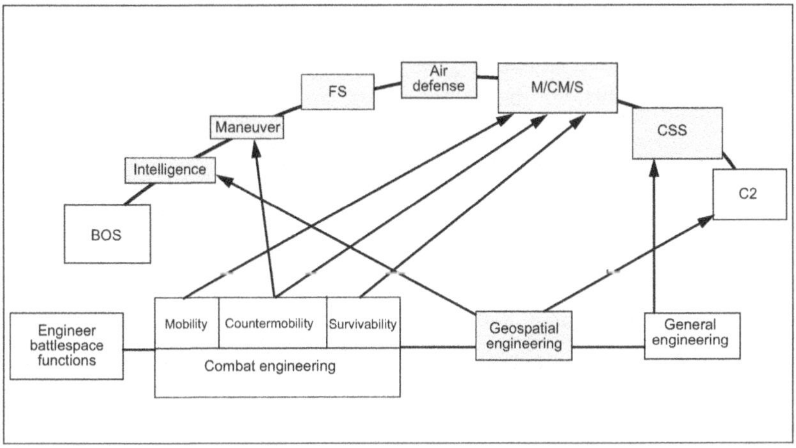

Figure 3-1. Engineer Primary Relationships to the Battlefield Operating System

COMBAT ENGINEERING

3-2. Combat engineering is an integral part of a combined arms unit's ability to maneuver. It is focused on the support of close combat forces. Combat engineers enhance the force momentum by physically shaping the AO to make the most efficient use of space and time to generate mass and speed while denying the enemy maneuver. Combat engineers accelerate the concentration of combat power, increasing the velocity and tempo of the force necessary to exploit critical enemy vulnerabilities. By reinforcing the natural restrictions of the AO, combat engineers limit the enemy's ability to generate tempo and velocity. These limitations increase the enemy reaction time and physically and psychologically degrade his will to fight. Combat engineering includes the subordinate functions of M/CM/S.

KEY FUNDAMENTALS OF COMBAT ENGINEERING

3-3. The following key fundamentals of combat engineering improve the probability of a successful operation.

Evaluate the Terrain

3-4. Engineers should evaluate how the threat may employ obstacles to shape the battlefield or create casualties given the terrain in the AO. They should also consider how the threat will use the terrain to shape his scheme of maneuver. Engineers should then integrate how the terrain will impact on friendly operations and how they can leverage their efforts to achieve the commander's intent.

Understand How the Threat Fights

3-5. Engineers must be familiar with the history of the AO and the conflict. Knowledge of threat doctrine and engineer methods and functions is critical. Engineers should consider patterns of obstacle employment—are they in compliance with mine employment? Not all armies or belligerents mark their minefields and many non-first-world countries rely on IEDs to mark minefield locations and hazard areas. Many local factions, militia, or military units lay mines and mark them with readily available materials rather than formal marking methods adhering to a doctrine. These markings are generally used to warn their own troops and local civilians of the presence of mines. Friendly units operating in these threat environments must know and understand these markings. This is very important because the Army has entered many areas of ongoing or previous conflicts where opposing forces exchanged terrain on multiple occasions leaving behind altered or unrecorded mines and booby traps. The zone of separation (ZOS) region in Bosnia-Herzegovina and areas fought over by the Taliban and the Northern Alliance are recent examples.

Use all Available Engineer Assets

3-6. Engineers seek to predict where obstacles are emplaced, detect or avoid them, and reduce or clear obstacle areas. Engineers use all of the friendly assets available to focus resources to find and avoid obstacle areas or to develop intelligence and fidelity about obstacle areas that must be traversed. At the operational level, engineers should seek OBSTINTEL collected by intelligence specialists, SOF, psychological operations (PSYOP), and civil affairs, and distribute this information to military planners and patrols to complete the COP. All OBSTINTEL should be reported and distributed quickly to enhance force mobility and protect the force.

MOBILITY

3-7. There is a great deal of confusion about the doctrinal definitions and use of the term mobility. DOD defines mobility as quality or capability of military forces which permits them to move from place to place while retaining the ability to fulfill their primary mission. This definition applies to all services and is very broad. The Army defines mobility as "those activities that enable a force to move personnel and equipment on the battlefield without delays due to terrain or obstacles." This definition applies to all branches as they seek to move throughout the AOs. Mobility refines the definition as facilitating the momentum and freedom of movement for the maneuver of forces by reducing or negating the effects of existing or reinforcing obstacles. It includes activities such as reconnaissance, countermine/counterobstacle operations, gap crossing operations, and combat roads and trails. *FM 3-34.2* is the primary reference for mobility planning. Mobility, as an element of the BOS, is broader than the engineer mobility function and includes reducing obstacles, providing bridge and raft support, increasing battlefield circulation, improving or building roads, and identifying routes and contaminated areas. It is the engineer function of mobility that we are addressing when we speak of mobility operations—obstacle reduction by maneuver and engineer units to reduce or negate the effects of existing or reinforcing obstacles. The objective is to maintain freedom of movement for maneuver units, weapon systems, and critical supplies *(see FM 101-5-1)*.

3-8. Mobility is a quality or capability of military forces that permits them to move in time and space while retaining their ability to fulfill their primary mission. A commander must be able to mass forces quickly, at a chosen place and time, to accomplish the assigned mission. The commander must be able to achieve superior tempo through a quick observe, orient, decide, and act (OODA) loop than the enemy. Mobility is critical to achieving this situation and maintaining it for extended periods of time over great distances. The combat-engineer mobility focus must support the requirements of the close combat force.

3-9. Sustained ground combat operations require coordinated movement and effective concentration of combat power against the enemy, in spite of enemy interdiction efforts. Air defense, air and ground transportation, reconnaissance and security, service support, and traffic control operations are among the chief concerns as these large movements occur. Ground combat operations have the best chance of success when they are synchronized with air superiority and air interdiction operations. The LCC or ARFOR commander directs the movement of subordinate forces, ensuring that by the end of a distinct phase of a major operation, decisive and shaping forces are positioned in a way that enables rapid transition to subsequent phases. Additionally, the ARFOR command capability to conduct sustaining operations may set the tempo of the overall joint operation. Within the framework of decisive, shaping, and sustaining operations, the ARFOR will designate and shift the main effort.

3-10. Operational mobility is linked closely to the concept of movement and maneuver operations, which includes the functions of providing mobility for operational forces and countering the mobility of enemy operational forces. Operational-level engineer units provide support for mobility and countermobility operations. Facilitating the maneuver of major formations without delays includes countering the effects of operationally significant obstacles, as well as enhancing operational movement, by preparing and improving facilities and routes critical to major operations. Operational countermobility delays hinder the movement of enemy operational formations, to include selecting and emplacing systems of obstacles for operational effect. Operational-level engineer units provide support for mobility and countermobility.

3-11. Terrain, natural and man-made, significantly influence operational mobility. It often dictates the capacity of LOCs, which can limit the size and composition of supported forces. In war, the operational-level commander considers the effect of terrain features on ground movement and the ability of air power to influence that movement by detecting ground forces and subsequently delaying, disrupting, and destroying these forces. In peacetime, the Army commander may consider how these features affect accomplishing missions that support peacekeeping or humanitarian operations.

ASPECTS OF MOBILITY

3-12. Mobility operations are intended to maintain freedom of tactical maneuver and operational movement through the following five functional areas:

- Countermine activities detect, neutralize (through a combined arms breach or bypass), mark, and proof mined areas.
- Countering employs tactics and equipment to breach or bypass and ultimately reduce obstacles other than mines.
- Gap crossing fills gaps in the terrain to allow personnel and equipment to pass.
- Constructing combat roads and trails expediently prepares or repairs routes of travel for personnel and equipment. This includes temporary bypasses for damaged roads and bridges.
- FACE prepares or repairs expedient landing zones (LZs), forward arming and refueling points (FARPs), landing strips, or other aviation support sites in the forward combat area.

Countermine Activities

3-13. Countermine operations are all efforts taken to counter an enemy's mine effort. Countermine operations are difficult because detection systems are imperfect and mine neutralization systems are only partially effective. Normally, countermine operations using explosive systems are conducted under enemy observation and fire. Countermine operations include mine detection, enemy minefield reconnaissance, combined arms breaching, and enemy mine operations prevention.

3-14. **Mine Detection**. The detection of mines is linked to proper reconnaissance techniques to include geospatial engineering assistance. A proper analysis of enemy techniques and devices lays the groundwork for effective reconnaissance. *FM 20-32,* Mine/Countermine Operations, is the primary reference for a discussion of mine detection.

3-15. **Enemy Minefield Reconnaissance**. Engineer reconnaissance and the specific considerations of minefield reconnaissance are discussed in *FM 5-170.* Proper reconnaissance creates the conditions necessary for successful obstacle breaching.

3-16. **Combined Arms Breaching**. As engineers plan for mobility operations, they may realize that they will have to conduct breaching operations. Enemy obstacles that disrupt, fix, turn, or block the force can affect the timing and force of the operation. Most obstacles can and will be observed by the enemy and will be protected with fires. They should be bypassed if possible. For those that must be breached, constant coordination and integration of all elements of the TF are vital for success. Combat engineers are key to the orchestration of the operation and are responsible for employing the tactics and techniques necessary to penetrate obstacles in the path of the force. Combined arms breaching operations are some of the most complex of modern warfare but are not an end in themselves. They exist only as a part of the maneuver forces' operation focused on the objective. The goal of breaching operations is the continued, uninterrupted momentum of ground forces to the objective; therefore, these operations should be planned and executed in support of the ground forces' needs to ensure that actions at the objective are supported by actions at the breach. Combined arms breaching is explained in depth in *FM 3-34.2*.

3-17. Successful breaching operations are characterized by applying breaching tenets. These tenets should be applied whenever an obstacle is encountered in the AO, whether during an attack or a route clearance operation. These breaching tenets include—

- Intelligence.
- Breaching fundamentals.
- Breaching organization.
- Mass.
- Synchronization.

3-18. Combined arms breaching operations require the constant application of the factors of METT-TC and the concentrated use of supporting arms. Fundamentals of combined arms breaching operations have evolved in concert with the fundamentals of ground combat and provide a logical and time-proven set of rules. These fundamentals include suppress, obscure, secure, reduce, and assault (SOSRA).

3-19. **Enemy Mine Operations Prevention**. The most effective means of countering a mine threat is to prevent mine laying. Proactive countermine operations destroy enemy mine manufacturing and storage facilities or mine-laying capabilities before the mines are laid. Planners must consider enemy storage and mine production facilities and assets for inclusion on the target lists. In addition to destroying mine manufacturing and storage facilities, units must consider targeting enemy engineers and equipment capable of laying mines. During the IPB process, engineers work with the Intelligence Officer (U.S. Army) (S2)/G2 to create an initial enemy obstacle template and variations of the template to show what the enemy is capable of creating over a period of time. Use of this information may place enemy engineers and equipment on the high-payoff target list (HPTL) with collection assets focused to find them.

Countering an Obstacle

3-20. Many issues encountered in counterobstacle operations apply to nonmine obstacles. Engineer reconnaissance should focus on collection efforts to detect the presence of enemy obstacles, determine their types, and provide the necessary information to plan appropriate combined arms breaching or bypass operations to negate the impact on the friendly scheme of maneuver. Reconnaissance also allows friendly forces to anticipate when and where the enemy may employ obstacles that could impede operations. It is prudent to incorporate plans whenever possible to deny the enemy the opportunity to establish effective obstacles. Achieving this goal can be accomplished by—

- Occupying the area before the enemy can exploit it.
- Preplanning artillery and close air support to deny or harass enemy units attempting to establish obstacles.
- Looking for or creating alternative routes for friendly force units.

- Using engineering knowledge of obstacles to create contingency plans for combined arms breaching or bypassing for quick neutralization of enemy obstacles.

Gap Crossing

3-21. Engineers focus on projecting combat power over gaps by tailoring the appropriate resources for the specific mission set. Engineer planners task-organize the appropriate bridge units to support gap crossing operations. Combat engineers can assist with an assault gap crossing using organic armored vehicle-launched bridges (AVLBs) and Wolverines or their heavy equipment to modify the existing gap or by using expedient bridging (rope bridges, small nonstandard bridging using local materials). Operational-level engineers resource subordinate units that do not possess the necessary organic standard bridging equipment. River crossing is a unique gap crossing mission that requires specific and dedicated assets from the BOSs. For a discussion of river crossing, refer to *FM 90-13*.

Combat Roads and Trails

3-22. The ability to move personnel and equipment is essential to maneuver warfare. This ability provides the commander with the means to increase tempo and speed, as well as concentrate mass at crucial times and places. The construction and maintenance of trails and roads are normally considered general engineering tasks and are, therefore, performed by engineering support units. However, locations near the forward line of own troops (FLOT) or time restrictions may require forward combat engineer units to perform these functions in an expedient manner or for short durations until support engineers are available. The two most likely scenarios that would involve this requirement would be bypass operations or support FACE operations. It is important for the engineer commander and staff to perform this function only in support of the maneuver plan. They should not allow engineering assets to be dissipated, rendering them unable to perform their primary role of supporting the commander's operational scheme of maneuver. Engineers should always strive to take full advantage of existing infrastructure and natural terrain features when constructing combat trails and roads.

Forward Aviation Combat Engineering

3-23. With the advent of airpower and its associated support requirements, engineers have acquired a mission to support aviation assets. This frontline support will normally take the form of creating LZs for helicopters and vertical and/or short take-off and landing aircraft or parachute drop zones for personnel, equipment or supplies. Engineers should always strive to take full advantage of existing infrastructure and natural terrain features when constructing expeditionary landing and/or drop zones. Airpower and helicopters are important to support maneuver warfare. The use of forward landing and/or drop zones can increase the speed and tempo of operations. For example, a closer proximity decreases turnaround time for aircraft and helicopters (for example, FARP sites); decreases personnel, equipment, and supply travel times from rear areas to the forward combat area; and decreases the response times of close air and helicopter support missions.

COUNTERMOBILITY

3-24. Countermobility is the augmentation of existing obstacles with reinforcing obstacles integrated with direct- and/or indirect-fire systems to disrupt, fix, turn, or block the enemy. The maneuver commander destroys enemy combat capabilities making use of the increased time for target acquisition. Once again, the countermobility function should not be confused with the countermobility BOS. The countermobility BOS is expanded to include not only obstacle construction but also smoke generation. The engineer commander focuses on proper obstacle integration with the maneuver plan, adherence to obstacle emplacement authority, and rigid obstacle control. *FM 90-7* is the primary reference for countermobility planning.

3-25. Countermobility is the physical shaping of the AO to alter the scheme of maneuver of the enemy. Countermobility operations block, fix, turn, or disrupt the enemy's ability to maneuver, giving the commander opportunities to exploit enemy vulnerabilities or react effectively to enemy actions. When planning countermobility obstacles, it is important to understand the commander's intent, timetable, and scheme of maneuver. These factors, along with available manpower, equipment, and materials, ultimately determine what is feasible to support the OPLAN. Two key actions in obstacle plans include—

- Avoiding obstacle plans that require so much material and the manpower they cannot be emplaced in a timely manner to provide useful support to the maneuver plan—for example, the enemy maneuver elements bypass the obstacle field before it is completed.
- Assuming that friendly forces are not impeded with friendly obstacles later in the operation.

3-26. Nonengineer units may need to augment engineer units with security and personnel to execute countermobility operations. The operational commander and various unit commanders must be aware of this possible requirement during planning.

SURVIVABILITY

3-27. Survivability encompasses developing and constructing protective positions such as earth berms, dug in positions, and overhead protection as a means to mitigate the effectiveness of enemy weapon systems. The survivability function is differentiated from survivability in the M/CM/S BOS since the BOS includes deception, camouflage, operations security (OPSEC), and NBC defense measures. Survivability is also often confused with force protection. *FM 3-0* states that force protection is one of four components of protection (force protection, field discipline, safety, and fratricide avoidance). Survivability operations are just one means to enhance force protection. *FM 5-103* is the primary reference for survivability planning.

3-28. Survivability is the ability of personnel, equipment, and facilities to continue to operate within a wide range of conditions faced in a hostile environment. It includes all aspects of protecting personnel, weapons, and supplies. Units must be able to reduce the exposure to threat acquisition, targeting, and engagement. Engineer-supported tasks such as the construction of field fortifications; hardening of command, communication, and combat training locations; improvements to weapons systems firing positions and infantry fighting positions, are critical to this effort.

Field Fortifications

3-29. DOD/NATO defines a field fortification as an emplacement or shelter of a temporary nature which can be constructed with reasonable facility by units requiring no more than minor engineer supervisory and equipment participation. Engineers construct fighting positions for combat vehicles, direct-fire weapons systems, artillery, and air defense artillery. Field fortifications provide a degree of protection from the effects of enemy weapons systems and a more stable weapons platform from which to sustain accurate volumes of fire. They sustain confidence in a soldier's ability to fight effectively where he otherwise could not survive.

Strong Points

3-30. DOD/NATO defines a strong point as a key point in a defensive position, usually strongly fortified and heavily armed with automatic weapons, around which other positions are grouped for its protection. Strong points are heavily fortified battle positions that cannot be overrun quickly or bypassed easily by enemy forces. They consist of an integrated series of well-protected fighting positions connected by covered routes and reinforced with extensive protective obstacles. They are designed to withstand artillery fire, air strikes, and mounted and dismounted assaults. The enemy can reduce them only by expending significant time, personnel, and equipment assets in overwhelming force.

3-31. It takes significant engineer effort to create a strong point. As a rule of thumb, it will require an engineer company to create a company-sized strong point. Altering a portion of an urban area to turn it into a strong point is generally easier than creating, but it still requires significant effort. Strong points smaller than company size are not usually discussed, except in urban areas.

GEOSPATIAL ENGINEERING

3-32. Geospatial engineering is collecting, developing, disseminating, and analyzing positionally accurate terrain information that is tied to some earth reference. These actions provide mission-tailored data, tactical decision aids, and products that define the character of the zone for the maneuver commander. Key aspects of the topographic mission are geospatial databases, analysis, positional control, and printed maps. These aspects provide the commander a common view of the terrain, which leads to a COP that he uses for C2. Engineer officers at theater, corps, division, brigade, and battalion levels are responsible for geospatial engineering. They provide commanders a clear understanding of the physical environment by enabling visualization of the terrain and explaining its impact on friendly and enemy operations. They identify terrain aspects that the commander can exploit to gain advantage over the enemy, as well as those that the enemy will exploit.

3-33. Geospatial engineering is a subjective evaluation of the terrain's physical attributes of the terrain and the physical capabilities of the vehicles, equipment, and people that must cross or occupy the terrain. Engineer officers as geospatial engineers also coordinate requirements for collecting and generating higher-resolution data and write geospatial annexes/appendices to OPLANs/OPORDs for their supported echelon. Geospatial engineering depends on topographic support for managing, developing, disseminating, and analyzing positionally referenced terrain information. The geospatial engineer must—

- Continually provide current and accurate digital information databases to maintain the commanders' vision of the AO.
- Analyze the effects of operations on the ground and weather factors contributing to maneuverability.
- Perform the critical task of maintaining a common operational background throughout all echelons.
- Provide a means for the commander to visualize the AO.

3-34. The geospatial function primarily supports the C2 and intelligence BOSs. *FM 3-34.230* and *JP 2-03* are the primary references for geospatial engineering.

GENERAL ENGINEERING

3-35. General engineering encompasses those engineer tasks that establish and maintain the infrastructure required to conduct and sustain military operations. Such tasks include the construction and repair of LOCs, MSRs, airfields, utilities, and logistical facilities. *FM 5-104* is the primary reference for general engineering planning. It is directly related to *JP 4-04.*

3-36. General engineering tasks are typically performed to the rear of the division boundaries but can also be performed in forward areas. This function is usually performed by engineer units above division level and consists of repair and construction tasks. Well-developed theaters have a preponderance of repair tasks and lesser developed theaters require more construction tasks. The principles of general engineering in an AO are speed, economy, flexibility, decentralization of authority and establishment of priorities.

3-37. A major requirement for general engineering is a suitable and sufficient supply of construction materials. Engineer commanders have the burden of locating or manufacturing many of these resources. All engineer units must be capable of exploiting available materials. Outsourcing may become an option to access materials as well. Quality control is imperative to ensure that materials meet the critical construction tolerances/standards.

SECTION II - INTEGRATING THE ENGINEER FUNCTIONS

3-38. For the current force and Stryker organizations, the engineer battlespace functions must be integrated to achieve synergy with the other BOSs. The engineer coordinator (ENCOORD) at each echelon is responsible for integration and synchronization of the engineer effort. The engineer battlespace functions are integrated across doctrine through the application of the two primary integration constructs of assured mobility and FFE.

ASSURED MOBILITY

3-39. Linking to the operational framework established in *FM 3-0*, the engineer battlespace functions support the maneuver commander's use of the elements of combat power for decisive, shaping, and sustaining operations within the full spectrum of operations. How these engineer functions are integrated across the elements of combat power is the question. Assured mobility as depicted in *Figure 3-2* integrates all of the engineer battlespace functions, however, assured mobility should not to be confused with the mobility function.

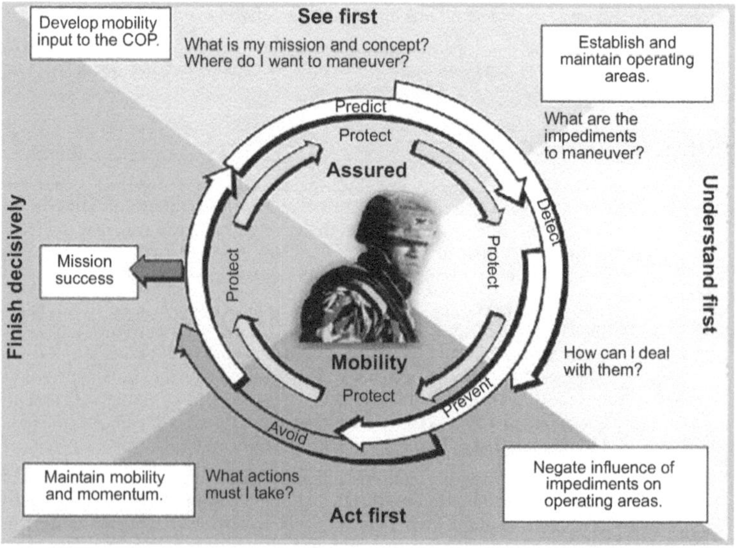

Figure 3-2. Assured Mobility and the Engineer Fundamentals

3-40. Assured mobility encompasses those actions that give the force commander the ability to deploy, move, and maneuver where and when he desires, without interruption or delay, to achieve the mission. A relatively new doctrinal framework, the imperatives and fundamentals of assured mobility are what enable friendly forces to exploit superior situational understanding (SU) and, therefore, gain unsurpassed freedom of movement. Put simply, this framework describes the processes that enable the commander to see first, understand first, act first, and finish decisively.

3-41. Assured mobility supports the maneuver commander's use of the elements of combat power to achieve decisive, shaping, and sustaining operations across the full spectrum of operations and conflict. The framework of assured mobility entails four imperatives that are linked to the elements of combat power *(see Figure 3-3)*. These imperatives are proactive, not reactive, and assure mobility only if integrated into the MDMP.

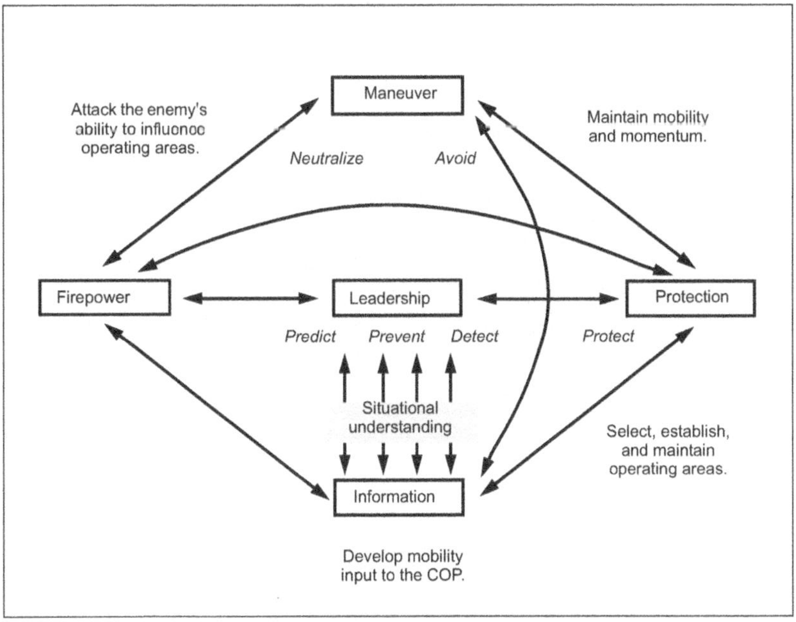

Figure 3-3. Imperatives of Assured Mobility and the Elements of Combat Power

3-42. The first imperative, develop mobility input to the COP, is the collection and integration of geospatial, cultural, and enemy information (aided by automated mobility planning tools) to establish mobility input to the COP for the entire AO. This information allows quick development of the initial and follow-on, real-time modified combined-obstacle overlay (MCOO) that enables the maneuver commander to select the focused operating areas within the AO that best provide positions of advantage. The operating areas are smaller areas designated within the AO that allow the commander to focus collection assets and efforts. The MCOO is defined by the desired endstate and is updated with new information to reflect real-time mobility aspects.

3-43. Knowledge of existing obstacles and monitoring of existing traffic patterns are two examples that allow us to see the battlefield in near-real time. Where obstacles are is as important as where obstacles are not. This information allows the maneuver commander to determine where he may maneuver, what resources will be required to get there, and how the enemy forces may attempt to influence this maneuver plan. The mobility input to the COP enables the maneuver commander to identify the operating areas in the AO and the associated mobility challenges. It is linked to ISR operations and continuously updates the commander and leaders with real-time mobility information. This imperative links the information element to the leadership element to provide SU.

3-44. The second imperative, select, establish, and maintain operating areas, means identifying the threat and restrictive terrain and location of a countermobility effort. This imperative is the linkage between seeing first and understanding first. With the aid of automated tools, critical mobility choke points within operating areas will be identified and a shaping plan. This plan includes predicting enemy actions and intents and the required sensor coverage to fill any information voids within the operating areas. Battlefield terrain reasoning and awareness (BTRA) allows us to template potential obstacles and locations of where the enemy might place obstacles. Sensors dedicated within the collection plan will be employed, or focused from other assets, on the critical areas to fill the information voids or improve our SU and maintain mobility input to the COP. The system allows the commander to track existing obstacles and catalog threat patterns to facilitate a predictive analysis of what the threat may do to terrain to impede friendly maneuver.

3-45. To solidify the COP, sensors are employed at critical areas to protect them from enemy influence. In coordination with these sensors, conducting standoff attacks on selected enemy capabilities to perform mine or countermine activities (or deceiving the enemy to focus his attention on other areas) will fix the current mobility SU. Sensors and forces may be emplaced at choke points, such as bridges, to counter enemy attempts to disrupt operations. Being able to control and monitor critical mobility areas is coordinating a mobility plan in conjunction with the scheme of maneuver. Simply put, "sensor staring" will enable the force to "own" the operating areas. By identifying the locations that may impede movement, the leadership element can resource where he wants to maneuver to avoid or neutralize these impediments while protecting the force and minimizing risks. This imperative supports the elements of protection and maneuver.

3-46. The third imperative, attack the enemy's ability to influence operating areas, is accomplished by interdicting the enemy's countermobility efforts and providing the maneuver commander with multiple avenue options. This imperative is a proactive means of employing standoff detection and obstacle neutralization systems to maintain mobility within the operating areas and assist with ensuring freedom of maneuver. The focus is on shaping the noncontiguous operating areas for decisive operations.

3-47. Standoff detection of obstacles, standoff neutralization, and attacking these obstacles in coordination with the maneuver plan, is critical to preserve resources. This is a proactive attack of the enemy's ability to employ obstacles. Attacking the enemy's ability to shape the AO against us is a key enabler for mobility. Current sensor systems and future intelligent munitions and antipersonnel (AP) land mine alternatives (APLA) will be used to proactively attack the threat.

3-48. The fourth imperative, maintain mobility and momentum, is accomplished by using each of the engineer battlespace functions to support firepower and maneuver. The first three imperatives are command, control, computers, communications, intelligence, surveillance, and reconnaissance (C4ISR) intensive. This imperative considers the fact that the threat is thinking and will adapt to our operations. The purpose of this imperative is to allow maneuver forces to neutralize the effects of obstacles in the context of their maneuver. Their goal is freedom of maneuver to seize objectives without delay along multiple and parallel routes. Detecting obstacles (including side-attack and wide-area mines) and neutralizing their effects without allowing them to adversely affect the maneuver plan or the momentum of the maneuver force are critical. As a last resort, an organic breaching capability exists to overcome enemy obstacles that may be encountered when it is necessary for the force to cross through the obstacles. Marking systems that provide visual, virtual and active identification of obstacles and cleared or safe areas are required.

3-49. In addition to supporting the assaulting force, maintaining mobility and momentum is an important role to ensure that the force can be supported. This includes maintaining and operating routes for pulse logistics as well as forward landing areas for air resupply. Engineers must be able to neutralize the effects of obstacles without disrupting our maneuver momentum in one of four ways: detect and avoid, detect and destroy from standoff, detect and reduce, or withstand the effects. FFE is used as an enabler for the engineer battlespace functions to support assured mobility and the elements of combat power *(see Figure 3-3, page 3-12)*, and the discussion of FFE.

3-50. The application of assured mobility is performed by all branches of the Army although the engineer branch has primary staff responsibility for it. As engineers support the full spectrum of operations, assured mobility is overlaid and integrated with those fundamentals of full-spectrum operations, as identified in *FM 3-0. Figure 3-4* visually represents the linkages and relationships between those fundamentals, assured mobility, the BOS, the engineer battlespace functions, and FFE. Each is related to and imbedded in those basic fundamentals.

3-51. Integrating new capabilities within the imperatives, engineers provide a combined arms proactive approach to mobility. To do this, engineers use information to maximize avoidance without committing into the threat's massed effects. Engineers support maneuver on multiple parallel routes and provide computing power used to identify numerous routes to the objective. They focus on attacking or neutralizing individual mines (just the ones that limit our mobility) instead of reducing lanes in minefields.

3-52. The fundamentals of assured mobility are predict, detect, prevent, avoid, neutralize, and protect. They are part of the full spectrum of operations that follow a continuous cycle of planning, preparing, and executing engineer operations that support decisive, shaping, and sustaining operations *(see Figure 3-4)*. Achieving assured mobility rests on applying the six

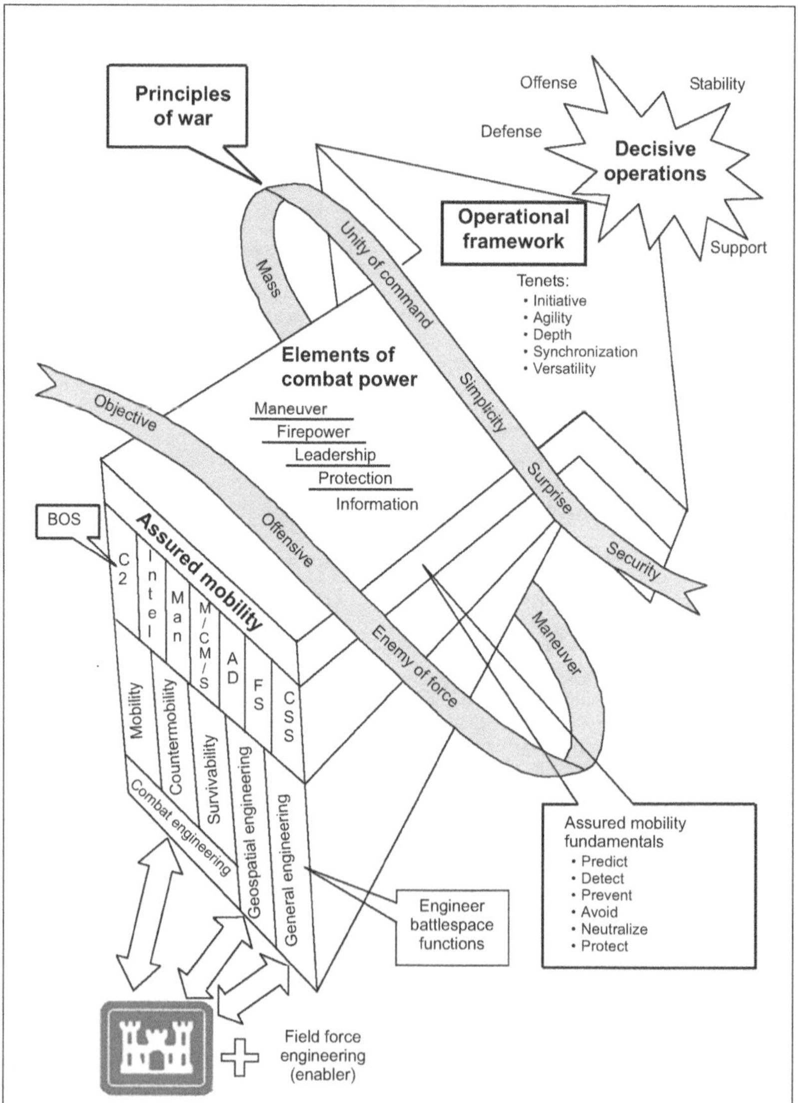

Figure 3-4. Fundamentals of Full Spectrum Operations and Engineer Battlespace Functions

fundamentals listed below. In essence, these fundamentals describe actions that sustain friendly maneuver ability and preclude enemy maneuver ability. They depend on superior SU, shared knowledge, and decisive execution.

- Predict. Predict actions and circumstances that could affect the ability of the force to maintain momentum.
- Detect. Detect early indicators of impediments to battlefield mobility and identify solutions through the use of ISR assets.
- Prevent. Prevent potential impediments to maneuver from affecting the battlefield mobility of the force by acting early. Political considerations and ROE may hinder our ability to apply the fundamental early in a contingency.
- Avoid. Avoid detected impediments to battlefield mobility of the force, if prevention fails.
- Neutralize. Neutralize, reduce, or overcome (breach) impediments to battlefield mobility that cannot be prevented or avoided. The breaching tenets and fundamentals apply when forced to neutralize an obstacle.
- Protect. Protect against enemy countermobility effects.

3-53. The most critical aspect of this framework is the linkage between the fundamentals. For instance, the linkage between predict and prevent, between detect and prevent, between detect and avoid, and between detect and neutralize are essential for success. A failure of any of these linkages will diminish a commander's ability to achieve decisive results.

3-54. The fundamentals of assured mobility are applicable from the strategic to the tactical level. *Table 3-1* provides examples of what resources might be used at various command levels, using the fundamentals against potential impediments. The list is not all-inclusive but intended as a tool to capture the distinctions of assured mobility at various levels.

Table 3-1. Potential Assured Mobility Resources

	Mobility Impediments	Predict	Detect	Prevent	Avoid	Neutralize	Protect
Strategic	Ground Routes	National assets	National assets	Blockades	Planning	Dive teams	Coalition support
Combatant Cdr	APOD	SOF	NGOs	Embargoes	Mine database	Construction units	HLS
	SPOD	FBI/Local police	HUMINT	Interdiction targeting of production facilities		TeleEngineering technical data	Key node force protection
	APOE	Request NIMA Data	UN demining data	PSYOPS	Alternate deployment routes	FBI/local police	Threat condition status
	SPOE	Schedule rapid terrain data acquisition	FBI/local police	FBI/local police	Research and Development Programs	USACE construction	FBI/local police
	Rail	Develop doctrinal template Research and development programs	Research and development programs	Research and development programs		Research and development programs	Research and development programs
Operational	Theater LOCs	DTSS—terrain analysis	ACR	Interdiction targeting	Mine database	EOD	Danger areas isolation
JTF/Corps	ISB's	Terrain data acquisition—terrain IR	UAV	PSYOPS	Movement control	Airfield repair teams	Security operations
	Theater log bases	Refine/develop doctrinal template	SOF	CA/CI activities	Campaign areas of attack determination	Fuel-air bombs	
	Theater airfields	Develop SITEMP	NGOs			TeleEngineering technical data	
			HUMINT			Construction units	
			SIGINT				

Table 3-1. Potential Assured Mobility Resources (Continued)

	Mobility Impediments	Predict	Detect	Prevent	Avoid	Neutralize	Protect
Tactical	Routes	DTSS—tactical terrain products	RSTA units	Preparatory fires	Mine database	Area clearance	Armored vehicles
Corps/ Division/ Brigade	MSR/ASRs	UAV	TUAV	Route security patrols	Movement control	RTE clearance	Security patrols
	FOB	Refine / Develop SITEMP	STAMIDS	Raids	Alt corridors	CA breaching	Local security
	Axis		PSS-12	Interdiction targeting		EOD	
	Operations bases		HUMINT	PSYOPS			
	Road network		SIGINT			River Crossing	
	Rail network						
	Obstacles						
	Gaps						
	Rivers						
	Defiles						
	Urban areas						

3-55. With the development of the COP, commanders can use the fundamentals of assured mobility to ensure freedom of maneuver. During planning, engineers predict areas that the enemy may use to impede movement, using geospatial terrain products and ISR capabilities. During preparation, engineers, in conjunction with the intelligence community, employ sensors to detect impediments in the AO. ISR provides SU in the AO and populate selected corridors with collection capabilities. Engineers apply the prevent fundamental by denying the enemy the ability to influence operating areas. We do so by employing detection capabilities, countermobility systems, and obstacle neutralization systems to attack enemy assets that attempt to influence corridors. During execution, the commander uses the COP to maneuver his force and avoid restrictions. If a force cannot avoid restrictions, a commander neutralizes the impediment using the breaching tenets (intelligence, breach organization, breach fundamentals, mass, and synchronization). Protection is the enforcement of force protection measures to deny the enemy the ability to harm the force as the force maneuvers and performs its engineer functions during all phases of planning, preparing, and executing a mission.

3-56. Assured mobility is built from relationships with the functions and the BOS. The maneuver support (MANSPT) cell is responsible for integrating the BOS contributions to the fundamentals of assured mobility. MANSPT representative ensures the maneuver force integrates the imperatives for execution.

FIELD FORCE ENGINEERING

3-57. Even though the Army is reducing in size and is becoming more expeditionary in nature, engineers are still required to perform all of the broad engineer functions in support of the force commander. No one engineer unit METL is broad enough to allow units to train in every functional area, *(see FM 7-0)*. Engineers must balance multifunctional capabilities with a limited internal degree of expertise. FFE is one method to access the specialized aspects of the engineer functions via TeleEngineering. FFE applies Engineer Regiment capabilities across the range of engineer functions (although primarily general engineering intensive) and in all phases and types of operations ODSS through reach and forward presence. (*See Appendix C* for more detail.) FFE recognizes the critical need for early integrated engineer participation in planning and optimizing engineer capabilities for mission development and execution. This virtual collaboration with the FFE integrates the strengths of Army engineers in table of organization and equipment (TOE)/table of distribution and allowance (TDA) units (AC/RC), USACE (services and technical engineer expertise), and USAES (tactical expertise with geographically dispersed units in the field). This capability allows engineers to develop feasible solutions in a timely manner with early elimination of ad hoc solutions. Based on the current operational environment with more complex requirements and operating conditions, FFE addresses concerns about engineer availability and synchronization, given the general limited technical engineering and environmental capability in most TOE engineer units. FFE allows engineers to reduce their footprint, maximize their capabilities by deploying only the necessary engineer skills (specialized

USACE teams and ENCOM modules) on the ground (reaching to access other technical expertise on a situational/as needed basis), and reduce the associated vulnerabilities and other negative aspects of deployment into the AO.

SECTION III - THE ENGINEER REGIMENT

3-58. The Engineer Regiment contains a wide range of organizations that contribute performing and supporting capabilities to engineer functions at every level. Determining the type and numbers of Engineer Regiment assets and capabilities to support any given operation will be performed by the combatant commander and/or the ARFOR, in conjunction with his engineer staff.

UNITED STATES ARMY CORPS OF ENGINEERS

3-59. The Chief of Engineers has aligned USACE divisions with combatant commanders as they reinforce and extend the capabilities of the Regiment *(see Figure 1-1, page 1-3)*. This relationship with the combatant commander allows direct access to USACE resources to support engagement strategies and wartime operations. The USACE mission covers the full spectrum of operations with five major functions:

- War fighting operations provide full-spectrum engineering and contingency support.
- Disaster relief operations respond to local, national and global disasters.
- Infrastructure builds and sustains the critical facilities for military installations and the public.
- Environmental operations restore, manage, and enhance ecosystems, local and regional.
- Water resource development creates synergy between water resources development and the environment.

3-60. USACE capabilities include access to the expertise of the seven Engineer Research and Development Centers (ERDCs) and all of the resources within the divisions, districts, and other sources. USACE is the primary proponent of FFE, which enables the engineer battlespace functions. FFE is provided through deployed tactical engineer units and USACE personnel (deployed and at their home station). The engineer commander maintains his flexibility and determines the mix of capabilities (troop, USACE civilian, and contractor) based on the tactical situation, time-phased requirements, required/available capabilities, funding, and force caps. The USACE division commander task-organizes division capabilities to meet the varying time-phased requirements. The division's capabilities depend heavily on reach through systems such as TeleEngineering. The FFE concept is applicable in joint and combined operations to provide a better engineer solution that can be implemented faster and with a smaller footprint. The Air Force and Navy have similar capabilities. The Air Force's FFE is called geo-reach and the Navy's FFE is called engineer recon teams (not to be confused with tactical Army engineer reconnaissance teams [ERTs]). The USACE objective of FFE is to more effectively execute its roles (engineering expertise, contract construction, real estate acquisition and disposal, and environmental engineering) in all operations. Additionally, the USACE FFE objective is to maximize use of reach to provide technical assistance and enable the engineer battlespace functions for the combatant commander. USACE accomplishes this by training, equipping, and maintaining specialized, deployable forward engineer support teams (FESTs).

ECHELONS ABOVE CORPS

3-61. At EAC there are engineer C2 HQ called ENCOMs and/or theater Army (TA) brigades and engineer groups that are assigned the responsibility of commanding and controlling how forces execute engineer functions in a theater AO. As the combatant commander's LCC establishes the theater backbone structure, the ENCOMs focus on operational issues such as enabling theater entry and supporting sustained forces. The ENCOM commander establishes a relationship with the Army or land component staff engineer to synchronize the engineer effort. CS assets task-organized to the component commander cannot be task organized to the staff engineer because he does not have command authority. This allows corps and division engineers (DIVENGs) to remain focused on the conduct of the tactical operations. The actual structure of the theater-level engineer force is not firm. The organization is tailored to the anticipated mission. Its capabilities are diverse and specialized. The theater force engineer uses the building block approach to determine the engineer structure based on the varying operational requirements. The theater force engineer has TDA and modified table of organization and equipment (MTOE) units from AC and RC organizations to tailor its capabilities. These unit capabilities include prime power, facility engineering, real estate, topography (and other geospatial capabilities), combat heavy, port construction, utilities, fire fighting, quarry, pipeline, diving, well drilling, and bridging. *Figure 3-5, page 3-22*, shows the existing operational engineer force structure for EAC and EAD.

3-62. The theater force engineer support spans more than just theater sustainment. He can support the maneuver force engineers by accomplishing missions within the combat zone. Control measures such as engineer work lines (EWL) are used to delineate engineer responsibilities between EAC and corps and below engineer units. ENCOMs and TA brigades are prepared to deploy capability modules to provide tailored capabilities to the ARFOR/ combatant commander. These modules are based on unique competencies that include C2, theater planning, facilities engineering, design and survey, and geospatial engineering.

CORPS

3-63. The corps engineer brigade commands and controls all engineer support to the corps and is assigned all engineer units that are not organic to divisions, separate maneuver brigades, and cavalry regiments. The brigade provides combat, geospatial, and general engineering support to the corps based on METT-TC. Corps geospatial engineering support normally is

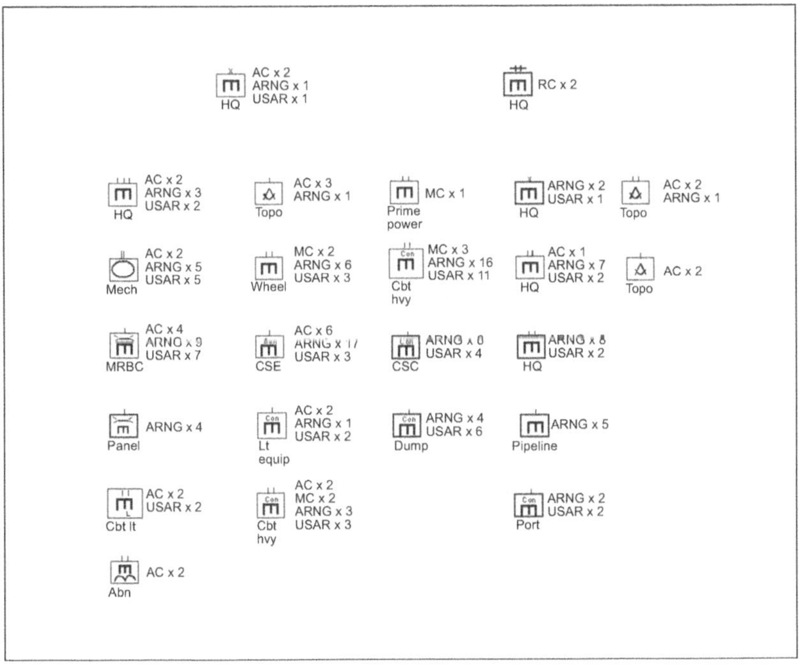

Figure 3-5. Engineer Units at Echelons Above Division and Echelons Above Corps

provided by a topographic engineer company placed in DS of the corps from the TA topographic engineer battalion. The corps engineer brigade augments engineers organic to divisions, separate maneuver brigades, and cavalry regiments. The corps engineer brigade may contain various numbers of engineer groups, corps engineer battalions (mechanized, wheeled, airborne, and light), combat-heavy engineer battalions, and separate engineer companies (fixed and assault float bridge, combat support equipment [CSE], light equipment [LE], and topographic). Other specialized engineer organizations will augment the corps engineer effort as the mission dictates. The brigade provides oversight and support for theater engineer forces from EAC units operating in the corps area. These units include prime power, construction support, pipeline construction, dump truck, and port construction, utilities, well-drilling, fire fighting, and other special teams and detachments.

3-64. The corps engineer brigade commander also serves as the corps engineer special staff officer. The corps staff engineer section (SES) provides engineer functional-area expertise to all corps staff elements. The SES provides recommendations to the corps staff on the use of engineer assets and ensures that the engineer battlefield functions are fully planned, integrated, synchronized, and executed to support the corps commander's intent and scheme of maneuver. The corps engineer tasks and prioritizes the work effort of the DS corps topographic company. In force projection theaters where no forward-based theater engineer structure exists, the corps engineer brigade could initially function as the theater engineer HQ and regional-contingency engineering manager (RCEM). To do this, the brigade needs special augmentation from an ENCOM and/or the USACE in the areas of construction management, real estate acquisition, and construction contracting support. The brigade would execute this function until an ENCOM, TA engineer brigade, or engineer construction group arrives in theater. In the absence of follow-on deployment of an ENCOM, TA engineer brigade, or engineer construction group, the corps engineer brigade (with the special augmentation listed above) may have to act as the theater engineer HQ indefinitely.

DIVISION AND BELOW

3-65. DIVENGs in an analog division serve two critical roles for the division. First, they provide engineer expertise at every echelon of command from the division to the company or team. Second, they provide the structure necessary to command engineer units at these echelons. Both of these roles involve all of the engineer battlefield functions: combat (M/CM/S), geospatial, and general engineering. As a combat multiplier, engineer units focus on maintaining division freedom to maneuver and attacking the threat's freedom to maneuver on the battlefield. As part of the division staff, the DIVENG focuses on integrating and synchronizing engineer missions to support the division commander's intent and scheme of maneuver.

3-66. In the Force XXI (FXXI) division, the engineer brigade HQ is replaced by an engineer cell in the division tactical (TAC) command post (CP), Assistant Chief of Staff, G3, Operations and Plans (G3) division main, and DIVENG section in the division main. This staff is tasked by the DIVENG to form an organic division staff engineer section (DSES). The DIVENG has no units task-organized directly to him. He provides the division commander with planning, integration, and synchronization of engineer operations in the deep, decisive, and sustaining fight. Integrating ISR requirements is critical to this end. The DIVENG's knowledge on maneuver and force integration allows for effective force integration of EAD engineer forces as mission enhancements. This includes acquiring and managing information, maintaining SU, and estimating the necessary direction of subordinate forces.

3-67. The M/CM/S element of the division TAC is comprised of chemical and engineer personnel. Their function is to enhance division mobility and degrade enemy mobility. This element is also mission with the survivability by avoiding or at least minimizing the effects of WMD. The M/CM/S element:

- Monitors, directs, and coordinates the activities of chemical and engineer units.
- Has an embedded engineer team that conducts reconnaissance to continuously update the division knowledge base and ensure that the

COP accurately reflects all obstacles (both friendly and enemy) and trafficability information concerning the AO.
- Has engineer and chemical elements that work closely with the G2 and FSE to track and plot known obstacles.

3-68. The G3 M/CM/S element in the division main consists of three distinct sections—engineer, chemical, and MP. The cell is focused on mobility (both friendly and enemy), countermobility (friendly and enemy) and survivability (protection) of the force. The engineer staff will also integrate the geospatial and general engineering requirements. The engineer section—

- Prioritizes, makes allocation recommendations, and task-organizes divisional engineer assets and other engineer units received from EAD.
- Assists the plans team as it develops OPLANs and OPORDs. Makes recommendations for the appropriate use of engineers and provides detailed knowledge of their capabilities and limitations.
- Assists the G2 section of the IPB cell and the G2 plans officer with the terrain analysis portion of the IPB.
- Maintains an accurate status and location of all friendly and known enemy obstacles, to include impediments to movement caused by the weather or terrain.
- Advises and coordinates with the DIVENG for planning and executing rear-area missions/tasks in support of sustainment operations across all five of the engineer functions. This includes providing terrain products through the use of the Digital Topographic Support System (DTSS) and providing terrain products used while developing or analyzing COAs, thus enhancing the division commander's ability to make informed battlefield decisions. CS assets task-organized to the FXXI division cannot be task-organized to the division staff engineer because he does not have command authority.

3-69. In the FXXI division, the engineer sections can electronically move and share information with subordinate and higher units. This allows for rapid synchronization of operations and dissemination of orders. Each process is complemented by the use of the Maneuver Control System-Engineer (MCS-ENG), FXXI Battle Command Brigade and Below (FBCB2), All-Source Analysis System-Remote Workstation (ASAS-RWS), DTSS and C2 personal computer (C2PC). As part of the division staff, the DIVENG section focuses on integrating and synchronizing all engineer missions to support the commander's intent and scheme of maneuver. Engineers of the FXXI and the analog divisions provide engineer expertise at each echelon of command and planning and synchronization for engineer operations in support of the maneuver commander's intent and the scheme of maneuver.

3-70. Each armored-cavalry regiment (ACR) has a regimental staff engineer and an organic engineer company. The regimental engineer provides engineer planning expertise to regimental and squadron commanders. Although the regimental engineer has no command authority, he remains the senior engineer advisor to the regimental commander. CS engineer units task-organized to the regiment work in conjunction with the regimental engineer to conduct parallel planning and battle tracking. The regimental engineer remains the primary engineer staff officer in the regimental staff even if the task-organized engineer commander is senior. CS units may be abruptly task-organized away; the regimental engineer provides continuity. The regimental engineer role involves all of the engineer battlefield functions: combat (M/CM/S), geospatial, and general engineering. The engineer company commander is a combat multiplier, focused on maintaining the Regiment's freedom of maneuver and mission accomplishment. As part of the regimental staff, the regimental engineer focuses on integrating and synchronizing engineer missions to support the regimental commander's intent and scheme of maneuver.

3-71. The SBCT has a brigade engineer and a single engineer company that is designed and organized principally for operating in SSCs, stability actions, support actions, and MTWs. The company provides combat engineering and limited general engineering (primarily sustainment), and embedded geospatial engineering to support the force in preventing, containing, stabilizing, or terminating a crisis. The brigade engineer is the primary advisor to the brigade commander and may also serve as the MANSPT coordinator. As the MANSPT coordinator, integrates and synchronizes all MANSPT assets into the planning process and conducts battle tracking. The brigade engineer has no command authority, but remains the primary advisor on engineer matters, even if a senior engineer commander is task-organized to support the SBCT.

SECTION IV - INTEGRATION INTO THE COMBINED ARMS TEAM

3-72. Each maneuver force echelon from corps down to battalion/TF level has an ENCOORD to integrate engineers into the combined arms fight. For some higher echelons, the ENCOORD is solely a staff officer; however, the ENCOORD is usually an engineer unit commander/leader with an associated staff officer. At the non-FXXI division level, for example, they are the DIVENG and assistant division engineer (ADE), respectively. In either case, the engineer is a special staff officer who is a member of the echelon battle staff and plays an integral part in developing plans and orders. Engineers work with all members of the battle staff and must understand their capabilities to integrate and synchronize the M/CM/S BOS and contribute to integrating the intelligence and CSS BOS.

3-73. The primary contribution of the staff engineer at the operational level is a solid engineer battlefield assessment (EBA). Without a good EBA, the staff engineer cannot properly contribute to developing a scheme of engineer operations (SOEO) and a task organization to support the plan. A good EBA distributed to subordinate engineers also facilitates the parallel planning process and integration of a BOS.

3-74. At the strategic and operational planning stages, engineer input makes an essential contribution to setting the conditions for operational and tactical success. It enhances the effectiveness of joint targeting by advising on the selection of targets within the enemy's national and military infrastructure. This is done to optimize the required effect on his forces and population while minimizing unwanted effects on nonmilitary targets and joint forces in subsequent operations. This allows attacks to be focused, reducing the effort required to achieve the desired strategic or operational effect and improving the joint capability to attack the enemy's war effort without causing disproportionate collateral damage to civilians and the environment. In DS to imminent operations, engineers must be included in targeting planning at all levels to prevent destruction of key infrastructure essential to future operations and to provide guidance on efficient achievement of the required countermobility effects. The engineer staff should also be involved in battle damage assessment (BDA).

3-75. Engineer support allows the maneuver commander to reinforce cohesion and HN cooperation by preserving HN infrastructure and relieving the effects of conflict on its environment and its population, including refugees. There are a wide variety of activities or tasks that could be conducted by engineers to support building routes and camps for displaced persons and refugees and mine awareness training for civilians. Engineers directly support preserving and protecting the force by hardening the C2 and other facilities, providing collective protection, and removing threats to troops and morale (UXO, mines, scatterable munitions, and IEDs). Engineers provide support for cover, concealment, and camouflage deception efforts, beyond the supported formation capabilities, to counter enemy intelligence operations and protect the force from the effects of enemy fires.

3-76. Engineers sustain the joint force directly through activities such as construction of fuel supply networks and production and supply of water, electricity, and accommodations. They are responsible for construction, maintenance, and post conflict restoration of services and facilities essential to operations. Engineers also provide essential support to logistic activities. In addition to assessing and recommending the most efficient use of existing infrastructure, they develop, construct, repair, and maintain a wide range of logistic facilities.

SECTION V - OPERATIONAL CONSIDERATIONS

FIGHTING AS ENGINEERS

3-77. Combat and CS engineers are the vanguard of the Army because they fight beside combat arms units. When conducting combat operations in close battle, they must be prepared to fight and employ their combat skills, using fire and maneuver to accomplish their engineer mission. On today's battlefield, the enemy can detect and engage engineers quickly, regardless of their location. Consequently, all combat engineers/sappers are organized, trained, and equipped to fight and destroy the enemy in addition to their primary responsibilities within combat engineering. This section addresses aspects of engineers in close combat, organized to fight as engineers.

3-78. Combat and CS engineers are organized, trained, and equipped to engage in close combat to accomplish their engineer missions and—

- Support a movement to contact (MTC) or attack as part of a maneuver formation in the movement to accomplish the formation mission.

- Fight as the reduction force during combined arms breaching operations.
- Assist the supported organization to defeat an unexpected attack.
- Protect a reserve obstacle that must remain passable until friendly forces are able to withdraw.
- Maintain security at a work site.
- Protect themselves in an AA or on the march.

3-79. The enemy will attempt to kill combat engineers as well as dismounted infantry or mechanized forces. It is imperative that engineers are trained to be physically aggressive and tactically competent. The combat engineers' secondary mission is to reorganize into infantry units and fight as infantry.

ENGINEER COMBAT ORGANIZATION

3-80. The 21B combat engineer is trained to accomplish the same basic tasks as the 11B infantryman, in addition to specialized engineer-unique tasks. Engineer squads/platoons/companies are trained to move rapidly and fight violently, either by themselves or in maneuver battalions as a part of a combined arms formation. The additional skills of the 21B make him a potent and versatile member of the combined arms team.

3-81. Mechanized (and the motorized SBCT) combat-engineer squads are organized around their armored vehicle and are armed with an array of rifles, squad automatic weapons, grenade launchers, light and heavy machine guns, and AT weapons. The squads also carry an array of demolition materials, configured into satchel and combat demolition charges, and are able to attack rapidly and violently with demolitions as well as fire. In the platoon, they carry a basic load of conventional and scatterable mines (SCATMINEs) sufficient to emplace a minefield quickly and defend it, if necessary.

3-82. Wheeled combat engineers are organized and equipped much the same as mechanized combat engineers. The major difference is the squad carrier—a 5-ton dump truck or family of medium tactical vehicles (FMTV). When dismounted, a squad and platoons are trained to function much like a dismounted infantry organization.

3-83. Light engineers move on foot, carrying critical tools and equipment as well as demolition materials. As squads or platoons, light engineers move as part of the light infantry formation. Capable of using fire and movement techniques, they also contribute employing demolition and fire to a close-combat fight.

3-84. Combat (heavy), topographic engineer units and specialized engineer companies/detachments are armed primarily with rifles and have a limited number of crew-served weapons. They are not organized to move within combined arms formations or apply fire and maneuver. They are capable of engaging in close combat with fire and movement primarily in a defensive role.

ENGINEER COMBAT CAPABILITIES

3-85. During offensive operations, combat engineer units are task-organized with maneuver units and are integrated into the combined arms formation.

The engineer unit is designed to provide demolition and reduction capabilities to the combined arms team. The engineer unit can also employ direct-fire weapons systems to aid in employing demolitions and reduction assets. Regardless of the mission, armored engineer vehicles are combat vehicles that provide a significant contribution to the combat power of the entire formation. To accomplish the mission, engineers fire and move under the direction of the formation commander, as necessary, using demolition skills when appropriate. Fire and movement techniques are based on rifle, automatic rifle, and grenadier covering fire, allowing the placement of demolition charges within striking range.

3-86. When involved in an assault, engineers fight dismounted on the objective. However, they focus on reducing the close-in protective obstacles as well as demolition tasks against positions and dug-in vehicles. Demolition charges produce significant shock and concussion effects on defenders, as well as destroy critical positions, munitions, and combat vehicles.

3-87. Combat engineers employed on reserve obstacles in the defense execute the technical procedures necessary to ensure target destruction. However, the engineer demolition party responds to enemy contact. They assist the demolition guard in securing the target by holding it open or gaining time to ensure for destruction. The engineer force may assist in target defense by installing obstacles to support the defensive scheme.

3-88. Engineer units engaged in emplacing obstacle systems provide their own local security. Within their capability, they employ close combat techniques against attackers to ensure that the obstacle system is completed. Construction, topographic, and other specialized engineer organizations also provide their own local security. In rear operations, they participate in base cluster defense. They install local protective obstacles and fight from defensive positions on the perimeter. They also form reaction forces that can expel or destroy enemy forces that penetrate a base cluster.

REORGANIZING TO FIGHT AS INFANTRY

3-89. Throughout history, engineer organizations have been required to fill the role of infantry as a secondary mission. The combat engineer/sapper is part of a well-armed and well-equipped organization capable of executing infantry tasks in conjunction with other combat units. Organizational deficiencies include the lack of organic fire control personnel, communications equipment, and medical personnel. If an engineer battalion has been designated to fight as infantry (a maneuver unit), then it requires the same support and integration of other maneuver elements (such as armor, AT, fire support, or air defense artillery [ADA] elements) into its task organization to accomplish its mission. While engineers always fight as engineers, reorganizing them and employing them as infantry requires operational considerations.

EMPLOYMENT CONSIDERATIONS

3-90. Reorganizing engineers as infantry is an operational-level decision that requires corps commander or higher authorization due to the nature and considerations involved with reorganization and the impact (resourcing and training) on the engineer unit and the higher organization. However, a commander must carefully weigh the gain in infantry strength against the loss of engineer support. Engineers provide far more combat power in their primary mission than when configured as infantry. Stopping the engineer work may reduce the combat power of a commander's entire force. Because of the long-term impact, a commander must notify higher HQ when employing engineers as infantry. A commander must carefully analyze infantry and engineer demands before deciding to employ an engineer unit as infantry. Employing engineers merely implies the higher commander using the engineers for a short period of time. On the other hand, reorganization is an action taken to shift internal resources within a degraded unit to increase its level of combat effectiveness. This requires resources, time, and training and is for an extended period of time.

3-91. An immediate requirement for the infantry may not require reorganization; engineers are simply committed to the fight. Reorganization occurs when time allows moving unneeded engineer elements and equipment from the battle area and augmenting the engineer structure with additional capabilities. A commander normally considers reorganizing when forecasting a shortage of infantry before a future operation or phase of an operation. The commander makes a decision after weighing METT-TC factors and determining an acceptable risk level. Reorganizing as infantry takes time and other assets.

Corps And Above

3-92. Corps combat engineer battalions working in either the division or the corps rear may be employed more easily as a separate infantry force. These units frequently work under the control of their battalion HQ and are not dispersed and integrated into other formations. They are also located to move forward and join the force in contact, form a reserve, or prepare and occupy blocking positions. The commander directing this employment should provide early warning to allow the unit time to assemble, reorganize, and prepare for the mission before commitment. The engineer unit must provide immediate liaison to the gaining maneuver command to facilitate planning and integration. This generally requires about 24 hours, unless the unit has previously prepared for a similar mission.

3-93. When an engineer unit is employed as infantry, one major consideration for the commander is to store engineer equipment, such as bulldozers, bucket loaders and road graders in tactical assembly areas. Equipment not used in the infantry role may be attached to other units for C2 purposes or to accomplish other engineer tasks. This decision is METT-TC driven and generally based on the overall concept of the operation.

3-94. The commander directing the employment should augment the engineer unit with additional medical personnel and vehicles, air defense teams, and fire support teams. The unit should also be augmented with armor, AT weapons teams, mortar teams, and other combined arms capabilities, as required, to perform the expected mission.

Division and Below

3-95. Generally, DIVENG battalions are task-organized throughout the division's area and are closely integrated with other maneuver arms. Engineers fight and conduct their operations in this configuration. Engineers in a combat vehicle or dismounted formation (with satchel charges or rifle fire) fight, as required, under the formation commander. Engineers who prepare defensive positions fight from those positions alongside the defenders, if attacked. DIVENGs use their close combat skills as infantry in an emergency while performing their engineer mission.

CONSIDERATIONS

3-96. Engineer units employed as infantry do not have the same capabilities as conventional infantry units. Because they normally operate the same as infantry organizations and have the same basic weapons, this is generally not a problem at squad and platoon levels. At higher levels, this is a concern.

3-97. The engineer company can effectively control other arms as a company/team because it normally works closely with them. The engineer company routinely maneuvers alone and is well suited to protect TF flanks during MTC operations or as the TF breach force during deliberate attacks.

3-98. To be fully effective at the battalion level, engineer units need to be augmented with heavy AT weapons and mortars, as well as the normal CS provided to any infantry unit. Engineer battalions frequently maneuver as battalions, so their training makes them effective in offensive operations and in a defensive role when employed as infantry.

3-99. Employing engineers as infantry is likely to occur when the reserve force has been committed and no other reserve force is possible. The engineer reserve force can be used in two ways: as a reinforcing force for units in contact with or as a blocking force against an attack or counterattack. It can accomplish this mission by building and occupying a strong point. Other uses of an engineer reserve force include—

- Augmenting an armor battalion with infantry to build a TF.
- Augmenting an infantry battalion with an additional company.
- Operating as the breach force commander for a brigade breaching operation.
- Operating separately in an economy of force role or as part of a brigade defense.
- Providing air-assault forces for seizing critical terrain.
- Providing a specialized assault force for urban operations (UO) or the attack of an enemy strong point.

Chapter 4

Planning Engineer Operations

A military engineer must have broad engineering experience between wars.

Major General Julian L. Schley
(Chief of Engineer, 1937-1941)

The Engineer Regiment is a large and diverse organization that provides support to commanders at every level. The lines of responsibility between USACE, MTOE, and TDA units have blurred with our involvement in full-spectrum operations and new threats. As echelons of responsibilities have blurred, we realize that every engineer unit is capable of supporting a maneuver unit at any level of war. For example, a combat heavy battalion may deploy a subordinate unit or capability to support a JTF at the strategic level or provide support for an Army corps at the operational level or even a mechanized battalion at the tactical level. Engineer planning is conducted at the strategic, operational, and tactical levels. The engineer staff should participate in all aspects of planning for operations, particularly during the early stages, to ensure the successful engineer preparation of the operational area. Engineer planners should consider a wide range of requirements.

SECTION I - INTRODUCTION TO PLANNING

4-1. Engineer doctrine addresses the range of full-spectrum operations, which include offensive, defensive, stability, and support operations. Engineers support these operations, or a combination of these operations in any environment. Versatility is a key principle for engineers participating in executing and transitioning between these operations. Essentially, any engineer organization is capable of operating and supporting the commander through the entire range of operations. This may be accomplished by tapping into the expertise, skills, and products available outside of their organization. Expertise may come from international and/or other service engineers. Internally, the Regiment has become a network-centric organization that allows every organization to access each other's unique skills via reach (backward and laterally), including TeleEngineering capabilities embodied in FFE. An MTOE engineer unit brings its core strengths and is reinforced by USACE, ENCOMs, USAES, or other assets within the Regiment for the specific operation in which it participates.

ARMY TRANSFORMATION AND THE ROLE OF ENGINEERS

4-2. The Army is currently engaged in a transformation to become a more strategically responsive force across the full spectrum of operations. The Army and its engineers are moving along three major paths- the Future Force, the current force, and the Stryker force. This FM supports the doctrinal requirements for current engineer operations, and it lays out the framework for operations during the Army's transformation. It addresses engineer operations in support of modern and digitally enhanced forces and Stryker force. By addressing these forces, we provide engineer doctrine that will pull us toward the future force and fulfill the role of being an engineer operational capstone manual.

4-3. The critical transformation path leads to the Future Force. It will be more strategically responsive, deployable, sustainable, versatile, agile, lethal, and survivable than the current force structure. The engineers enabling the Future Force will have these characteristics in order to provide proactive support and imbed combat (M/CM/S), geospatial, and general engineering capabilities. Engineers are a primary enabler of assured mobility. Today, the science and technology community is working hard to achieve advanced capabilities such as stand-off mine detection and neutralization for the close fight. Engineers will develop the systems and techniques necessary to stay ahead of the potential asymmetric capabilities of our adversaries.

4-4. At the same time, the Army will retain those forces (the current force) as we know them today. These forces are being recapitalized through the continuation of existing modernization programs, such as digital technologies. This will be part of the force used for war should they be required anytime in the next 10 years or so. Engineers will continue to field digital capabilities, and provide geospatial support, and in some cases, Bradley engineer fighting vehicles (BEFVs), Wolverines, and other necessary systems to provide modern capabilities to the current force.

4-5. The Stryker force is bridging the gap in capabilities between today and the Future Force. The Stryker force will validate an organizational and operational model for the Future Force. This organization is based on a medium-weight combat vehicle, the Stryker, that allows a more responsive and deployable unit. The engineer squad vehicle (ESV) Stryker variant requires adjustments in the tactics, techniques, and procedures (TTP) involved with SBCT breaching operations. The organic terrain team in the SBCT provides improved geospatial capabilities to the brigade. ISR capabilities and sensors are distributed to the lowest-level units and contribute to the collection plan and OBSTINTEL. Engineers in this structure will adhere to their base doctrine in *FM 3-34.221*.

ENGINEER PLANNING FOCUS

4-6. The planning process and considerations for the engineer are very similar, regardless of whether involved at the joint or the Army level. The omission of engineer planning in any phase of an operation adversely affects the entire plan. In many cases, the engineer commanders may be required to educate supported commanders on engineer capabilities, methods, and depth of the Engineer Regiment.

4-7. Engineer operations require engineer planning at all levels across the range of military operations. Engineer planners must determine the mobilization, deployment, employment, sustainment, and redeployment requirements of the combatant commander or senior Army commander's concept of operations. Operational planning merges the operation plan of the joint force, specific engineer mission, and available engineer forces. Tactical planning occurs primarily at the level of the unit specified to accomplish the tactical task or mission. The other service components can also accomplish tactical planning. When planning joint/Army operations, engineer planners should consider the wide range of diverse requirements for engineer support operations. Successful engineer support to the joint force requires the early involvement of engineers in the planning process and in all phases of joint operations. Understanding terrain and how engineers support the joint force and enhance air, land, and sea operations provides the essential background for planning engineer operations. Early development of a comprehensive plan for engineer operations ensures the availability of engineer forces, equipment, and materiel in support of joint/Army operations.

4-8. The challenges of planning successful engineer operations in support of joint operations within diverse theaters are vast and varied *(see Figure 4-1, page 4-4)*. The engineer staff must be involved in planning from the initial stage of the process by providing geospatial products to the commander and staff. Understanding how engineers affect air, land, and sea operations equips the planner with the background to form a comprehensive plan of engineer actions. This universal application of engineers is crucial at all levels of war.

4-9. Early in the planning process, the engineer staff produces an EBA. During operations, engineer planners should consider such things in their EBA as—

- Geospatial terrain analysis.
- Intelligence requirements.
- Topographic support.
- Construction support (including construction safety requirements and safety criteria).
- Countermine operations.
- Force protection.
- HN forces.
- HN infrastructure.
- Multinational operations.
- Interagency operations.
- Contractor support.
- Materiel acquisition.
- Operational phases.
- Environmental considerations.
- Funding requirements.
- Resource management.
- Other critical considerations, including threat engineer capabilities, terrain, and weather conditions having an impact on particular AO.

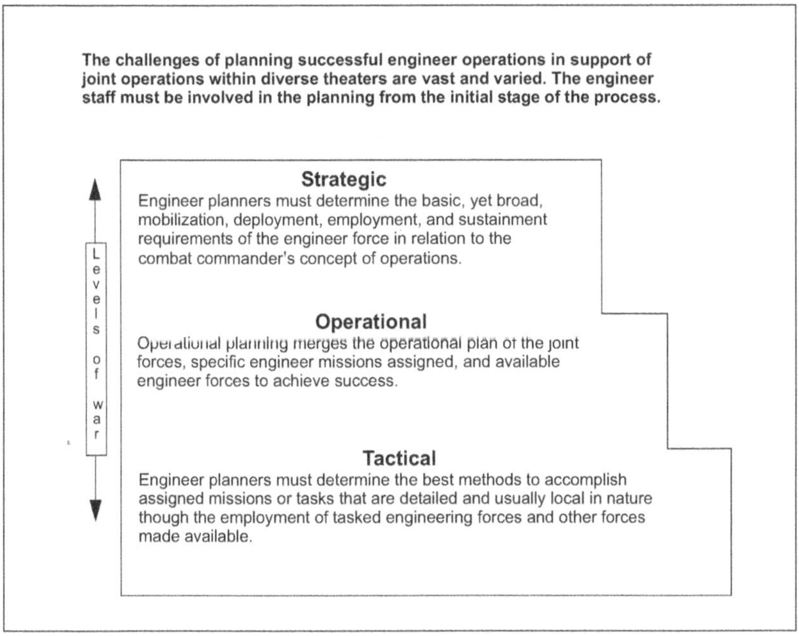

Figure 4-1. Engineer Planning

4-10. *Planning* is the means by which the commander envisions a desired outcome, lays out effective ways of achieving it, and communicates to his subordinates his vision, intent, and decisions, focusing on the results he expects to achieve. Planning is the first step of the operations process that leads to preparation and execution. Throughout the planning process, commanders focus on the results they expect to achieve. This is true at all echelons and levels of war.

4-11. Planning is a dynamic and continuous process. It applies the elements of operational design to a specific situation to accomplish a mission. Commanders assess enemy and friendly centers of gravity, decisive points, and the associated objectives that allow them to attack or defend. The engineer commander's vision, as summarized within the intent, expresses the military conditions that achieve the end state. The concept of operations describes how the various elements of the command will operate together and how they are nested in purpose, space, and time to support the higher commander's or HQ's intent.

4-12. When engineers plan for operations (threat manifested as an opponent or a natural disaster), they must—

- Understand the higher commander's or HQ's intent and be an integral member of the planning staff.
- View the battlefield/AO; have extensive knowledge of the terrain and geospatial products available, OBSTINTEL, and threat capabilities; and anticipate enemy COAs (ECOAs). ISR assets are integrated into finding and answering combat, geospatial, and general engineering intelligence requirements to predict, detect, and prevent threat capabilities. Engineers should visualize the end state and all engineer tasks and purposes from start to finish and transition to future operations.
- Use engineer systems to the best advantage. There is no single solution to engineer problems. Engineer commanders know their equipment and configure/package it in the best combination to address impediments to the friendly maneuver plan. They should resource equipment and organizations to accomplish the identified tasks within the time allotted, identify reduced engineer capabilities, and identify methods to mitigate the risks and leverage all capabilities in the Regiment.
- Concentrate overwhelming engineer capabilities. Engineers mass capabilities to accomplish all missions to facilitate maneuver operations and impede threat movement. During defensive operations, engineers leverage all capabilities to isolate the threat and protect the force.
- Shock and overwhelm the threat. Engineer units move quickly and synchronize to facilitate maneuver operations and counter all threat impediments. They should deny the threat time to prepare and execute their plan. During stability and support operations, engineers may look to assist the population recover from the shock and damage of a threat, such as a natural disaster.
- Provide continuous engineering support. Engineers cannot just look at the tip of the spear but must consider the depth and ensure engineer capabilities are available for freedom of movement from the rear boundary forward. This includes the integration of environmental considerations.
- Man, arm, fuel, fix, move, and sustain the sappers. Engineers ensure that all logistical requirements are thought through to the end state and resourced to accomplish the mission and facilitate future operations.

4-13. Typically, the engineer commander is dual-hatted commander and senior engineer staff advisor or ENCOORD. At the TA level and above and currently in FXXI divisions, the senior staff engineer may no longer be the commander. However, it is still critical in either case that the commander or senior staff engineer conduct parallel planning using the MDMP with task-organized or subordinate engineers to facilitate a realistic assessment. The engineer commander's intent and planning guidance direct the activities of the staff and subordinate commanders. The staff assists the commander with the coordination and detailed analysis necessary to convert the planning guidance and commander's intent into a plan. It becomes a common point of reference for operations.

4-14. Operational and tactical planning complement one another but have different aims. Operational planning prepares the way for tactical activity on the most favorable terms (proper resourcing) and continually seeks to foster and exploit tactical success. Major operations depend on the creative use of tactical action to accomplish a strategic or operational purpose within a specific situational context against an adaptive opponent. Tactical planning emphasizes flexibility and options. Comprehensive planning may be feasible only for the first event or phase in an operation; succeeding actions depend on the enemy response and circumstances and require sound branches and sequels.

4-15. Scope, complexity, and length of planning horizons differ between operational and tactical planning. Campaign planning coordinates major actions across significant time periods and distances to achieve operational objectives. Planners mesh service capabilities with joint and multinational formations as well as interagency and NGOs. Tactical planning has the same clarity of purpose as operational planning but has a shorter planning horizon. The plan guides subordinates as they progress through each phase of operations. Comprehensive, continuous, and adaptive planning characterizes successful operations at the operational and tactical levels of war.

4-16. The joint combatant commander or senior Army commander's engineer planning concepts focus on the relationship of geography and force projection infrastructure to the concept of operations. Engineer planners must determine the basic, yet broad, mobilization, deployment, employment, and sustainment requirements of the combatant commander's concept of operations. At all levels of planning, the senior engineer commander and/or the ENCOORD at each echelon must support the supported commander's OPLAN/OPORD and the internal OPLAN/OPORD for the engineer organization. In selected circumstances, the senior engineer commander may also be the supported commander.

ENGINEERS AND THE LEVELS OF WAR

STRATEGIC LEVEL

4-17. In strategic level of war, art and science are used to develop and employ armed forces as part of the President's instruments of power (diplomatic, information, military, and economic). The instruments of power are synchronized to secure national and multinational objectives. Combat, CS, and CSS engineer capabilities available to the JFC are described in *JP 2-03, JP 3-34,* and *JP 4-04.* Engineers provide unique capabilities derived through USACE and AC and RC engineer forces to allow the JFC the greatest flexibility when developing a strategy to achieve the LCC's objectives. At the strategic level, the availability of ports, roads, airfields, and other infrastructure affects the sequencing of units and the tempo of entry operations. Engineer support to force projection is essential to operational mobility. The USACE and the Army's ENCOMs are the critical strategic resources that support these force projection missions.

4-18. The ENCOMs have established relationships in each major theater of war. The Army's two ENCOMs (412th and 416th) are aligned to the LCC for specific combatant commanders and have forward cells co-located with the supported LCC. Additionally, subordinate organizations of the USACE are aligned with the combatant commanders and provide a bridge of support and expertise to every level, including tactical engineers. The Regiment aligned four USACE divisions with the four overseas geographic combatant commands at the direction of the CSA. Since the Transatlantic Programs Center was aligned to support Central Command (CENTCOM)/United States Army Forces Central Command (Third US Army) (ARCENT) but is neither a division nor a general officer command, the Chief of Engineers aligned the Southwestern Division to CENTCOM/ARCENT, with the Transatlantic Programs Center in support. These strategic support relationships are shown in *Table 4-1* and further articulated in a CSA letter to the Chief of Engineers dated 28 July 2000. Note that strategic engineer support to Northern Command (NORTHCOM) is to be determined.

Table 4-1. Strategic Engineer Support Relationships

COCOM	ASCC	ENCOM	USACE
EUCOM	USAREUR	412th	North Atlantic Division
CENTCOM	Third United States Army/ Army CENTCOM	416th	Transatlantic Programs Center
PACOM	USARPAC and Eighth Army (Korea)	412th	Pacific Ocean Division
SOUTHCOM	USARSO	416th	South Atlantic Division
NORTHCOM	TBD	TBD	HQUSACE

4-19. Engineers strategically respond to the combatant commander's requirements to deploy both AC and RC capabilities, ranging from modularly configured capabilities to entire organizations with complete C2. In the past, the RC was mobilized, en masse, to support the large military that opposed the Soviet threat. Today, the Army has become more expeditionary and the AC and RC have had to become more responsive in order to provide specific required capabilities not available locally/organically to a combatant commander. Often, individual RC units with specific skill sets are activated for short periods to augment the AC forces. In an effort to minimize the force footprint and reduce sustainment requirements, engineers deploy the minimal force necessary as outlined in strategic restrictions. Military engineers have many efforts underway to modularize their expertise and skill sets into deployable C2, ASCC augmented, facility engineering, troop construction, and geospatial teams. Reach is another method engineers use to reduce their AO footprint. Reach is a form of FFE *(see Appendix C)*. In peacetime, engineers participate in improving HN infrastructure to support the combatant commander's engagement strategy. Engineers also provide assistance to HLS and international disasters. Engineers are assigned, apportioned, allocated, and used to augment the Army and component commander's operations. The civilians and engineers of the USACE provide support to the nation at the strategic level by managing the civil works and military construction projects to improve the facilities used by our armed forces.

OPERATIONAL LEVEL

4-20. Operational planning merges the OPLAN/OPORD of the joint force, specific engineer missions assigned, and available engineer forces to achieve success. At the operational level, engineers contribute to battles by prioritizing limited assets and mitigating risks. Engineers seek ways to contribute to decisive, shaping, and sustaining operations by setting the conditions for success and facilitating the component or operational commander's objectives. Engineer commanders anticipate requirements and resource them with engineer units from their command the request engineer support from outside of their organizations. Engineer commanders and staffs also leverage resources available to the ENCOMs, USAES, and USACE ERDC through FFE teams and contractors. Engineers seek ways to leverage other service engineer and interagency capabilities. Engineer organizations may also act as the controlling HQ for certain focused missions.

4-21. Many times operational engineering is general engineering, but even tactical-level combat engineering tasks achieve operational results. For example, clearing an airfield of mines and other hazards is a tactical engineering mission that provides new operational-level capabilities. This tactical task facilitates the rapid flow of light forces and supplies into a forward area by air instead of over unsecured land routes. Operational-level engineering and tasks are not limited to aerial port of debarkation (APOD) and seaport of debarkation (SPOD) and COMMZ areas.

4-22. The defense planning guidance (DPG) directs forces to develop strategies to overcome antiaccess or area denial efforts by a threat that prevents forces from performing entry operations and gaining access to areas where they can operate from a tactical- or operational-level distance. This requires an operational-level engineer analysis that is linked closely to higher- level logistics and movement planning. Operational-engineer considerations include—

- Coordinating with logisticians to leverage existing infrastructure and identify restrictions that can be widened by engineer effort (such as increasing the maximum on the ground (MOG) for airfields, improving marshalling/staging areas.
- Examining methods to facilitate rapid intra-theater repositioning for forces/assets (such as converting stretches of highways into airfields).
- Using reverse planning from the objective to the entry point to identify how best to support operational maneuver (such as reconstruction of major bridges).
- Planning for success—post-hostility stability operations demand infrastructure requirements beyond military needs (contracting for rebuilding, moving civilians, and performing reconstruction).
- Coordinating early and synchronizing with other interagency organizations (such as prioritizing efforts and defining roles and responsibilities between the UN, governmental organizations, NGOs, USACE, other military organizations).

- Emphasizing that operational and strategic-level engineer and logistician sustainment planning efforts merge.

TACTICAL LEVEL

4-23. The commanders or planners of the unit or units specified to accomplish the tactical task or mission normally perform tactical planning; however, the other service components can also accomplish tactical planning. Engineer planners determine the best methods to accomplish assigned missions or tasks that are detailed and usually local in nature by employing tasked engineering forces and other forces that are made available. The JFC should ensure that engineer forces are placed properly in this arrangement and employed to influence the joint force AO.

4-24. The tactical level employs units within the full spectrum of operations. For combat operations, engineers are employed as part of a maneuver force to achieve tactical objectives. Engineers are resourced by the operational commander to accomplish M/CM/S, geospatial, and general engineering tasks as part of an engagement or battle. Engineers are task-organized to provide combat engineer capabilities, sustain combat power with their CSS, and deal with uncertainty. Engineers may also be called upon based on the METT-TC, to function as a maneuver TF or company team commander in special situations. In stability or support operations, engineers may become the priority of effort to provide general engineer support such as constructing life-support areas (LSAs) and improving force protection measures.

SECTION II - ENGINEER RESPONSIBILITIES AND FORCE PROJECTION

FORCE PROJECTION

4-25. Force projection is the ability to project the military element of national power from the continental United States (CONUS) or another theater, in response to requirements for military operations. Force projection operations extend from the mobilization and deployment of forces to the redeployment to CONUS or home theater *(see JP 3-35).*

4-26. The most important characteristic of force projection is synchronizing assets at all levels of war and projecting forces rapidly in response to a crisis or other military requirement. Force projection operations usually begin as contingency operations, involving imminent or actual involvement during war, or as conflict on a regional scale or stability or support mission. Force projection also applies to rapidly deploying forces to respond to a HLS requirement or national emergency or disaster. In combat operations, theater aims may be achieved faster by committing a smaller forward presence force rather than waiting for a larger, but less timely, response option. In this case, US forces could be opposed; however, force projection may occur unopposed. Unopposed operations could afford forces time to continue to build combat power, train, and acclimate after they arrive in theater. The engineer will conduct force projection as part of the overall joint and, possibly, multinational force operation. Engineer support efforts require close coordination with joint and coalition military engineer forces and other agencies to meet force projection requirements. Operational requirements for force projection enablers may require creating or upgrading an intermediate staging base (ISB), a rapid port enhancement (RPE), or similar support. These missions would require extensive use of engineer support in the earliest stages of force projection.

4-27. Force projection operations are undertaken in response to a crisis. That crisis can occur in isolation, as is the expected case in stability operation. However, a crisis also can occur during a major operation. Force projection operations can be received as a series of processes that can occur in an overlapping or sequential manner. When the contingency occurs during a major operation, these processes assist in both resolving the crisis and in returning the contingency forces back into the ongoing operation as rapidly as possible.

4-28. Force projection encompasses five processes—mobilization, deployment, employment, sustainment and redeployment—that normally occur in a continuous, overlapping and iterative sequence for the duration of the mission *(see Figure 4-2* and *FM 100-7).*

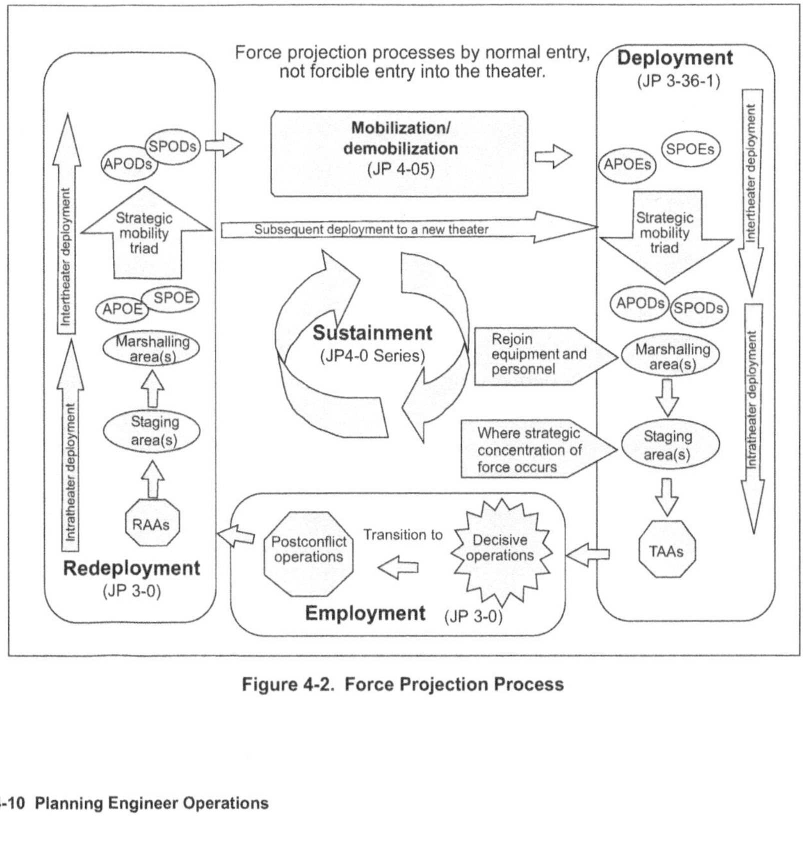

Figure 4-2. Force Projection Process

4-29. Deployment activities, for example, might be so closely followed by decisive operations that they are indistinct from one another. Decisive operations may begin well before the force has completely deployed. At a minimum, commanders and staffs must consider the:

- Coordination of sequencing and phasing of forces (maneuver, operational support, operational protection, SOF, and C2/intelligence and surveillance [IS]/ISR).
- Requirement and time frame to establish and build up the theater infrastructure.
- Environmental issues and the need to reduce the risk of environmental impacts on force health protection.
- Protection of forces, to include rear area security operations (tactical combat force [TCF]) and ADC.
- Preparation time for deployment and operational readiness—types of units and their readiness.
- Combatant commander's critical items list in the time-phased force and deployment data (TPFDD) flow.
- Requirement and level of in-theater stocks.
- HN capability and availability.

4-30. Any particular force projection operation may not include all of the processes. For example, a force projection operation may be the first phase of an evolving major operation. Redeployment of all forces may not begin until the end of the subsequent phase, of which the force projection was a single phase.

FORCE PROJECTION CONSIDERATIONS

4-31. Force projection operations challenge all leaders. Early critical decisions made under uncertain circumstances will be required at all levels. These decisions can greatly affect future conditions for successful mission accomplishment. During the initial planning stages, consideration should be given to the entire deployment cycle, including postconflict operations. It is imperative for engineer leaders to understand the anticipated end state to conduct reverse planning and properly plan the ordered flow of units and the resources to properly take them through reception, staging, onward movement, and integration (RSOI). Unit mobilization and deployment can occur at the same time, or sequentially, and are based on force requirements and strategic aims. When an engineer unit deploys, it will do all that is necessary to meet the demands of the overall mission. *Figure 4-2, page 4-10* shows the deployment cycle as stated in *FM 100-7*.

4-32. Operational design and unit execution must be capable of overcoming any unforeseen obstacles. In most force projection scenarios, combat engineers will be unable to cope with the requirements for general engineering, real estate support, and related technical services required to develop and maintain the operational support base. However, they can provide some of the initial C2 and planning until additional engineer support becomes available. Typically, ENCOMs, with a combination of CS and CSS engineers, respond to these operational-level engineering requirements—with a mix of military and contractor capabilities, integrating tailored organizations, and elements from USACE. *FM 3-0* describes several key considerations that apply to force projection operations. The following are engineer mission capabilities that may arise during the force projection process:

LETHALITY FOR THE DEPLOYING FORCE

4-33. An important strategic/operational consideration for planning contingency operations is to introduce credible lethal forces early. The early entry force must possess the required lethality to accomplish the mission and protect the force the moment it arrives in theater. The lethality of combat forces securing operational objectives can be enhanced through a combination of the engineer battlespace functions. Engineers may contribute to the lethality of the early combat force by emplacing minefields and other obstacles. They protect lodgments by constructing secure C2 nodes, logistics bases, and other necessary fortifications and survivability positions.

ANTICIPATION AND INTELLIGENCE

4-34. Force projection anticipation is the expectation of being alerted and deployed. To properly anticipate mission requirements, engineer commanders should make every effort to ensure that equipment and personnel shortages are identified before being alerted. If faced with a unique support mission (non METL related), anticipation implies the possibility of fielding new equipment. Rapidly introducing US forces to an area requires accurate, detailed, timely, and continuous intelligence. If units have been assigned a region of focus in peacetime, planning can occur before an alert and deployment. Commanders/operators need direct access to engineer assessments of the theater infrastructure for developing COAs and evaluating the infrastructure. Engineers provide geospatial products of likely contingency areas to support the IPB process for all current and potential operations. When assigned as the ENCOORD, engineers advise the G2/S2 of significant environmental factors and ensure that these impacts are integrated into the IPB process. They also assess the available infrastructure for possible general engineering requirements, which include airfields, MSRs, ports, utilities, and logistics facilities. Engineers must anticipate requirements for construction, construction standards, and critical resources to support deployed forces. Other intelligence and preparation might include learning the enemy engineers' capabilities and preparing for them accordingly.

FORCE TAILORING AND TEAMWORK

4-35. Force tailoring is the process of determining the right mix and sequence of the following:

- AC/RC component engineers.
- Combat engineers.
- CS engineers.
- CSS engineers.
- Contractors and contracted engineers.

4-36. Rapid deployment forces may have little opportunity to tailor forces. Follow-on forces can be tailored to meet the specific concerns of the long-term mission. Proper planning should give the operational commander the resources and dispositions to deal with anything that might jeopardize the mission of protecting the force. Commanders consider METT-TC, strategic lift, pre-positioned assets, civilian-contractor support, and HNS when tailoring forces. Deploying units must be very flexible and versatile, to perform missions beyond the unit METL but within the broad scope of the engineer functions.

BATTLE COMMAND

4-37. During force projection, engineer commanders must develop an appreciation for the types of roles/missions that the organization may support across the spectrum of operations. They must quickly visualize the battlefield environment through to the end state and articulate clear guidance that meets the intent of the supported maneuver commander. The commander's intent and other guidance provide the impetus for staff planning and set the conditions for anticipating engineer requirements.

4-38. Commanders must deal with deployment, entry, and potential combat, concurrently, while adjusting to the evolving conditions of each. Deployment can cause the physical separation of units in space and time and the separation of the unit from the next higher HQ. Additionally, units may be placed under unfamiliar organizations. Therefore, simplicity and the ability to adapt and adjust are key considerations. Engineers must support—

- Deployment of themselves and other forces.
- Lodgments and base camps with construction and leasing.
- Maneuver operations with combat engineering.
- Federal, state, and local agencies.

4-39. Engineers will accomplish these tasks at the small-unit level. Engineers at all echelons are challenged to ensure that they understand the mission due to separation in time and space. Engineer commanders at all levels should describe and direct their intent to influence the situation. Moreover, commanders must demonstrate confidence in their subordinates and stimulate a level of initiative and motivation that accomplishes the mission with very little supervision.

LOGISTICS

4-40. Like the initial entry forces in TOs, logistics must be tailored and flexible. The availability of ports, airfields, roads (infrastructure), and other assets will affect the sequencing of units and the tempo of entry operations, as well as the overall logistics planning. ENCOMs classically support force projection logistics operations by leasing or constructing forward support bases, ISBs, and lodgments; improving APODs and SPODs; and providing engineer support for joint logistics over-the-shore (JLOTS) operations. Engineers must work with logistics planners to synchronize the flow of engineer logistics with the flow of engineer units into the TO, including working closely with HN and contracted logistics support.

MEDIA IMPACT

4-41. Emerging information communication technologies, the evolving global information environment, and the media's ability to provide live coverage from anywhere in the world to everywhere throughout the world bridges the gap between the tactical, operational, and strategic levels. Media coverage of Army operations can influence public opinion, political decisions; and the direction, range, and duration of operations. In the emerging information environment, working effectively with the media is a critical element of mission success.

4-42. Engineer operations, especially those executed in support of contingency operations, are likely to attract significant media coverage. Engineers at all levels—

- Must be prepared to operate in a media-intense environment.
- Need to understand that the Army has a vital interest in facilitating media coverage and communicating the Army's perspective.
- Must be prepared to support open and independent reporting and access to units as early as possible.
- Should be trained to interact with media representatives and confidently provide complete, accurate, and timely information.
- Must know that public affairs support is available.

POST CONFLICT ACTIVITIES

4-43. Issues related to the strategic end state, postconflict activities, and transition to peace are considered throughout the force projection operational planning and execution. This is also applicable to post crisis or disaster operations. Engineers tend to play a significant role in conducting postconflict activities due to the long-term military presence required for force protection and general engineering requirements.

4-44. Engineer planners must plan for success in postconflict planning if they are to be prepared to support postconflict operations. The major consideration is the renewal of traffic and the demands on the civilian support infrastructure to keep the AO stabilized after combat operations. Title X and the Army's minimum military requirement focus for posthostilities are often at odds, as engineers consider the transition to stability or support operations. Engineer planners must consider in advance how C2 and resources and military engineer effort and materials will be used in the transition vis-à-vis any international or other US agency effort. This is most significant during the transition from combat to stability operations where critical infrastructure support is destroyed or damaged from the war but other agencies are not yet available or in position to assume the mission. Military engineer operational planning for this transition period is particularly important to the stability of the postconflict theater.

SECTION III - THE MILITARY DECISION-MAKING PROCESS

4-45. There are two Army doctrinal planning procedures defined in *FM 101-5*. In units with a formally organized staff, the military decision-making process

MDMP helps commanders and staffs develop estimates, plans, and orders. It provides a logical sequence of decision and interaction between a commander and his staff. The MDMP provides a common framework that supports the maximum use of parallel planning for all staffs. However, at the lowest tactical echelons, commanders do not have a staff. Consequently, they (and their subordinate leaders) follow troop leading procedures (TLP). Both procedures hinge on the commander's ability to visualize and describe the mission/operation. Both are means to an end, and their value lies in the result, not in the process. Each process can be performed in detail, if time permits, or in an abbreviated fashion in a time-constrained environment. TLP will not be addressed in this manual, but rather in the respective engineer unit manuals (such as *FM 3-34.221*) focused on the specific TLP associated with that unit and echelon of command.

4-46. The different methods of planning are described in *FM 5-0*. For joint planning methods refer to *JP 5-0* and *JP 5-00.2*. The particular procedures used in joint planning depend on the time available to accomplish them. However, engineering considerations are similar for both deliberate and crisis action planning. Engineering considerations in the deliberate planning steps encompass the same engineering considerations as in the crisis action planning steps. This method applies to the full spectrum of operations. A correlation of these steps is provided, and considerations are outlined in the phases of the deliberate planning process.

4-47. The different methods of planning are described in *FM 101-5*. For joint planning methods refer to *JP 5-0* and *JP 5-00.2*.

4-48. Engineer commanders or the senior ENCOORD should develop a parallel planning process with their task-organized engineer units to facilitate realistic assessments. This near parallel process feeds into the force commander's MDMP *(see Figure 4-3, page 4-16)* and provides input for an engineer OPORD or annex to be published, nearly simultaneously, maximizing the time available for execution. This helps to ensure that the engineer capabilities (and included specialized areas such as environmental considerations) are integrated into the MDMP.

RECEIPT OF MISSION

4-49. During this phase, the Joint/Army force engineer assembles the resources required to support the mission analysis and COA development tasks for the concept development phase. Engineers must immediately begin development of mission specific geospatial products after the receipt of the mission in order for them to be available during mission analysis. As a general rule, commander allocates a minimum of two-thirds of the available time for subordinate units to conduct their planning and preparation, leaving one-third of the time for the commander and staff planning. Issuance of a warning order (WO) facilitates parallel planning, allowing subordinates the maximum time to conduct their own planning.

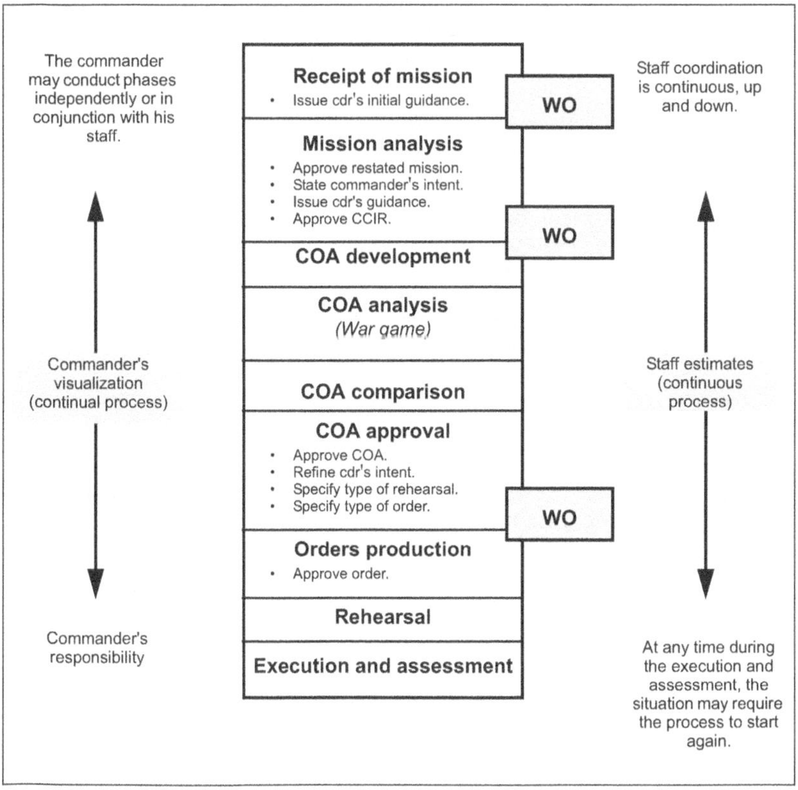

Figure 4-3. The Military Decision Making Process

MISSION ANALYSIS

4-50. A mission analysis has seventeen subordinate steps that provide the framework for success in the MDMP. The engineer has a role to play in each of these steps, although some are more critical than others. *See FM 101-5.*

4-51. The end product of the mission analysis process is a restated engineer mission statement and the development, preparation, and issuance of planning guidance to the staff and subordinate commands. Developing an engineer end state and a clear understanding throughout the chain of command contributes to the mission accomplishment and achievement the desired objectives. During the EBA, the ENCOORD identifies the specified and implied engineer tasks (may be more than M/CM/S) and their associated purposes. From these tasks, the ENCOORD will recommend essential mobility survivability tasks (EMSTs) to the higher commander. Engineer considerations of the EBA during this step of planning include identifying the threat, the terrain, and the friendly forces, to include—

- Threat engineer capabilities, order of battle, tactics, and TTPs.
- Threat engineer templates eventually linked to the collection plan.
- Existing mines, UXOs, and so forth, including access to the databases.
- A terrain analysis in support of battlefield visualization.
- An infrastructure and facilities assessment (bulk fuel storage, ports, airfields).
- Environmental considerations, to include environmentally sensitive areas and associated impacts.
- The availability of HNS.
- Digital mapping requirements.
- The capabilities of assigned engineer forces.
- The capabilities of multinational engineers.
- All engineer capabilities.
- The identification of EMSTs.
- The engineer end state.
- The recommended information requirements (IR).

PLANNING GUIDANCE DEVELOPMENT

4-52. During this phase, the ENCOORD receives the maneuver commander's guidance and priorities for engineer operations. The ENCOORD also seeks to gain the maneuver commander's approval of his proposed EMSTs. The engineer commander assesses all available information derived from the mission-analysis process to provide the commander and staff with the input required to develop and support the initial COAs. The maneuver commander communicates the desired engineer tasks to achieve the desired purpose and effects. The joint/Army force engineer uses this combined assessment to identify the resources required to support each COA, make recommendations based upon available time and resources, and recommend force tailoring to best support the combatant commander's intent. This assessment is the linchpin of successful engineer integration into operations.

COURSE OF ACTION DEVELOPMENT

4-53. COA development is the critical phase of the planning process. In this phase, engineers fully integrate and synchronize combined arms operations as part of the maneuver plan. The ENCOORD synchronizes the SOEO and confirms the EMSTs required to achieve the desired purpose and effect. He then assigns resources and recommend priorities to accomplish the SOEO.

4-54. During COA development, the engineer may consider the following:

- All identified engineer tasks and resources required.
- The detailed reverse planning process.
- River crossing planning.
- Situational obstacle planning.

- The options for joint force operational movement, maneuver, and protection.
- The C2 options to best employ the engineer capabilities of the joint force.
- Recommendations for ISBs, FOBs, forward logistic support sites, and avenues of approach (AAs).
- MSRs and available LOC facilities.
- The development of EMSTs.

COURSE OF ACTION ANALYSIS (WAR GAME)

4-55. COA analysis allows the staff to synchronize the BOS for each COA and identify which COA best accomplishes the mission. Through war gaming, the staff attempts to foresee the action-reaction-counteraction dynamics of the battlefield that would result from each COA. Engineers should develop and recommend the criteria for evaluation of EMSTs during this phase to subjectively assess the war game. Engineers should focus on the impact of the threat engineer functions and the friendly engineer functions with respect to the impact of the terrain on the operation. The ENCOORD synchronizes the methods to achieve the desired effects and ensures that they are accomplished. The process of war gaming allows the staff to understand the advantages and disadvantages of each COA and forms the basis for the commander's course of action comparison and decision. COA war gaming may identify branches and potential sequels that require additional planning by the engineer and the staff.

COURSE OF ACTION COMPARISON

4-56. Based on the identified criteria for evaluating the war gamed COAs, the engineers highlight the advantages and disadvantages from a SOEO perspective for each COA. The actual comparison may follow any technique that results in a recommendation. An effective technique for comparing COAs is a comparison matrix. Using specific criteria, each COA is compared to the others. While comparing COAs, engineers determine which SOEO best supports the mission.

COURSE OF ACTION APPROVAL

4-57. The staff recommends the best COA to the maneuver commander based on an outline of the COA and its advantages and disadvantages. The force commander considers the staff recommendation presented by the Operations Directorate (J3), G3, or Operations and Training Officer (U.S. Army) (S3); approves the SOEO to include EMSTs; and announces a decision and concept/ intent for the approved COA. At this point, the engineer commander can issue another WARNO using the force commander's updated information. This facilitates the planning for engineer subordinate units. The engineer staff officer makes a recommendation to the commander during the decision brief. The type and amount of detail that the engineer briefs depends on the needs and preferences of the individual commander. In general, it covers the—

- Concept of engineer support.
- Engineer mission priorities.
- Critical engineer events/actions.
- Task organization and command/support relationships.
- Obstacle overlay (including SCATMINE employment authority and concept for use by the system type).
- Survivability estimate and priority.
- Critical tasks directed to subordinate units.
- Engineers' work/construction time line.

4-58. Other members of the battle staff brief information that the engineer provides input for during the estimate process. This is particularly true of the intelligence portion of the decision brief and the comparison of COAs. Once the commander makes a decision, the estimate provides the bulk of the information needed to prepare the maneuver force's OPLAN or OPORD.

ORDERS PRODUCTION

4-59. During the planning development phase, an approved COA is expanded into a complete OPLAN, OPLAN in concept format (with timed-phased force and deployment data [TPFDD]), or functional plan. The process is the same for all plan types. To support this phase, the engineer provides input for the appropriate annexes and appendices of the plan to include the SOEO and the EMSTs, as found in *FM 101-5* and *Chairman of the Joint Chiefs of Staff Manual (CJCSM) 3122.03A*. Engineers prepare Annex F and its associated appendices for Army plans and Annex D (Appendix 6) and Annex L for joint plans. Besides the civil engineering support plan (CESP) and Annex L, the joint/Army engineer staff usually provides input, depending on the mission and combatant commander's intent. Refer to *JP 4-04* for further information on the CESP and Annex L.

JOINT ORDERS/PLANS

4-60. Although not included in *CJCSM 3122.03A*, engineers should seriously consider preparing an engineer appendix to Annex C. Major operational engineering considerations may affect the success of a joint operation. The emphasis should focus on the overall priority of the engineer effort by phase, as well as by any unique C2 considerations and support relationships between combat engineers and supporting theater-level engineers.

4-61. Engineers should review the entire order/plan for adequacy, feasibility, acceptability, and consistency with joint and Army doctrine for all combat, geospatial, and general engineering considerations. Special attention should be given to the following areas:

Task Organization, Annex A

4-62. Refer to Annex A to ensure sufficient capability to meet identified requirements. Additionally, planners should provide input to the flow of the engineer force, as detailed on the TPFDD.

Operation, Annex C

4-63. Engineers review and participate in the writing of Annex C. Their participation may be required in many of the subordinate appendices. Although not listed in *CJCSM 3122.03A* as a designated appendix, engineers may also find it appropriate to prepare a supporting engineer appendix to the appendix to *Annex C*. The format for this appendix would be similar to Annex F in *FM 101-5*.

Air Base Operability, Appendix 8, Annex C

4-64. Engineer considerations may factor heavily into the overall air base operability (ABO) concept of Appendix 8, Annex C, including the concepts of the five basic functions to—

- Defend (installations).
- Survive (provide expedient protection).
- Recover (assess the damage and effect repairs).
- Generate (alternatives for damaged systems).
- Support (the recovery effort).

Counterattack, Appendix 12, Annex C

4-65. The engineer will be a critical contributor to Appendix 12, Annex C. Any counterattack will be directly affected and, in some cases, defined by the availability and capabilities of combat, topographic, and general engineering support.

Explosive Ordnance Disposal, Appendix 13, Annex C

4-66. Appendix 13, Annex C should provide an outline of expected EOD operations in support of the plan. EOD personnel assist in identifying and removing UXO to support mission requirements and/or return facilities to a usable status. At the joint level, EOD is considered an engineer responsibility rather than a logistic one.

Force Protection, Appendix 15, Annex C

4-67. The combatant commander may require the joint force engineer to provide a list of the forces available to support the protection plan for Appendix 15, Annex C. Based on the commander's intent, activities of the enemy, and the predominant threat, the joint force engineer may provide appropriate assistance, including—

- Facility hardening.
- Revetments.
- Berms.
- ADC.
- Installation security improvements (barriers, perimeter fencing, monitors, and cameras).

Critical Infrastructure Protection, Appendix 16, Annex C

4-68. The combatant commander may require the joint force engineer to provide engineer expertise to support the critical infrastructure protection plan. The joint force engineer may provide assistance in the writing and implementation of the plan.

Mobility and Transportation, Appendix 5, Annex D.

4-69. Engineers provide input to this appendix to support the given OPLAN/ CONPLAN. Most of the support at this level will be general or topographic engineering related. All three tabs will likely require engineer input.

Civil Engineering Support Plan, Appendix 6, Annex D

4-70. Engineers provide Appendix 6, Annex D as an action plan for the six broad categories of facilities necessary to support the given OPLAN/ CONPLAN. The appendix defines the general character and magnitude of the civil engineering support (synonymous with general engineering for this appendix).

Civil Affairs, Annex G

4-71. Engineers should focus on areas affecting dislocated civilians, humanitarian demining operations (HDOs) efforts, government stability, and the destruction or degradation of civilian infrastructure. Engineers play a key role in providing shelter (tent cities) and performing essential repairs to facilities and utilities (water, sanitation, and power production). Engineering personnel can also help identify local resources to assist in civil-military operations (CMOs).

Joint Operation Planning and Execution System Operation Plan/operation Order/ contingency Plan, Annex L

4-72. Annex L, which contains three appendices, is the equivalent of Appendix 2, Annex F (environmental considerations). It is the parallel document for a joint staff and is generally written by the joint force engineer.

Geospatial Information and Services, Annex M

4-73. Annex M, which contains three appendices, identifies the geospatial forces assigned or attached, their manner of employment, and the required geospatial products and services. Engineers assist the combatant commander by identifying available geospatial assets to support the plan and making recommendations for additional support. In a Joint Operation Planning and Execution System (JOPES) OPLAN/OPORD/CONPLAN, this is the equivalent of Appendix 3, Annex F (geospatial information and services). It is the parallel document for a joint staff.

ARMY ORDERS/PLANS

4-74. Engineers should review Annex A to ensure sufficient capability to meet the identified requirements. The engineer staff officer lists the engineer units under the proper control HQ with the correct command or support relationships. Additionally, planners provide input to the flow of the engineer force as detailed on the TPFDD.

4-75. Engineers review Annex C and ensure the inclusion of obstacle effects or other graphics to assist in conveying the commander's intent on the overlay.

4-76. In Annex D engineers work with the fire support officer/staff to integrate obstacles with fire. Of particular interest are SCATMINEs and confirming that all obstacles are covered by fire.

4-77. Annex F is the principal means through which the engineer defines engineer support to the maneuver commander's intent, the SOEO; EMSTs, and coordinating instructions to subordinate commanders. It is not intended to function as the internal order for an engineer organization, where the engineer commander will articulate intent; the concept of operations; and coordinating instructions to subordinate, supporting, and supported commanders.

4-78. An integral component of the deliberate decision making process is the production of Annex F. The preparation of the annex seeks to clarify the engineer support to the OPLAN/OPORD and includes the—

- Priorities of work to shape the theater/AO (not in a tactical level engineer annex).
- Operational project planning, preparation and execution responsibilities (not in a tactical level engineer annex).
- Engineer organization for combat.
- The SOEO to include EMSTs.
- EMSTs for subordinate units.

NOTE: Guidance to maneuver units on obstacle responsibilities should be listed in the body of the basic order under paragraph 3a, Tasks to maneuver units, not here.)

- Allocations of Class IV and Class V (obstacle).

4-79. For Appendix 1, Engineer Overlay, Annex F, the engineer/ENCOORD may produce this overlay in conjunction with Annex C to highlight obstacle information (to include water obstacles) or breaching operations. A deliberate river crossing operation may require a separate annex as part of an order.

4-80. The engineer/ENCOORD performs as the staff integrator and advisor to the commander for environmental considerations. Appendix 2, Annex F, is a parallel document to Annex L in a joint OPLAN/OPORD/CONPLAN. *See FM 3-100.4* for an example of this appendix. When specific command procedures dictate, other staff officers include some environmental considerations in logistics and medical annexes. Unit planning at the regiment or brigade level and below will normally include only those elements required by the higher HQ orders or plans that are not already included in a unit standing operating procedure (SOP). If this appendix is not written, appropriate material will be placed in the coordinating instructions (paragraph 3, Execution) of the basic order.

4-81. The engineer/ENCOORD performs as the staff integrator and advisor to the commander for geospatial information and services. Appendix 3, Geospatial Information and Services, Annex F, is a parallel document to Annex M in a joint OPLAN/OPORD/CONPLAN. *See FM 3-34.230* for an example of this appendix.

SECTION IV - OPERATIONAL CONSIDERATIONS

4-82. The MDMP is valid at the operational level. At the operational level, the supported OPORD/OPLAN may be either joint or Army-specific, depending on the forces involved. *See CJCSM 3122.03A* for the formats that go with the JOPES process.

4-83. The engineer support planning process at the operational level focuses on the following interrelated activities:

- Engineer facilities study. The study is derived from the Joint Engineer Planning and Execution System (JEPES) computer model, which analyzes data. The study is used to develop a CESP and becomes part of Appendix 6, Annex D of the joint OPORD/OPLAN.
- Engineer annex. Annex F to the OPLAN is prepared using the results from the engineer facilities study and the CESP. The annex provides instructions for executing the engineer part of the OPLAN.

ENGINEER FACILITIES STUDY

4-84. According to *CJCSM 3122.03A*, the combatant commander's engineer planners use the JEPES computer model to prepare estimates of theater-level wartime engineer requirements for the following items in support of an OPLAN:

- Facilities.
- Engineer man-hours.
- CESP.
- Engineer annex.
- Environmental annex/appendix.

4-85. The primary purpose of JEPES is to assist the combatant commander and service-component engineer planners in determining whether the OPLAN provides the correct amount of engineer capability at the right place and is timed correctly to support deploying forces.

4-86. The JEPES is one of several tools the commander has to assess the validity and the accuracy of an engineer plan. The JEPES data, along with the engineer analysis and command guidance, provides a commander with another means to check the supportability of the engineer plan for a specific OPLAN.

4-87. In deliberate planning, the combatant commander includes a CESP within the logistics annex of the OPLAN. Independent of the combatant commander's plan, the Army's service-component engineers routinely develop their service plan as a means of detailed, deliberate planning.

4-88. The TPFDD is the primary driver of the JEPES model. This model extracts information such as the unit type, the destination location, the arrival time, and the population from the TPFDD. Given this input, the JEPES estimates construction man-hour and facility-type requirements to support the bed down of forces deploying into a theater. The JEPES also computes the engineer assets (man-hours) that are available to meet the estimated requirements. The JEPES provides Class IV output in the form of long tons and short tons. The results from the analyzed JEPES data are gross estimates that are used in the deliberate planning process for analyzing COAs for engineer support to the OPLAN. Because of the integral relationship between the JEPES model, the OPLAN, and the TPFDD for a theater, the JEPES does not readily lend itself to crisis planning in theaters where an OPLAN and TPFDD have not been prepared.

4-89. The JEPES model algorithms are based primarily on the support facilities necessary for the RSOI of all inbound forces. The model calculates facility requirements for a unit's final destination on the TPFDD but does not compute other engineer missions and support requirements within the theater. Aspects of the estimate that are not automatically calculated by the JEPES include—

- The construction of forward logistics bases and enemy prisoner of war (EPW) and displaced persons camps.The construction, maintenance, and repair of MSRs.
- The survivability of command, control, and communications (C3) nodes.
- The construction, expansion, or maintenance of port activities.
- Support for JLOTS operations.
- The construction of attack aviation strips, theater ammunition storage points, and fuel pipelines.
- Support for tactical elements in combat engineering (M/CM/S).

4-90. The JEPES has a capability for manually inputting specific requirements such as EPW camps; petroleum, oils, and lubricants (POL) pipelines; MSR construction; and other requirements, specified or implied in the mission analysis and planning guidance. This data is entered into a user's input file, which the JEPES combines with the other TPFDD requirements.

4-91. The output from the JEPES is a gross estimate, reflecting engineer capabilities, as depicted by the TPFDD. The accuracy and reliability of the information generated by the JEPES model is directly affected by the—

- Accuracy of the unit data on the TPFDD.
- Level of accuracy of the assets in the joint operations area (JOA).
- Level of specificity on the TPFDD.
- Assumptions for HN-provided facilities.

4-92. The product of the JEPES output analysis is the engineer facilities study. Engineer planners at the combatant commander and component levels use this study to prepare their CESP and the engineer annex to the OPLAN. The study also becomes Tab C of the engineer annex in joint orders.

CIVIL ENGINEERING SUPPORT PLAN

4-93. The CESP is used in preparing the engineer annex to the OPLAN. It is the primary planning document in which the engineer staff considers the minimum essential facilities and the construction capabilities required to support the commitment of military forces. When developing the CESP, the engineer staff should consider the—

- Engineer facilities study.
- Engineer intelligence of the theater.
- HN capabilities and HNS agreements.
- Construction contract capabilities.
- Mission of other US and allied forces.
- War damage estimates.
- Facility engineering responsibilities.
- Logistics support plans.
- Subordinate units.
- Any other aspect of the operation that impacts general engineering support.

4-94. In deliberate planning, the combatant commander includes the CESP within the logistics annex of the OPLAN. Independent of the combatant commander's plan, the Army's service-component engineers routinely develop their service plan as a means of detailed, deliberate planning. The CESP differs from the engineer facility study by considering other planning aspects that the JEPES was not designed to estimate. The CESP usually addresses the following engineer aspects in addition to the JEPES data:

- Information on available resources, facilities, and characteristics within the region relevant to the construction mission and construction capabilities.
- Restrictions imposed on the use of bases and installations.
- Major construction resources and their allocations.
- Future construction standards to be used as the theater matures.
- Responsibility for construction management among components.
- Responsibility for determining the facility's use in light of competing requirements from the components.
- Priorities at different phases during the conflict.
- Provisions for withdrawal, such as base denial and the movement of residual assets and stored critical Class IV supplies.
- Contract construction agent (CCA) missions and responsibilities and their relationship to engineer assets.
- Engineer support guidance and agreements for support commands and the area support group (ASG).
- Class IV construction materials availability.

4-95. Supported by the JEPES data analysis or other studies, the CESP states the priorities, programs, and general policy when seeking general engineering support. The format for the CESP is governed by *CJCSM 3122.03A* and included as Appendix 6 of Annex D in joint orders.

ENGINEER ANNEX

4-96. In JOPES, the engineer annex is primarily covered by the engineer responsibility to write Annex L, Annex M, and Appendix 6 to Annex D, although the engineer has significant participation in other portions of the OPORD and OPLAN as identified in *paragraphs 4-59 to 4-69* above. A separate engineer appendix to Annex C may be necessary or desirable in certain situations. In Army orders, Annex F will include Appendix 1, Appendix 2, and Appendix 3.

Chapter 5

Operational Principles of Engineer Operations and Organization

I would desire to have companies of Sappers formed—they should be instructed in every thing that relates to the construction of Field works—how to dispose of the Earth—to cut the Slopes—face with turf or sods—make fascines—arrange them properly—cut and fix Palisades, etc.

Louis Duportail
Chief of Engineers, Continental Army
January 18, 1778

Engineers must understand and be knowledgeable of basic Army tasks first, but this responsibility is closely followed by having a firm grounding in the fundamental Army missions, roles, and functions that are supported by the engineer battlespace functions. They must not only have an appreciation for Army engineer organizations, but also have a working knowledge of the organizations and capabilities of each of the other elements that make up the Engineer Regiment. This knowledge must then be applied to the tactical, geophysical, and political environments in which they are operating. The engineer branch contains all three of the force structure categories: combat, CS, and CSS. As such, it is a threefold branch with diverse capabilities that interact across all areas of the Army. When you add the other elements of the Engineer Regiment, the Army engineer must be multifunctional and versatile.

ENGINEER OPERATIONAL PRINCIPLES

5-1. Engineer operational principles are extracted chiefly from the principles of war and the Army tenets. There are a number of principles that engineers adhere to in order to optimize the engineer battlespace functions and to achieve success in the context of today's technologies and organization. The principles apply to all types of engineer units and to the full spectrum of operations. The eight operational principles associated with engineer operations are—

- Mass.
- Versatility.
- Security.
- Economy of force.
- Momentum.
- Simplicity.

- Synchronization.
- Initiative.

MASS

5-2. Engineers mass the effects of their combat power in time, space, and purpose to overwhelm the enemy or gain control of a situation. Engineer assets may be massed in conjunction with maneuver forces or independently. Mass does not necessarily equate to more engineer assets, it refers to the massing of engineer capabilities and effects. Effects can be massed by requesting special capabilities and assets (to include, but not limited to divers, pipeline construction, bridging organizations, FFE, and geospatial engineering capabilities) from higher HQ for special or unique requirements. Massing of effects is facilitated by accessing expertise through reach and synchronizing them to achieve the desired effect(s) at the times and places required to facilitate the maneuver commander's success. The most critical massing of engineer capability may be that of massing engineer expertise and "brain power" from a variety of locations other than the mission site. Note that the main effort of the maneuver force may not be where the engineer effort is massed. Engineer resources may be massed in a sustainment role supporting disaster relief in a foreign country or performing specialized combat missions within the maneuver plan, such as a river crossing.

5-3. Engineers must begin their work early to complete it on time. To effectively mass engineer effects, they have to anticipate future missions and reposition their unit, if necessary, while accomplishing current missions. Engineers must share the commander's vision and be proactive in anticipating mission requirements. As the operation transitions through the subsequent phases of the operation, engineers must be able to synchronize their actions to meet mission requirements and achieve operational and tactical success. Engineers focus on the success of mission rather than on habitual support to particular organizations. This does not mean that habitual combat engineer associations are not vital or retained whenever possible. Improved SU allows engineer commanders to anticipate requirements and mass M/CM/S, geospatial, and general engineering effects over simultaneous targets in the AO. Assured mobility and the selected massing of engineer assets depend on the ability to draw information from the COE and develop them into a shared SU.

5-4. Engineers organic to maneuver forces do not stay with those forces when they are held out of the fight. They remain out of action only long enough to refit or reconstitute after a major action before they are recommitted. Plans must ensure timely return to the parent maneuver force before their commitment to combat to ensure adequate integration and preparation time. Combat engineer forces can be positioned to support the combined arms reserve or counterattack forces and are considered to be committed, not held in reserve. There are time and distance considerations that affect how to plan and organize the massing of tactical engineer effects. Unlike fire effects, which can be reprioritized and massed quickly, tactical engineer effects are often ground-based and subject to time-distance factor limitations.

VERSATILITY

5-5. Engineers are capable of operating in multiple environments and performing a myriad of tasks that include M/CM/S, geospatial, and general engineering. These missions can occur during offensive, defensive, stability, and/or support operations under the C&S of a combined arms, joint, coalition, or interagency organization. It is the engineer's ability to transition quickly between these operations that reveals their versatility. (As an example, both combat and combat heavy engineers may be required to transition between conducting combat operations and repairing utilities or facilities while simultaneously providing support to civilians during UOs.) It is not unusual for engineers to have a battle-focused METL and be alerted to support a fire fighting mission. Engineers may also perform additional missions with augmentation and the approval of the higher commander. Engineers can organize and perform C2 operations during defensive attack, or stability and support operations. In these roles, engineers typically require other BOS elements to be task-organized with them to achieve the proper C2 structure.

SECURITY

5-6. Engineers never permit the enemy to acquire an advantage while performing their engineer battlespace functions. Army engineers are well integrated into the ISR plan and attempt to have SU at all levels. They embrace all measures available to protect themselves and conserve their assets. They integrate themselves into maneuver formations when possible for maximum security. When operating independently in remote parts of the AO, engineers provide their assets to maintain local security at all times. The same tactical security requirements apply in a support operation. Equipment and personnel are vulnerable to crime and hazardous materials (HAZMATs).

ECONOMY OF FORCE

5-7. Understanding the principle of mass, engineers know that they cannot provide troops or assets to every identified task in the AO. Assets and/or effort are task-organized to mass the effects and yet reduce the risks for the commander throughout the AO. Engineers recognize that risk may be required in military operations and that resources are based on priorities. They may reduce the risks by improved SU and task-organizing limited engineer units or assets to accomplish the mission as an economy of force. The higher commander should always understand the risks associated with engineer missions. Economy of force implies that the senior engineer commander must prioritize engineer effort. In a support mission, combat heavy engineers may have to mass effort in one region and minimize construction efforts in an economy of force with specific priorities, providing the minimum essential needs of a populated area.

MOMENTUM

5-8. Engineers ensure that the force maintains offensive momentum regardless of the enemy, the terrain, or any other impediment within the fundamentals of assured mobility and in line with the commander's intent. They are proactive, not reactive, according to the commander's intent and scheme of maneuver. Momentum implies that engineers work through the problems they encounter using resources (such as FFE) at their disposal or improvising other means to prevent the mission from stalling or failing. Engineer operations enable the momentum of the force throughout the AO in support of decisive, shaping, and sustaining operations.

SIMPLICITY

5-9. Engineers achieve simplicity by providing clear, uncomplicated plans and unambiguous, concise orders to ensure thorough understanding. Whenever possible, engineers resource subordinate units with substantial assets to prevent complicated movement of formations/assets/materials across the battlefield/AO and to simplify the plan. Orders are simplified and direct to alleviate confusion. Rehearsals are used to identify confusing areas and simplify them for subordinates. Combat engineer elements may, at times, remain with a given maneuver organization rather than be shifted to another organization or mission in the interest of simplicity. CS and CSS engineers use SOPs to provide simplicity when task-organizing between units. Combat heavy engineers maintain simplicity for construction projects using primarily basic theater construction standards.

SYNCHRONIZATION

5-10. Engineer commanders and his staff members serve as the senior member of the M/CM/S BOS and the senior coordinator of MANSPT to ensure that they synchronize MANSPT functions. The commander's staff synchronizes the plan and the commander synchronizes the execution. The staff promotes synchronization through SU and a COP of all friendly and enemy engineer effort in the AO. Additionally, engineers integrate engineer efforts across BOSs operating as combat, CS, or CSS forces.

5-11. Engineers are closely integrated with maneuver and fires. The scheme of maneuver governs the engineer plan. Fire, maneuver, and mobility and countermobility form a triad. Neither fire nor maneuver is truly effective if the combat formation cannot move at will and deny battlefield maneuver to the enemy. Engineers operate well forward to integrate mobility and countermobility into the triad.

INITIATIVE

5-12. Engineer leaders and soldiers at every level will operate with the commander's intent and common sense to use initiative in the absence of orders. Leaders know their missions, including reconnaissance, breaching, obstacle emplacement, and airfield and port construction and repair. They set the terms on their own, according to the commander's intent, when they are in the absence of specific orders. Initiative is enabled when engineers are able to develop and sustain a good SU.

ENGINEER ORGANIZATIONAL PRINCIPLES

5-13. Engineers are organized and equipped to support linear and nonlinear operations *(see Chapter 6)*. Engineers are a portion of a larger BOS and, as such, are required to task-organize to accomplish specific missions. To leverage their capabilities for the maneuver commander, engineer units are organized for success to perform all of the battlespace functions to support full-spectrum capabilities. The following principles apply to combat, CS, and CSS engineers:

- Deployability.
- Modularity.
- Multifunctionality.
- Span of C2.
- Sustainability.
- Commercial integration.

DEPLOYABILITY

5-14. To perform their missions, engineer units must be organized and prepared to rapidly deploy to support the unit they are task-organized to support. AC and RC engineer units are well-rehearsed and prepared to deploy with configured basic loads for a variety of potential missions ranging from combat operations to HLS support. Soldiers are qualified according to Army regulations (ARs), FMs, and unit guidance on their weapons and individual and collective tasks. Units that require contractor support to operate have a plan for autonomous operations or a plan to maintain contract support in the AO. Concepts for future engineer equipment is primarily designed for rapid deployment and ease of loading and unloading onto railcars, aircraft, and sealift. These concepts are balanced with design for M/CM/S, geospatial, and/ or general engineering capabilities. Deployability is directly enhanced by the modularity of units.

MODULARITY

5-15. Engineer units have a baseline standard suborganization with the ability to quickly deploy and be task-organized. Multifunctional soldiers provide the flexibility required in full-spectrum operations. Flexibility must exist at all levels of the organization and within each module. This enables rapid force tailoring before deployment and during employment. Modular design provides the versatility and agility to configure into the smallest unit building blocks to deploy the minimum capabilities for assignment to higher C2 HQ (either engineer or maneuver forces). The mission accomplishment should not be delayed or adversely affected by an unnecessary footprint of larger engineer units.

5-16. Mission requirements drive the size and composition of the engineer force. A mix of different units and organizations is often necessary to achieve the proper balance of capabilities. Modular engineer support capabilities may be provided from various sources other than Army engineer units. Engineers from other services, USACE, other military branches, and outside agencies can all contribute to ensuring the right mix of engineer support. This mix may change as the operation progresses through its various phases. These modular capabilities must be organized to maintain responsiveness, flexibility, and effectiveness. The use of reach capabilities (to include TeleEngineering) within the Engineer Regiment makes it possible to minimize the footprint of engineers in theater while still supplying the necessary engineer support to committed forces.

MULTIFUNCTIONALITY

5-17. Engineer units must be capable of performing multiple engineer functions and should anticipate these requirements. Units are organized to perform specific engineer functions but are prepared to conduct other engineer tasks outside of their unit METL, but within the broad range of engineer battlespace functions, when missions are anticipated. As units support full-spectrum operations, they may have to quickly transition from combat operations to support operations requiring a completely different set of tasks. Units operating in noncontiguous areas are capable of performing limited engineer functions outside their area of proficiency when empowered and enabled with reach capabilities and expertise.

SPAN OF C2

5-18. Engineer operations are often characterized by supporting a large geographic area and having many diverse organizations under its control. Therefore, engineer C2 capabilities strive to be interoperable with all Army, joint, and multinational organizations. The C2 is capable of scaling larger or smaller in size and capability, as required by the subordinate or geographic requirements. Engineers are prepared to provide C2 to other nonengineer units when METT-TC requires stability or support operations. Timely and accurate information is power; therefore, C2 facilities must function vertically and horizontally, monitoring and tracking combat operations and sending accurate reports to the right person at the right time. Monitoring and reporting the status of engineer missions are critical to C2 decision making as is adjusting engineer forces and assets to meet planned or unanticipated mission requirements. Mission control and effective communication down to the lowest organizational level of engineers is critical to achieving this goal.

SUSTAINABILITY

5-19. Engineer units are organized with the appropriate logistical support to move, maintain, and fuel, or they coordinate the appropriate service, HN, or contractor for support. Engineer resources are always limited. Materiel, transportation assets, and time restrictions limit the engineer's ability to execute missions. Engineer unit sustainment and the supporting logistical structure must be planned in detail. Logistical limitations may restrict the mission accomplishment by the engineer force. Engineer operations are often resource-intensive. Engineer commanders should exhaust unit capabilities before tapping into external logistical support from the HN or LOGCAP.

COMMERCIAL INTEGRATION

5-20. Engineer units are capable of operating commercial engineer equipment and using available resources to expand their versatility. This allows for smaller deployment requirements while improving efficiency and, in some cases, the overall effectiveness of the engineer unit. Engineer deployments require coordination with contractors, ENCOMs and the USACE to facilitate access to commercial equipment, assets, and real estate. Engineer resources (labor, services, equipment, and materials) belonging to the HN, other services, and multinational forces are present in every theater. At all levels, but especially at the tactical level, improvising may be necessary to convert on-site materials and equipment for military use. HN resources may be well-suited for general engineering support, especially in the rear areas. Local resources may augment available engineers, releasing more engineer units for other missions. Through the processes of communication technology, engineers also have the capability to reach from nearly any location worldwide to obtain products, specialized expertise, and other necessary information. Reach capability may help to minimize the footprint of engineer forces in a given forward AO.

Chapter 6

Battlefield Framework

The battlefield is the epitome of war. All else in war, when war is perfectly conducted, exists but to serve the forces of the battlefield and assure success on the field.

BG S.L.A Marshall, 1947

The Army's warfighting focus enables a diverse (full-spectrum) force to meet the needs of the JFC in war, conflict, and peace. In war, Army forces form the nucleus of the land component—imposing their will on enemies and causing their collapse. In conflict, Army forces deploy quickly into an AO to deter adversaries and potential enemies from establishing their forces and to preclude them from gaining an operational advantage. If deterrence fails, Army forces defeat the enemy, terminate conflict to achieve national objectives, and establish self-sustaining postconflict stability. Early movement of Army forces retains initiative and freedom of action by providing the JFC complementary means to conduct decisive offensive operations at a time and place of the commander's choosing. If theater circumstances require it, Army forces provide the means to block the enemy's offensive and deliver the counteroffensive blow necessary to win as rapidly as possible. In peace, Army forces train for war. They also help shape the international security environment through engagement activities and support civilian authorities both at home and abroad in response to natural or man-made disasters. Regardless of the type of commitment of Army forces, the degree of engineer participation is likely to be high.

BATTLEFIELD ORGANIZATION

6-1. Within the discussion of the operational framework *(see FM 3-0)* is the concept of battlefield organization. Battlefield organization is the allocation of forces in the AO by purpose *(see FMs 3-0* and *3-07 [100-20])*. It consists of three all-encompassing categories of operations: decisive, shaping, and sustaining. The purpose unifies all elements of battlefield organization by providing common focus for all actions. Commanders organize forces according to purpose by determining each unit's operation will be decisive, shaping, or sustaining. These decisions are the basis of the concept of operations. When circumstances require a spatial reference, commanders describe the AO in terms of deep, close, and rear areas. These spatial categories are especially useful in operations that are generally contiguous and linear and feature a clearly defined enemy force. Combat engineering tends to support the categories of decisive and shaping operations. Geospatial engineering supports all categories, with a primary focus on decisive and shaping operations. General engineering may also support categories. While it is not a separate engineer battlespace function, sustainment engineering is the term often used to describe those general engineer missions that support the category of sustaining operations.

LINEAR OPERATIONS

6-2. Linear operations are how the Army has traditionally fought, with ground forces sharing boundaries and oriented against a similarly organized enemy force. Linear operations employ deep, close, and rear areas within the battlefield framework of decisive, shaping, and sustaining operations.

6-3. Operations in deep areas are conducted in a designated area forward of the close area, with the purpose of shaping enemy forces before their arrival into the close area. The deep area is defined in terms of time and space. Engineers in the offense use reconnaissance assets to predict and detect the enemy to prevent and avoid obstacles that may impede the friendly force. When required to neutralize, engineers may coordinate for fires where they anticipate the enemy is overwatching obstacles and to set the SOSRA conditions for the eventual close fight; breaching, or gap crossing operations. In the defense, engineers may emplace scatterable munitions or conventional obstacles covered by direct or indirect systems to disrupt the enemy's maneuver in the deep fight.

6-4. Operations in close areas occur where the commander envisions close combat as imminent. The maneuver commander masses his combat capabilities to achieve a decisive action. In the offense, engineers balance their limited assets between the requirement to mass with redundancy and the risk with an economy of force. Engineer effort is focused on avoiding impediments then on neutralizing. In the defense, engineers enhance force protection providing survivability positions. Engineers emplace obstacle groups to reinforce the maneuver commander's plan and increase his target acquisition time in the EA.

6-5. Sustaining operations are conducted behind the commander's rear boundary. They assure freedom of action and continuity of operations, sustainment, and C2. Forces in this area are often prone to attack and take enhanced force protection measures to protect forces and facilities. Engineers provide support to enhance force protection by providing survivability to protect vulnerable assets. Engineers are integrated to provide rear commanders with assured mobility support to assist in predicting, detecting, preventing, avoiding, neutralizing, and protecting in the rear area.

NONLINEAR OPERATIONS

6-6. Nonlinear operations are how the Army fights with ground forces in noncontiguous and simultaneous operations while allocating forces in the AO by purpose. Engineers contribute to decisive, shaping, and sustaining operations by establishing the imperatives of assured mobility. Developing a COP leverages information by using geospatial tools to combine terrain data and an integrated ISR collection plan to attain a level of understanding within the elements of combat power.

Decisive Operations

6-7. Decisive operations are those that directly accomplish the task assigned by the higher HQ. Decisive operations conclusively determine the outcome of major operations, battles, and engagements. There is only one decisive operation for any major operation, battle, or engagement for any given echelon. The decisive operation may include multiple actions conducted simultaneously throughout the AO. Commanders weight the decisive operation by economizing on combat power allocated to shaping operations. In the offense and defense, engineers normally focus on freedom of maneuver through assured mobility. Decisive operations integrate all aspects of the assured mobility imperatives, with emphasis on maintaining mobility and momentum. This requires the senior engineer to synchronize the M/CM/S BOS capabilities to protect and sustain the ability to maneuver when and where the maneuver commander desires, enabling him to maintain pressure and lethality upon the enemy. It also requires synchronizing those engineer tasks that are embedded in the other BOSs.

6-8. In stability operations, decisive operations are usually those that achieve and maintain stability, protect lives and property, or promote peace. Unlike decisive operations in offense or defense, decisive operations in stability operations do not always have immediate impacts; sometimes results take years to achieve. In support operations, decisive operations normally prevent or mitigate the effects of natural of man-made disasters. They relieve or reduce conditions such as disease, hunger, or privation. Decisive operations could be stabilizing areas by providing security for personnel, facilities, or capabilities; rendering certain services to populations; or reestablishing critical infrastructure. Engineers play a significant role in both stability and support operations, and engineer actions may contribute to or actually be the decisive operation, especially during support operations.

Shaping Operations

6-9. Shaping operations at any echelon create and preserve conditions for the success of decisive operations. Shaping operations include lethal and nonlethal activities conducted throughout the AO. They support the decisive operation by affecting enemy capabilities and forces, or by influencing an adversary's decisions. Shaping operations use all elements of combat power to neutralize or reduce enemy capabilities. They may occur before, concurrently with, or after the start of the decisive operation. They may involve any combination of forces and occur throughout the AO. Typically prior to decisive operations, engineers contribute to shaping operations by establishing and maintaining operating areas. They gain control of friendly and enemy EAs, areas of interest (AOIs), operating areas, and mobility corridors that connect these areas. Engineers may then attack the enemy's ability to influence friendly operating areas. This proactive shaping allocates combat power and ISR capabilities to secure friendly maneuver areas and/or attack threat operating areas to prevent the threat ability to influence friendly maneuver operations. In stability operations, shaping operations often convert temporary gains into long-term political successes. The capabilities required to exploit stability often differ from those needed to achieve stability. This may require such measures as rotating different types of units in and out of the AO as the operation progresses through different stages. In support operations, shaping operations may include influencing perceptions, ideas, and information, as well as maintaining legitimacy. Shaping operations entail transferring tasks to civilian agencies or the local government. For military engineers, this may also include transferring responsibilities to USACE, civilian contract engineers, or HN organizations. Regardless of the type of operation, commanders may designate a successful shaping operation as a decisive operation. In that case, commanders weight the new decisive operation with combat power from other shaping operations. The concept of operations clearly describes how shaping operations support the decisive operation.

Sustaining Operations

6-10. Sustaining operations generate and maintain combat power. They are operations at any echelon that enable shaping and decisive operations by providing CSS, rear-area and base security, movement control, terrain management, and infrastructure development. Engineers contribute to sustaining operations by providing general engineering support to establish the theater backbone structure. Sustaining operations include rear-area and base security measures taken by military units, activities, and installations to protect themselves from acts designed to impair their effectiveness. Sustaining operations has four components: intelligence, base and base cluster self-defense, response force operations, and combined arms TCF operations. Engineers provide geospatial products that are distributed throughout the force to develop the COP from the FLOT to the rear most boundaries. FESTs from USACE and theater level ENCOMs provide design and labor capabilities to construct staging bases, logistic bases, FOBs, and LSAs. Engineer support to infrastructure development applies to all fixed and permanent installations, fabrications, or facilities that support and control military forces. Infrastructure development focuses on facility security modifications and includes ADC and repairs. Engineer operations provide mobility support throughout all LOCs for unimpeded sustainment support to forward forces. Engineering performed in support of sustainment operations may be referred to as sustainment engineering. (This is not another category or addition to the engineer battlespace functions.) The vast majority of this support will be general engineering and include at least some geospatial engineering.

FRAMEWORK CONSIDERATIONS

THEATER ECHELON

6-11. The ARFOR or ASCC engineer tailors the engineer structure to meet the requirements the JFCs plans or orders. While focused on operations in the CZ, these engineers also support the theater with general engineering support. The ARFOR or ASCC engineer must be closely tied into current and future operational planning and have his own C2 structure to ensure the timely and proper execution of the intent and scheme of maneuver. He must also anticipate and plan for future mission requirements. Engineer mission responsibilities at the operational level include constructing, maintaining and rehabilitating the theater support base, and providing support to other services, agencies, and other multinational forces in the TO. This may include support of the initial force projection requirements to create or improve the force projection enablers. The ability of CSS units to conduct sustainment operations, as well as moving and sheltering combat and CS forces, depends on adequate, responsive engineer support.

6-12. The number and type of operational-level engineer organizations depends on the operational environment, mission requirements, HN infrastructure, and availability of existing engineer support to the TO. Operational-level engineer units provide—

- Geospatial information and services support to the theater.
- Construction management for the ARFOR or ASCC.
- Troop construction and repair to all US elements in the COMMZ.
- Contract construction support.
- LOC development and sustainment support.
- Infrastructure enhancement and sustainment support.
- Real estate and facility acquisition, management, and maintenance support.
- General, geospatial, and combat engineering support to tactical-level organizations on an area or mission basis.

6-13. Establishing EWLs is a standard engineer C2 method used by the ARFOR or ASCC engineer, in conjunction with the task organization, to designate AORs for subordinate engineer organizations. EWLs are generally established from the corps rear to the rear of the COMMZ in a TO. EWLs may also be established to designate engineer support boundaries in operations other than war. These lines generally match ASG and HN military and political boundaries. EWLs will be adjusted as an operation transitions. Because engineers focus on mission requirements rather than area support, EWLs may be independent of other control measures (for example, to permit operational-level engineers to concentrate forward in the corps area to free up corps engineer assets or to conduct specific missions on a task basis forward of the EWL.)

6-14. Engineers are nearly always required in a TO. This force must conduct a carefully tailored, well-planned, and well-rehearsed mission. Contingency operations require a greater proportion of engineers to support the force.

6-15. Requests for engineer support (whether Army, other service, or HN) in the COMMZ are received and passed from an ASG to the supporting engineer group or from a TSC to a supporting engineer brigade or the ARFOR or ASCC engineer. The ENCOORD assigns missions based on ASCC priorities, drawn from the combatant commander or JFC priorities. Engineer units in the COMMZ execute the missions under control of the ASCC task organization. Unfulfilled requests for COMMZ engineer support are passed by dual routes to the higher ENCOM and the ASCC engineer, as well as through ASG channels to the TSC for resolution at the ASCC and joint levels.

6-16. Theater construction management often spans multiservice requirements. The combatant commander may establish a theater regional contingency engineering management (TCEM) or RCEM cell to prioritize engineer construction activities and augment the joint force staff. Army representation in this cell is normally drawn from the ARFOR/ASCC engineer staff. Other service component engineer personnel are also part of this cell.

6-17. The combatant commander may also establish joint boards to address engineer issues outside of normal operations and ensure coordination among the services. Such boards include the Joint Facilities Utilization Board (JFUB) and the Joint Environmental Management Board (JEMB). Each is discussed in *JP 3-34* and *JP 4-04*. Again, Army representation is normally drawn from the ARFOR or ASCC engineer staff and includes other service component engineers.

6-18. The TCEM or RCEM cell and joint boards are used to coordinate other military and non military resources. They synchronize and integrate the total engineer effort. Other military resources available to the combatant commander and the JFC include USACE teams, Navy construction engineers (Seabee teams), Rapid Engineers Deployable Heavy Operations Repair Squadron, Engineer (RED HORSE) units, or multinational engineer units. HN support, including local contractors and other engineer and logistical contracting options such as LOGCAP, provides a vast variety of engineer capabilities.

Combat Engineering

6-19. Theater and operational-level engineer units typically only perform combat engineer (M/CM/S) missions when operating in the division area as reinforcement to divisional engineers. Operational mobility and countermobility tasks are included under the general engineering function. Combat engineering tasks may also be performed in the theater rear area but tend to be associated with support of the TCF in its rear-area mission. *See FM 3-90* for a discussion of this mission.

Geospatial Engineering

6-20. The theater topographic battalion HQ is typically located in the COMMZ and will generally fall under the ENCOM, if an ENCOM is in theater. With the topographic planning and control team at the theater HQ, the battalion coordinates with the Topographic Engineering Center (TEC), HN, NIMA, and the National Ground Intelligence Center to fulfill theater mapping and satellite imagery products. They may also integrate USACE products to provide a more complete product in some cases. The general support (GS) topographic company supports noncorps units at the theater level and assists the DS companies in corps areas. They are normally collocated with the battalion HQ and may send elements to other locations, to include an EAC intelligence center.

General Engineering

6-21. Theater construction management often spans multiservice requirements. The theater commander may require the establishment of a TCEM or RCEM to prioritize and control engineer activities, with particular emphasis on construction activities. This management responsibility may be performed by the senior engineer staff and leadership (usually an ENCOM) on behalf of the ASCC. The engineer brigades and groups plan, coordinate, and supervise their assigned work in their respective AORs.

6-22. New construction, maintenance, and repair work in the COMMZ support and sustain combat operations forward. This includes roads, railways, pipelines, bridges, airfields, ports, buildings, utilities, EPW camps and installations, displaced persons camps, and any other requirements to support or take the load off combat operations.

6-23. Combat heavy engineer battalions have a variety of equipment, tools, and skills to do all types of construction. Some capabilities, such as asphalt and paving, require pairing with separate engineer companies and teams to accomplish the mission. Construction support companies operate and maintain equipment to augment combat engineer battalions, heavy, and other units as well.

6-24. Dump truck companies provide additional haul assets. Pipeline construction support companies provide technical personnel and specialized equipment to support pipeline construction and related facilities. Port construction companies provide technical personnel and specialized equipment for constructing and restoring ports, to include JLOTS facilities, inland waterway facilities, and POL marine terminals.

6-25. It is not uncommon for the ASCC to draw on other military and nonmilitary resources to integrate into the total engineer effort. Other military sources include USACE teams and support, Seabee teams, RED HORSE units, or other similar multinational engineer units. HNS, civilian contractors, and civil works agencies/businesses from outside the AO also provide a vast variety of potential augmentation capabilities.

6-26. While not exclusively a general engineering concern, environmental considerations *(see Appendix K)* come under general engineering for AUTL purposes. Engineers proactively advise the commander on environmental issues which promote awareness in the unit. Commanders protect the natural environment in which US military forces operate, to the greatest extent possible, consistent with operational requirements. Commanders must ensure compliance, to the greatest extent possible and as far as practicable within the confines of mission accomplishment, with all applicable environmental laws and agreements, including those of the HN. For more information on environmental regulations, laws, and treaties, *see FM 3-100.4*. The goal of compliance is to minimize the potential adverse impacts on human health and the environment while maximizing readiness and operational effectiveness. *FM 3-100.4*, and *JPs 3-0, 3-34, and 4-04* each highlight the critical integration of environmental considerations into all operations and training.

6-27. The ASCC has overall responsibility for real-property maintenance activities (RPMA). The TSC, through its ASGs, normally provides the needed RPMA support. Principal RPMA functions in a TO include operating, repairing, and maintaining facilities and utilities; fire prevention and protection; and refuse collection and disposal. The TSC provides technical RPMA guidance to the ASGs. ASGs provide RPMA support to all Army facilities in their AOs, to include leased facilities, unless HNS is or has available leased facilities. RPMA may also be provided for other service component facilities within the AO.

6-28. The theater's organizational RPMA requirements that exceed the TSC and ASG capabilities are forwarded to the supporting engineer group or brigade for execution according to the theater priorities. Configuring engineer units to support the ASGs is largely based on the expected RPMA workload. They are tailored to the specific ASG that they will be supporting to accomplish RPMA missions.

6-29. RPMA may be administered on a centralized or decentralized basis. If decentralized, the TSC, through its ASGs, uses supporting engineer teams, facility engineer detachments (FEDs), USACE, other CCA contracted RPMA services', or HNS to accomplish the mission. ASGs have small cells to receive, prioritize, and coordinate local engineer support requirements. They assess workloads, sequence critical requirements, and request assistance for the engineer group in GS.

6-30. In operations where a robust TSC/ASG structure doesn't exist, but base camps, logistic bases, and other operating bases have been established, the FEDs have taken on the larger task of conducting Directorate of Public Works (DPW) bare-base facilities management services. Leveraging their internal capabilities with other engineer resources in the theater, the FEDs become the focal point of administering RPMA to their individually assigned bases.

6-31. Many specialized engineer teams can be tailored to meet the needs of a particular ASG and support it under the C2 of an engineer HQ team. With the exception of the specialized team assigned to the ASG, the engineer group typically controls all of the engineer units in its assigned area. The ENCOM may also administer RPMA on a centralized basis. *FM 5-116* contains additional details on engineer operations in support of RPMA.

6-32. Contingency real estate support teams (CRESTs) are called upon early in a contingency operation to support acquiring real estate and facilities in support of reception, staging, RSOI bed down, and unit operations. Generally, requirements are centralized and front loaded to meet the anticipated theater level operational needs.

CORPS ECHELON

6-33. The corps generally fights at the tactical level of war; however, it may fight at the operational level during SSC operations. Although the corps has an engineer brigade assigned to it, the number and type of engineer units assigned to the brigade depend primarily on the mission of the corps, the number and type of their divisions and separate brigades, and the operational environment. Normally, each committed division of the corps will include elements of an engineer group from the corps engineer brigade. *See FM 5-100-15* for more information on corps engineer operations. *See Figure 6-1* for a notional corps engineer brigade organization.

6-34. Engineers perform battlespace functions (combat, geospatial, and general engineering) and have missions in all parts of the corps area. The general engineer mission in the corps AO is to construct and maintain LOCs and tactical march routes. This mission is continuous due to the effects of enemy actions, heavy traffic, and weather.

Engineer Corps Brigade Supporting Three Heavy Divisions

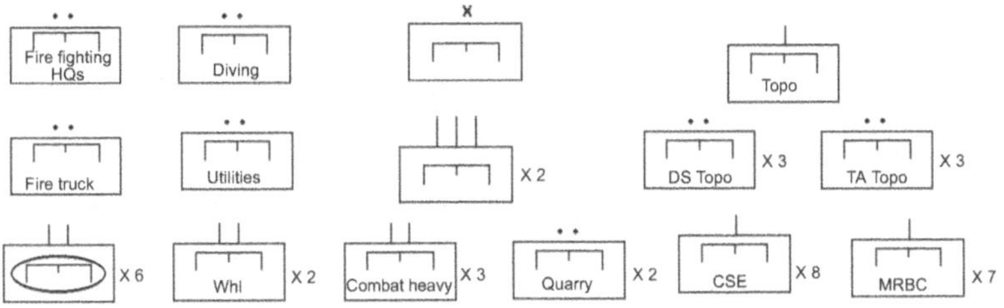

Engineer brigade competencies—

- Control corps engineer units not organic to divisions, separate maneuver brigades and ACR.
- Provide combat (M/CM/S), geospatial, and general engineering support to the corps.
- Augment engineer support to the divisions and other subordinate units.
- Require a major commitment of corps units to keep the LOCs and tactical march routes open to sustain the committed forces, shift other forces, and implement corps-directed attacks. Corps combat battalions perform those tasks with CSEs companies augmenting other units with horizontal equipment to maintain roads and airstrips.
- Provide wet and dry bridging multirole bridge company (MRBC) for use in LOC and other routes. A diving team performs scuba diving in forward combat areas (such as in river and far-shore reconnaissance), clears underwater obstacles, and supports assault float-bridging operations.
- Provide construction assets (combat heavy battalions and quarry teams) for operational construction requirements such as forward log bases, heliports, airfields, bed downs, and MSRs.
- Provide geospatial engineering for the corps; terrain teams provide geospatial (topographic) support for each assigned division.
- Provide fire fighting assets to support aviation units and petroleum units.
- Require utilities teams to support the corps support command (COSCOM) with RPMA support in the corps area.

Figure 6-1. Notional Corps Engineer Brigade Laydown

6-35. Corps engineer organizations provide guidance on protective shelters and camouflage measures to units and support corps units with geospatial information and services, as directed by the mission. Corps engineer units operating in the forward area reinforce divisional engineers in combat engineering roles. Many of the tasks performed by corps battalions in the corps rear area are also done in the forward area while performing missions in support of division operations.

Combat Engineering

6-36. Corps engineer units are typically combat support engineers. They usually perform (M/CM/S) missions when operating in the division area as reinforcement to divisional engineers. These tasks may also be performed in the corps rear area but tend to be associated with support of the TCF in its rear-area mission. *See FM 3-90* for a discussion of this mission.

6-37. Separate corps brigades and ACR have an organic engineer company, which is usually not sufficient to handle all required engineer tasks when these units are committed. These companies are designed and focused for supporting the mobility and countermobility portions of combat engineering. The corps engineer brigade reinforces these organizations with additional combat battalions and separate companies based on the mission and the situation.

Geospatial Engineering

6-38. A DS topographic company of the theater topographic battalion typically provides all forms of geospatial support for a corps. It is positioned in the corps rear area, providing a terrain analysis team to the corps CP.

General Engineering

6-39. General engineering activities occur throughout the theater and are not limited to the COMMZ. Within the corps area, operational construction requirements, such as forward logistics bases, heliports, and MSRs are needed to ensure the continuity of support for combat operations. These activities are in addition to the corps-generated construction requirements (bed down, logistics bases, and rehearsal ranges) that keep corps construction assets fully engaged. The nature of corps operations generally limits construction to the austere, essential facilities needed to sustain the current fight or support near-term major operations in offensive, defensive, and selected stability operations.

6-40. Corps combat engineer battalions perform general engineering tasks in the division area. CS equipment companies augment the combat battalions with equipment to move earth and maintain or create horizontal surfaces such as roads and airstrips. Combat battalions may also assemble tactical bridges provided by panel-bridge companies or allocated from theater stocks for use on LOCs and other routes. MRBC erect their own tactical bridges (fixed and floating) to support river crossing operations.

6-41. Light corps combat battalions and light equipment companies reinforce light DIVENGs, particularly during their initial deployment. The airborne corps battalion can parachute into an operational area and construct an assault airstrip. The light equipment company augments light forces with additional air-transportable earth-moving equipment.

6-42. Utility teams support the COSCOM and provide RPMA and base operations (BASOPS) support throughout the corps area. Additional support may be available through HNS or corps engineer brigade assets. The ENCOM provides additional support and technical guidance, as necessary. Frequently, in stability or support operations, USACE contractors (or other contracting agents) provide RPMA engineer services under the general oversight of Army engineers (the FED, which provides DPW bare-base facilities management).

DIVISION ECHELON AND BELOW

6-43. Divisions perform major tactical missions and can conduct sustained battles and engagements. The engineer forces organic to each division are tailored specifically to support that type of division. The corps engineer brigade provides additional engineer units based on the division's specific mission and tactical situation. The engineer infrastructure at the higher echelons makes it possible to commit and sustain divisions in combat.

Combat Engineering

6-44. The division combat engineer battalion is designed to perform the combat engineering function (M/CM/S) and to participate in the close fight. Corps combat battalions also participate in the division close fight. A corps commander will usually place at least one corps combat battalion in a command relationship to a division or engineer group supporting a division to reinforce the close fight. Corps and DIVENG elements often mix and cross attach elements to enhance the relative capabilities and strengths of each. The HQ of the habitually associated engineer battalion generally commands all engineers supporting the major effort of a brigade. A notable exception to this may be in a light division where only an engineer company is habitually associated to a maneuver brigade. There is also only a single organic engineer company in the SBCT. Each committed legacy heavy maneuver brigade normally needs the equivalent of an engineer battalion or one company per battalion TF. These levels of engineer support are adjusted based on the METT-TC analysis. Additional engineer capabilities may be task-organized from within the division or from EAD and EAC engineer units.

6-45. Additional corps engineer battalions operate in the division on an area or mission task basis. Separate engineer companies, especially bridge companies, operate in support of the division as required. When a division has the priority and need for a large number of corps engineers, it will likely also have an engineer group in support to control the activities of these engineers. When allocated to a division, the group HQ controls all engineer operations within the division's rear area. It also allocates and controls engineer organizations and resources to perform specific or selected tasks within maneuver brigade areas. This group HQ, or another group HQ may also function as the planning and control HQ for engineer tasks affecting the division as a whole (as in divisional assault river crossing of a breach by one or more maneuver brigades of the division).

Geospatial Engineering

6-46. The DS topographic company in the corps area also provides a terrain analysis team to the division. The team typically locates at the DIVENG HQ, with DTSS teams allocated to the assistant brigade engineers (ABEs). The DIVENG team works with the division's G2. TerraBase™ and other newly-developed programs are valuable tools to assist engineers at brigade and TF levels. The engineer cell of the SBCT contains an organic terrain team.

General Engineering

6-47. Combat engineer units rarely perform general engineering tasks within the division since their focus is on combat engineering tasks. General engineering tasks within the division are typically performed by corps level or higher CS engineer units. Tasks performed by combat engineers that obstruct and channel enemy maneuver (barriers and obstacles), protect friendly forces (defilade positions and shelters), and enhance forward mobility (combat trails and roads) are considered to be combat engineering functions. These missions are incorporated into the OPLAN and OPORD of the division and support the maneuver brigade/TF in accomplishing their respective missions and the commander's intent.

6-48. RPMA and BASOPS needs are generally very limited at the division echelon and below, although certain contingency operations will heighten their importance. Support is provided by the corps engineer brigade and higher echelon ENCOMs.

INTEGRATING OTHER ENGINEER ORGANIZATIONS AND CAPABILITIES

6-49. Integrating the variety and special capabilities of engineer organizations requires an understanding of the various capabilities and limitations of the engineer assets available for any given mission. Besides Army engineers, there are a variety of other organizations that may be available to support the overall effort.

6-50. The first of these are the other members of the Engineer Regiment, primarily USACE. They may have assets directly integrated into the military C2 structure and linked to the ENCOMs or be already operating under contract in theater. Whether providing construction contract and design support in the AO or outside of the contingency area, USACE can obtain the necessary data, research, and specialized expertise not present in the theater through TeleEngineering and other reach capabilities. *See Appendices I* and *J* for additional USACE capabilities.

6-51. Other US and allied military service engineers may also have engineer units and capabilities assigned to a given mission or operation. *See Appendices D, E, F, G, H,* and *I* for more information on other US and multinational military service engineer capabilities. HN engineer capabilities may be available if an adequate infrastructure exists. Potentially, this could include a wide array of civil and public works organizations. *See Appendix H* for a discussion of potential HNS.

6-52. It is increasingly common to contract a wide range of engineer services with local or third party national organizations and civilian contractors. These assets are typically used to free up military assets, minimizing the military footprint in a theater, when requirements exceed military capabilities or when the engineer operations and requirements are to be conducted in areas that are relatively safe from active combat.

Chapter 7

Battle Command for Engineers

In the profoundest sense, battles are lost and won in the mind of the commander.
Captain Sir Basil Liddell Hart, 1934

Battle command applies the leadership portion of the elements of combat power. It is based on the commander having a profound understanding of the conduct of full-spectrum operations and being able to visualize, describe, and direct the actions of subordinates. An engineer commander has a difficult responsibility in performing dual roles as the organic engineer commander and the senior engineer advisor to the maneuver commander. In today's operating environment, the engineer commander's leadership is essential to ensuring mission success. Understanding *FM 6-0 (101-5)* and how battle command relates to C2 operations is essential for all commanders.

BATTLE COMMAND AND COMMAND AND CONTROL OPERATIONS

7-1. The JFC and ARFOR organize assigned forces to meet mission requirements and use the resources available to them in the most effective manner. Engineer forces are extremely adaptable and can be tailored to best meet mission requirements. Each contingency operation will be unique, incorporating any of a range of different combinations of engineer entities and supporting relationships. Additionally, the structure and requirements that developed for engineer support will almost certainly vary as the operation transitions. Engineers will generally support the mission effort. However, they may also be called on to lead or command the operation, particularly when the mission is focused on disaster relief or nation building.

7-2. For most missions, the senior engineer commander in theater will C&S engineer forces as a specialized command with an independent command structure, maintaining technical and operational ties to strategic-level agencies and establishing working relationships with their ARFOR staff counterparts. Command authority is delegated by the ASCC or ARFOR commander, as is any command relationship with the TSC.

7-3. C2 operations are the exercise of authority and direction by a designated commander over assigned and attached forces to accomplish the mission *(FM 6-0).* The engineer commander must leverage available systems to most effectively achieve his intent. These systems include FBCB2 and MCS-ENG. As stated in *FM 3-0,* the goal of C2 is mission accomplishment. The engineer commander seeks to accomplish this by understanding the end state and ensuring that his assets are postured for transitions and future operations. He typically exercises authority and direction over engineer forces made available to him through establishing command or support relationships. However, he must also work within specific command or support relationships as well. Within these constraints, he is responsible for organizing resources and directing their operations. *See FM 6-0* for a more in depth discussion of battle command and all of the other primary aspects of C2.

7-4. *Figure 7-1* represents C2 visually and depicts the organizing principle of Army C2 doctrine. The focus of C2 is the commander. He determines what to do and directs actions. C2 is a continuous process; the staff must diagnose what is happening, even if unexpected, and develop appropriate solutions for the commander's decision.

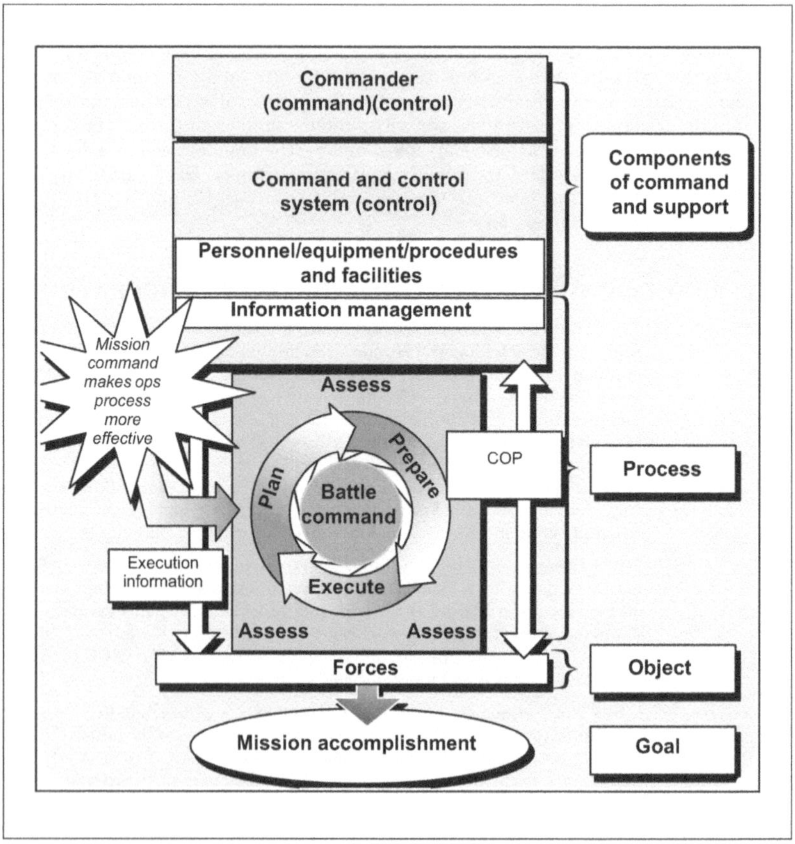

Figure 7-1. Command and Control Organizing Principle

LEADERSHIP ASPECTS—VISUALIZE, DESCRIBE, DIRECT

7-5. Commanders at all levels provide skilled judgment based on study, training, experience, and creative thinking. *FM 3-0* describes the critical aspects of leadership—visualize, describe, and direct *(see Figure 7-2, page 7-4).*

7-6. Upon receipt of the order, the engineer commander begins to formulate (visualize) a plan. He assesses his capabilities weighed against enemy capabilities on the terrain or support mission requirements. The commander conducts an informal engineer assessment after he reads the order, with input from his staff estimate. His staff prepares an EBA as part of the staff estimate. Using METT-TC, Army tenets, principles of war, engineer principles, and operational framework the commander visualizes the potential engineer tasks and the kind of information he needs to finalize a plan. The force commander may request products from geospatial engineering assets and FFE for technical support to improve his ability to visualize the combat, geospatial, and general engineering requirements for the operation. Engineer commanders may use sketches of the AO and locations of all the potential engineer tasks with purpose to visualize how the operation will occur.

7-7. The commander identifies (describes) what each subordinate commander, two levels down, must understand in the absence of orders. The commander uses the framework of the battlefield as a common reference to describe his vision to subordinates. Descriptions are provided of how he envisions supporting the operation regarding time, space, resources, purpose, and action for each type of operation—decisive, shaping, sustaining. He concludes with the commander's intent for engineer operations and planning guidance. The engineer commander's intent expresses the conditions for success and the desired end state of engineer forces on the objective.

7-8. The commander uses his focused intent (direct) as the foundation for this staff to develop the SOEO and supporting engineer concept of operations. This effort is synchronized by the senior engineer, chemical, and MP commanders within the M/CM/S BOS and through the MANSPT cell coordinator in the SBCT. The engineer commander directs his efforts by resourcing engineer mission requirements with personnel, time, and equipment.

ROLES OF ENGINEER COMMANDERS AND STAFFS

7-9. Whether performing as a commander or a supporting engineer staff element (referred to as the ENCOORD), engineers at each echelon monitor current operational activities, execute and control engineer missions, and plan future engineer operations in conjunction with the supported commander and forces. The depth and breadth of engineers involvement depends on the role the engineer plays within the military organization and the nature of the overall operation. His responsibilities may range from planning combat engineer operations to planning and coordinating environmental protection, critical areas, and protection levels. Success as an ENCOORD is critical to the success of the engineer unit. Commanders are challenged to generate a force that is tailored to current and anticipated missions, flexible in changing circumstances, and able to meet the constraints of lift capabilities. The factors of METT-TC drive the initial tailoring of the force. Prioritizing and leveraging available assets for acceptable degrees of risk guide the commander in

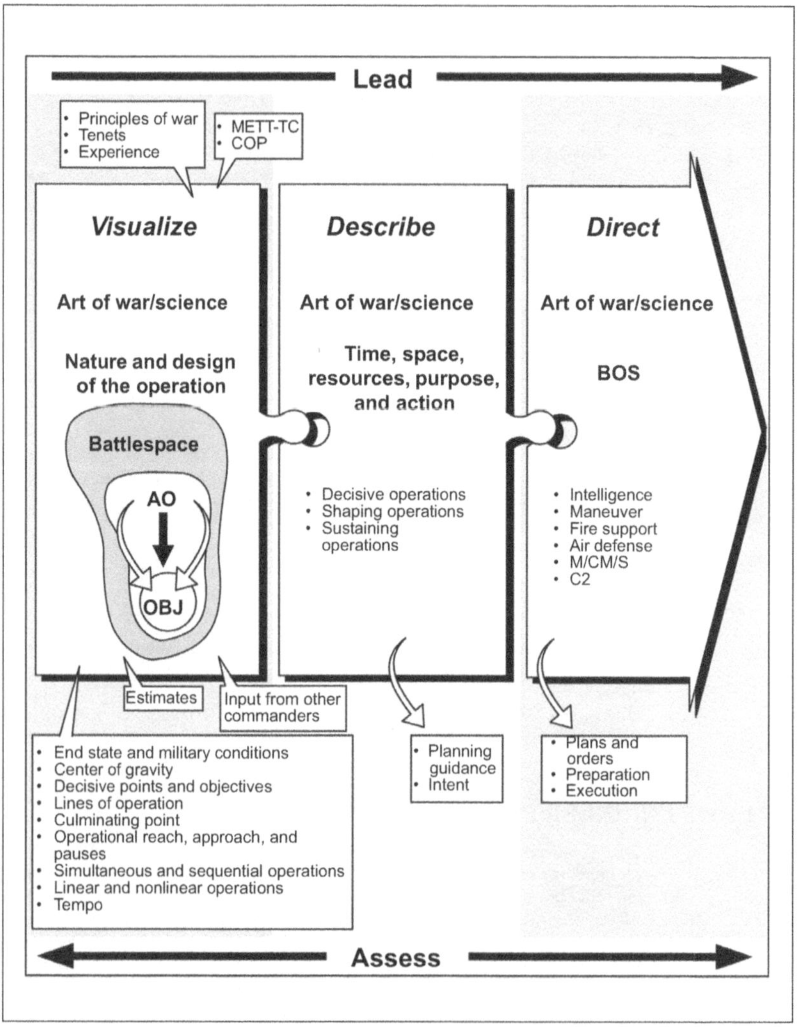

Figure 7-2. Leadership Aspects

determining the timing, number, and type of units to employ. Engineer units that make up the architecture of Army engineer forces are diverse and possess highly specialized capabilities. *(See Appendix D.)* Because the demands of any given operation vary, there is no set engineer structure; however, certain maneuver organizations do have a specific engineer element organic to them, and a building block approach based on need is used to determine the ultimate force refinement.

7-10. *Figure 7-3* is a useful depiction of the multiple engineer organizations that may comprise the theater engineer task organization. This organization lays the foundation to meet mission requirements. This mature structure develops incrementally, over time during a contingency response. *Appendix D* catalogs the deployable MTOE units within this structure as well as TDA organizations associated with the Engineer Branch. A discussion of other service engineer organizations is included in *Appendices E* through *G*. The discussion of potential contract engineering and other potential engineering support is included in *Appendices H, I,* and *J*.

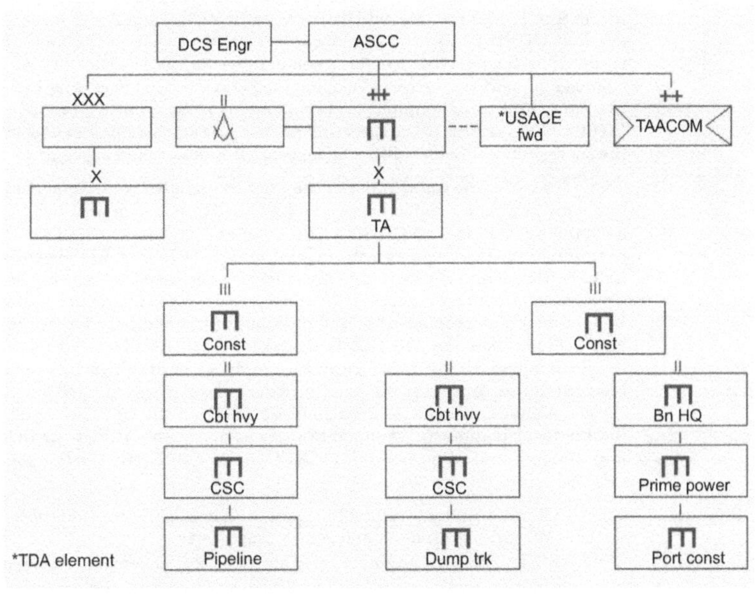

Figure 7-3. Notional Army Service Component Commander Theater Engineer Laydown

7-11. Each military operation is unique in itself, possessing varying degrees of the four basic operational components—ODSS. Engineers have a key role in each of these components, and they execute combat (M/CM/S), geospatial, and general engineering missions. The nature of the missions executed, both sequentially and simultaneously, determines the engineer's role.

7-12. As operations change and move across the spectrum of conflict, many of the products that engineers deliver at the unit level do not change. They continue to improve roads, upgrade infrastructure, lay pipeline, dig, and build. Force protection support is an engineer mission at every point on the spectrum. C2 structures and hierarchies are similarly organized, although at the lower end of the spectrum, engineers tend to play a proportionately greater role and may even be the controlling senior HQ in some instances. The prominence of the engineer role will be particularly high in austere environments. Additionally, it is in these environments and situations where USACE plays a larger role outside of its traditional boundaries and extends the participation of the entire Engineer Regiment.

THE ENGINEER COORDINATOR

7-13. The ENCOORD is the senior engineer officer within the force. However, in the SBCT, FXXI organized DIVENG staffs, ACRs, and Army level staffs, the ENCOORD is also the senior engineer commander in the force. He is responsible for advising the force commander on the best use of available engineer support resources, developing the SOEO, issuing necessary orders in the name of the commander, and implementing the approved SOEO. The ENCOORD is critical to the targeting process as well. His input is essential to the success of assured mobility and logic predict, detect, and prevent.

7-14. Typically, the commander of the organic engineer unit supporting the command or a senior staff officer is the ENCOORD. When an organization has a commander as the ENCOORD, the assistant or deputy ENCOORD is a permanent staff officer, representing the ENCOORD in his absence. An ENCOORD is normally located at every level of command within the theater, from the combatant commander or theater level to the maneuver battalion. Depending on the composition and missions of the engineer forces and the levels of operation, the ENCOORD can take on numerous forms at respective levels, each with its own scope and span of control and associated responsibilities. Whatever the Army echelon or level, the ENCOORD becomes the commander's means for accessing support from the various elements of the Engineer Regiment. When joint functions apply, the ARFOR ENCOORD may be designated as the joint ENCOORD. The ENCOORD's specific responsibilities include the following:

- Planning and controlling the engineer battlespace functions of combat (M/CM/S), geospatial, and general engineering.
- Recommending engineer organization for operations, to include combat.
- Advising on and obtaining engineer services and support through the reach capabilities of the Engineer Regiment.
- Maintaining the engineer COP of current operations.
- Advising on and developing policies for real estate acquisition, management, and disposal.
- Planning and coordinating with the S3, G3, J3, and fire support coordinator (FSCOORD) on integrating obstacles and fires.

- Planning and coordinating with the FSCOORD using artillery-delivered SCATMINEs.
- Advising the commander on the use of organic and nonorganic engineer assets.
- Advising the commander on employing and reducing obstacles.
- Advising the commander on environmental issues (to include the command environmental program), coordinating with other staff officers to determine the impact of operations on the environment, and helping the commander integrate environmental considerations into the decision-making process.
- Planning and coordinating environmental protection, critical areas, and protection levels.
- Providing a terrain mission folder to determine its effect on friendly and enemy operations.
- Managing the digital terrain data storage device (coordinating with the S2, G2, Intelligence Directorate [J2] for planning and distribution).
- Planning and supervising the construction, the maintenance, and the repair of camps and facilities for friendly forces, EPWs, and civilian internees.
- Assisting the S2, G2, and J2 in the IPB, including preparing the EBA portion of the engineer estimate.
- Participating in the targeting meeting.
- Participating in the JFUB, generally as the chair.
- Participating in the Joint Civil-Military Engineering Board (JCMEB), generally as the chair.
- Participating in the JEMB, generally as the chair.
- Providing information on the status of the engineer assets on hand.
- Making recommendations based on technical information to the appropriate operations and logistics staffs for the route of the MSRs and the logistics areas.
- Planning the reorganization of engineers to fight as infantry combat units when the appropriate commander deems their emergency employment necessary.
- Coordinating with interagency department engineers, such as the Federal Bureau of Investigation (FBI).

ASCC ENGINEER

7-15. The ASCC staff normally includes the ASCC engineer as a member of the commander's special staff. He integrates engineers into the ASCC's plan to sustain Army forces and support other services or allied forces. The senior engineer commander may be dual-hatted as the ASCC engineer or his deputy may assume this position as a primary role.

7-16. In a developed contingency operation that includes a COMMZ or rear-area support structure and boundaries, the senior engineer commander would most likely be an ENCOM commander with C2 of numerous engineer units, capabilities (topographic, design, and RPMA and assets (contract construction and HN). As such, the ENCOM commander executes the ASCC's goals of operational sustainment and mobility in synchronization with the roles already identified above. This is done through operational, planning, and technical staffs that are integrated within the ASCC as well as those organic to staff within the ENCOM. The ENCOM typically functions as a separate and specialized command under the ASCC. A theater engineer brigade or engineer group could also provide this support.

CORPS ENGINEER

7-17. The commander of the corps engineer brigade is the corps engineer or ENCOORD at the corps level. He plans and executes engineer operations in the corps area. As the engineer brigade commander, the ENCOORD commands all nondivisional engineer units in the corps. As the engineer special staff officer, he is responsible to the corps commander for all engineer-related matters in the corps AO.

7-18. The engineer brigade commander has a staff element located in the corps CPs. This staff is under the direction of the assistant corps engineer (ACE), who integrates engineers into the corps planning process. The ACE conducts planning with the corps staff and provides advanced warning of future corps operations through engineer channels to the corps engineer brigade, DIVENG brigade, separate brigades, engineer groups, and (ACR) engineer companies. He receives reports from these levels to keep the corps staff informed on current engineer operations.

DIVISION ENGINEER

7-19. The commander of the division (armor/mechanized and light) organic engineer brigade or engineer battalion is dual-hatted as the commander and the special staff DIVENG or ENCOORD. The DIVENG or ENCOORD plans and executes engineer operations in the division area. The division order often gives control of the elements of the engineer brigade to the maneuver commanders, yet the DIVENG commander retains command responsibilities and employs those elements left under his direct control. As the engineer special staff officer, the DIVENG is responsible (to the division commander) for all engineer-related matters in the division AO. He remains responsible, regardless of the rank of the senior corps or EAC engineer unit commander in the division AO. Nondivisional engineer units in the division AO provide liaison to the DIVENG.

7-20. To assist the commander in the role as a special staff officer, the DIVENG has a staff element located in the division CPs. It is under the direction of the ADE, who integrates engineer operations during the division's planning process. Through engineer channels, the ADE provides advance warning of future division operations to the division and supporting corps and EAC engineer battalions and maneuver brigades. He also receives reports from these levels to keep the division staff and the higher HQ ENCOORD informed on current engineer operations in the division. With staff assistance, the DIVENG controls corps and EAC units in the division area. He should task EAD engineer units in his role as the engineer brigade (heavy)/battalion (light) commander.

7-21. The senior engineers within FXXI divisions and ACRs are not commanders. They are not resourced with a full staff and CP and are not capable of independent C2 operations. In FXXI and ACRs, the senior engineers are special staff officers imbedded within the maneuver CP. EAD engineer units may not be task-organized directly to the special staff officer.

BRIGADE ENGINEER

7-22. In an armored or mechanized division, the organic engineer brigade generally provides an engineer battalion to each armor/mechanized maneuver brigade. In a light, air assault, or airborne division, the organic engineer battalion provides an engineer company to each maneuver brigade. The respective engineer battalion commander and company commander assume the brigade engineer (or ENCOORD) responsibility. In the SBCT, the brigade engineer is a special staff officer, not a commander, and functions as both the ENCOORD and BOS integrator for the MANSPT cell. The brigade engineer integrates engineers into the brigade planning process and coordinates current engineer operations in the brigade AO. As the brigade engineer, he commands all engineer units in the brigade. As the brigade engineer special staff officer, he is responsible to the brigade commander for all engineer-related matters in the brigade AO. Separate maneuver brigades and ACRs have a staff engineer organic to the brigade or regiment.

BATTALION AND TASK FORCE ENGINEER

7-23. The battalion or TF engineer often establishes a normal association between an engineer company and a heavy maneuver battalion or a light engineer platoon with a light maneuver battalion. Maintaining that association is one factor to consider in the tactical planning process, since there are advantages to the company commander or platoon leader functioning as the TF engineer. The company commander or platoon leader normally associated with a battalion/TF is also the staff engineer and advisor to the TF commander. The company commander has a small HQ section to assist the TF commander in integrating engineers into the TF planning process and in executing the engineer portion of the operation. The light engineer platoon leader will not have this luxury. The battalion/TF engineer makes operational reports through the TF S3 and provides other required reports through engineer channels, as necessary.

7-24. Since the SBCT has only one organic engineer company, engineer platoons will not be habitually task-organized to maneuver battalions/TFs. As a result, METT-TC will determine the level of engineer support task-organized to each maneuver battalion/TF, if any. In most cases, engineers will be organized to weight the main effort. In that instance, the company commander will serve as the TF engineer. When allocated engineers, the other maneuver battalions/TFs will usually have a platoon leader as the TF engineer.

7-25. When additional engineers operate with the TF, the associated company commander or platoon leader normally remains the TF engineer. However, the reinforcing engineer unit commander/platoon leader and staff task-organized to the maneuver unit or organic engineer will assist with the detailed planning.

COMMAND AND SUPPORT RELATIONSHIPS

7-26. Engineers have a number of specialty units with unique capabilities. Commanders task-organize these units to subordinate maneuver commanders to provide these capabilities. The commander can task-organize them in a variety of command support relationships. Based on the hierarchy, a subordinate commander may not impose a more restrictive command/support relationship if the unit is task-organized further down.

JOINT RELATIONSHIPS

7-27. At the theater level, when Army forces operate outside the US, they are assigned under a JFC *(see JP 0-2, JP 3-0,* and *FM 100-7).* A JFC is a combatant commander, subunified commander, or JTF commander, authorized to exercise COCOM or OPCON over a joint force. At the theater level, the combatant commander provides strategic direction and operational focus to forces by developing strategy, planning the theater campaign, organizing the theater, and establishing command relationships for effective unified action. The JFC plans, conducts, and supports the campaign in the theater of war, and during subordinate theater campaigns, major operations, and battles. The four joint command relationships are COCOM, OPCON, TACON, and support *(see Table 7-1).*

Table 7-1. Joint Command Relationships and Inherent Responsibilities

Inherent responsibilities are:	If relationship is—		
	COCOM	OPCON	TACON
Has command relationship with—	Gaining combatant commander, gaining service component commander	Gaining command	Gaining command
May be task organized by—	Gaining combatant commander, gaining service component commander	Gaining command	Parent unit
Receives logistic support from—	Gaining service component commander	Service component command; parent unit	Parent unit
Assigned positions or AO by—	As required by gaining component commander	As required by gaining command	As required by gaining command
Provides liaison to—	As required by gaining component commander	As required by gaining command	As required by gaining command
Establishes and maintains communications with—	As required by gaining component commander	As required by gaining command	As required by gaining command and parent unit
Has priorities established by—	Gaining component commander	Gaining command	Gaining command
Gaining unit can impose further command relationship/ authority of—	OPCON, TACON, DS, mutual support, GS, close support	OPCON, TACON, DS, mutual support, GS, close support	DS, mutual support, GS, close support

7-28. COCOM is a nontransferable command authority, exercised only by unified or specified combatant commanders, unless otherwise directed by the SECDEF. Combatant commanders exercise authority over assigned forces. COCOM provides full authority to organize and employ commands and forces the combatant commander considers necessary to accomplish missions. The combatant commander exercises COCOM through subordinate commands, to include subunified commands, service component commands, functional commands, and JTFs.

7-29. OPCON is inherent in COCOM. It is the authority to perform those functions of command over subordinate forces involving organizing and employing commands and forces, assigning tasks, designing objectives, and giving authoritative direction necessary to accomplish the mission. OPCON may be exercised at any echelon at or below the level of the combatant commander and can be delegated or transferred. Army commanders use it routinely to task-organize forces. The transfer of OPCON of units between two combatant commanders must be approved by the SECDEF. ENCOMs typically operate in an OPCON status.

7-30. TACON is the authority normally limited to the detailed and specified local direction of movement and maneuver of the tactical forces to accomplish an assigned task. It allows commanders below COCOM level to apply force and direct the tactical use of logistic assets but does not provide authority to change organizational structure or direct administrative or logistic support. The combatant commander uses TACON to limit the authority to direct the tactical use of combat forces. TACON is often the command relationship established between forces of different nations in a multinational force. It may be appropriate when tactical-level Army units are placed under another service HQ. TACON does not provide organizational authority or authoritative direction for administrative and logistic support. The commander of the parent unit continues to exercise those responsibilities unless otherwise specified in the established directive. Army commanders make one Army force TACON to another when they want to withhold authority to change the subordinate force organizational structure and leave responsibility for administrative or logistic support with the parent unit of the subordinate force.

7-31. Joint doctrine establishes support as a command authority. Commanders establish it between subordinate commanders when one organization must aid, protect, or sustain another. Under joint doctrine, the four categories of support are general, mutual, direct, and close *(see JP 0-2* and *JP 3-0)*. GS and DS describe the supporting command's focus. Mutual and close support are forms of activity based on proximity and combat actions. Army doctrine establishes four support relationships: direct, reinforcing, general, and GS reinforcing *(see FM 3-0)*.

ARMY RELATIONSHIPS

7-32. Commanders build combined arms organizations using C&S relationships. Command relationships define command responsibility and authority. Support relationships define the purpose, scope, and effect desired when one capability supports another. *Table 7-2* summarizes Army C&S relationships.

Table 7-2. Army Command and Support Relationships with Inherent Responsibilities

If relationship is:		Inherent responsibilities are:							
		Has command relationship with—	May be task organized by—	Receives CSS from—	Assigned position or AO by—	Provides Liaison to—	Establishes/ maintains communica-tions with—	Has priorities established by—	Gaining unit can impose further command or support relationship /authority of—
COMMAND	Attached	Gaining unit	Gaining unit	Gaining unit	Gaining unit	As required by gaining unit	Unit to which attached	Gaining unit	Attached, OPCON, TACON, GS,GSR, R, DS
	OPCON	Gaining unit	Parent unit and gaining unit, gaining unit may pass OPCON to lower HQ. *See note 1.*	Parent unit	Gaining unit	As required by gaining unit	As required by gaining unit and parent unit	Gaining unit	OPCON, TACON, GS, R, DS
	TACON	Gaining unit	Parent unit	Parent unit	Gaining unit	As required by gaining unit	As required by gaining unit and parent unit	Gaining unit	GS, GSR, R, DS
	Assigned	Parent unit	Parent unit	Parent unit	Gaining unit	As required by parent unit	As required by parent unit	Parent unit	Not applicable
SUPPORT	DS	Parent unit	Parent unit	Parent unit	Supported unit	Supported unit	Parent unit, supported unit	Supported unit	*See note 2*
	Reinforcing (R)	Parent unit	Parent unit	Parent unit	Parent unit	Reinforced unit	Parent unit reinforced unit	Reinforced unit then parent unit	Not applicable
	General support reinforcing (GSR)	Parent unit	Parent unit	Parent unit	Parent unit	Reinforced unit and as required by parent unit	Reinforced unit and as required by parent unit	Parent unit then reinforced unit	Not applicable
	GS	Parent unit	Parent unit	Parent unit	Parent unit	As required by parent unit	As required by parent unit	Parent unit	Not applicable

NOTE 1: In Nato, the gaining unit may not task organize a multinational unit (see TACON).
NOTE 2: Commander of units in DS may further assign support relationships between their subordinate units and elements of the supported unit after coordination with the supported commander.

7-33. The following are C&S relationships:

- Assigned.
- Attached.
- OPCON.
- DS.
- GS.
- TACON.

7-34. ADCON is the direction or exercise of administration and support authority over subordinate or other organizations, including the organization of service forces, resources and equipment control, personnel management, unit logistics, individual and unit training, readiness, mobilization, demobilization, discipline, and other matters not included in operational missions of the subordinate or other organizations. ADCON is synonymous with administration and support responsibilities identified in Title 10 USC. Unless otherwise specified by the SECDEF, ADCON of Army forces remains within the Army chain of command, from the lowest levels to the ASCC and back to the Secretary of the Army. ADCON is always subject to the command authority of the combatant commander.

COMMAND RESPONSIBILITIES

7-35. The engineer commander leads his engineers with the authority placed in his position. The commander balances his effort as a commander and senior advisor and integrator of the M/CM/S BOS for the combatant or force commander. In this unique role as a dual-hatted commander and staff integrator and coordinator, the engineer commander is first and foremost responsible for integrating and synchronizing the M/CM/S BOS and geospatial and general engineering into the concept of operations. It is this approach to support the bigger picture that allows engineers and their units to achieve success. Planning is part of the broader field of C2 and is a fundamental responsibility of commanders. While the staff completes much of the detailed analysis and preparations of plans and orders, the commander plays a central role in planning through the commander's intent and planning guidance. These guide the activities of the staff and subordinate commanders.

STAFF RESPONSIBILITIES

7-36. The senior engineer commander or staff engineer is first and foremost responsible for integrating and synchronizing all combat (M/CM/S), geospatial, and general engineering functions. This senior staff coordinator assists the maneuver commander with the coordination and detailed analysis necessary to convert the commander's intent and planning guidance into a plan. When another engineer organization arrives with a commander of equal or junior grade to the ENCOORD or the organic senior staff engineer, the organic ENCOORD will remain the senior advisor to the maneuver commander.

Chapter 8

Full-Spectrum Operations

Every age has its own kind of war, its own limiting conditions and its own peculiar preconceptions.

MG Carl von Clausewitz, 1832

The Engineer Regiment has a very challenging task to support its METL for a broad range of military operations ranging from decisively winning wars to keeping the peace and responding to other requirements within HLS. The Regiment, as a whole, is organized and equipped to respond to the broad range of full-spectrum operations. However, it is the challenge of the commander to plan, prepare, and execute thes ODSS missions. It is the operations process of planning, preparing, and executing with continuous assessment that the commander uses as a means to the end.

PLANNING, PREPARING, AND EXECUTING

8-1. The operations process of planning, preparing, and executing has proven successful to all commanders. It is a means for the commander or ENCOORD to break down responsibilities to subordinates and the staff. The engineer will face many challenges as he thinks through the potential for engaging in the full spectrum of operations. He needs to consider the complex tasks associated with transitions between these types of operations as well. This chapter provides some planning, preparing, and executing considerations for the spectrum of operations the engineer will face. It is the commander's ability to continuously reassess through the use of the operations process that allows him to dictate the terms of the operation and support the force commander or LFA in an HLS operation.

8-2. Planning for operations leads to an engineer commander making decisions before and during execution. At the core of decision making is knowing if, when, and what to decide. It includes understanding the consequences of decisions. Decisions are the means by which the commander translates the vision of the end state of the operation. Decision making is both science and art. At the higher echelons, art tends to govern over science and vice versa at the lower levels. Many aspects of engineer operations are quantifiable, such as the number of mines available or the materials required to construct a life support area. This is the science of war. Other aspects such as leadership, understanding the complexity of the operation, and anticipating the best place to breach a threat's defense in spite of the uncertainty of the enemy's actions are part of the art of war.

8-3. Preparing for operations includes activities conducted by engineers before executing to improve their ability to conduct an operation. At a minimum, these activities include plan refinement, rehearsals, reconnaissance and surveillance, coordination, inspections, and movement. Reconnaissance and rehearsals are of critical importance to engineers. Engineers do not have just one solution for all the tasks they may encounter. Therefore, it is essential to conduct reconnaissance rehearsals to identify the means and methods to accomplish known engineer missions and build flexibility into the plan for possible contingencies.

8-4. Executing is putting a plan into action by applying engineer capabilities to accomplish the mission. Inherent to executing operations is deciding when to execute the plan and when to adjust the plan based on the changing situation. Commanders use judgment and initiative to assess the situation and make decisions based on the model of assess, decide, and direct.

8-5. Because of the complexity of full-spectrum operations, engineers are critical in the transition between operations. As operations transition from defense to offense to stability or support, engineers have to show their versatility by rapidly resourcing and moving assets and capabilities to maintain momentum. Engineer capabilities are massed for periods of time and then redistributed to support the new operation, while continuing to support existing operations.

OFFENSIVE OPERATIONS

8-6. Surprise, concentration, tempo, and audacity all characterize offensive operations. Assured mobility provides a framework for the commander to examine using the operations process. Effective offensive operations capitalize on accurate intelligence and relevant information regarding enemy forces, weather, and terrain. The commander maneuvers his force to advantageous positions before contact. Security operations and defensive information operations keep or inhibit the enemy from acquiring accurate information about friendly forces. Contact with enemy forces before a decisive operation is deliberate, designed to shape an optimal situation for a decisive operation. A decisive operation is a sudden, shattering action that capitalizes on subordinate initiative and a COP to expand throughout the AO. The commander executes violently without hesitation to break the enemy forces' will or destroy them. *FM 3-0* discusses the four characteristics of offensive operations.

8-7. Army forces conduct offensive operations simultaneously with defensive, stability, and support operations while performing joint, multinational, and interagency operations. Similarly, while engineers may be focused on one or more of the engineer functions, all five functions operate simultaneously in support of offensive operations. *Figure 8-1* shows that relationship.

8-8. For offensive operations, prediction is a critical engineer contribution. By gathering the products available from geospatial engineering sources, engineers contribute to the commander's ability to see the terrain and frame the COP.

8-9. Engineers then describe the threat in terms of his subelements' functions on the terrain with weather effects. The threat is identified in terms of—

- How and where the threat is able to conduct engineer related functions such as assaulting and fixing?

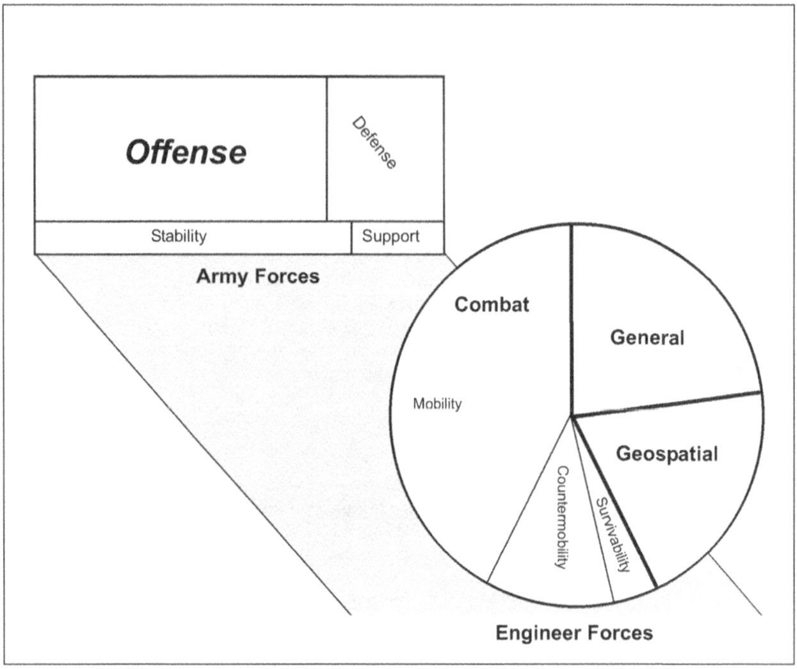

Figure 8-1. Generic Offensive Operations Relationships

- What assets are available to him?
- Is the threat in range to transport mines with assets available?
- How many mines can he carry?
- What terrain optimizes their use and impedes friendly maneuver?

8-10. Engineers should use sensors to observe these areas in the intelligence collection plan. It is during the planning phase that the commander identifies the links that predict must have with prevent and detect.

8-11. During the preparation phase, engineer assets are postured with their task-organized C2 HQ. The focus is on preventing impediments from impacting the friendly forces scheme of maneuver. This is accomplished by identifying viable routes around them or proactively planning for the removal of the impediments if an alternate route is not possible or desirable by the maneuver commander. There must be established links from the collection plan that are focused on preventing the threat from interdicting the friendly scheme of maneuver. Engineers also prevent enemy impact by participating in combined arms rehearsals to facilitate rapid execution to avoid or neutralize an obstacle. The threat actions that cannot be prevented must then be detected. Prediction efforts are linked to identifying and detecting impediments to the maneuver forces' freedom of movement. The engineer commander and his staff coordinate with the maneuver commander for ISR assets to detect potentially hazardous areas. Engineers also prevent enemy impact by conducting combined arms rehearsals to facilitate rapid-capability execution to avoid or neutralize an obstacle.

8-12. When executing offensive operations, the maneuver force uses its COP to link its detection to avoid obstacles along the route and provide assured mobility. The maneuver force can actively avoid an obstacle by interdicting threat countermobility before emplacement or passively avoid by identifying, marking, and bypassing an obstacle. If the friendly force commander is compelled to neutralize the obstacle to his front, the force will employ the breach tenets of intelligence, breach organization, breach fundamentals, mass, and synchronization. While supporting the maneuver commander, the engineer contributes to protecting the force by interdicting threat ability to influence friendly maneuver, constructing life support areas, or by digging in assets such as radars.

FORMS OF MANEUVER

8-13. Assured mobility supports all the forms of maneuver identified in *FM 3-90*. Unimpeded maneuverability is essential to the commander to achieve decisive operations. The five forms of maneuverability are specified by the higher commander. However, his guidance and intent, along with the mission that includes the implied tasks, may impose constraints such as time, security, and direction of attack that narrow the forms of offensive maneuver to one alternative. The fundamentals of assured mobility allow the commander to focus his assets and effort to maneuver on the battlefield within these constraints. Engineers synchronize their capabilities across the BOS to support assured mobility for each of the forms of offensive maneuver.

Envelopment

8-14. An envelopment is a form of maneuver in which an attacking force seeks to avoid the principal enemy defenses by seizing objectives to the enemy's rear to destroy him in his current position. It avoids the enemy's strengths and obstacles. At the tactical level, envelopments focus on seizing terrain, destroying specific enemy forces, and interdicting enemy withdrawal routes. A force seeking to execute an envelopment must have a substantial mobility advantage over the enemy.

Turning Movement

8-15. A turning movement is a form of maneuver in which the attacking force seeks to avoid the enemy's principle defensive positions by seizing objectives to the enemy's rear and causing him to move out of his current positions, or divert major forces to face the threat.

Penetration

8-16. A penetration is a form of maneuver in which an attacking force seeks to rupture enemy defenses on a narrow front to disrupt the defensive system.

Destroying the continuity of that defense allows the enemy's subsequent isolation and defeat, in detail, by exploiting friendly forces. The penetration extends from the enemy's security area through his main defensive positions and into his rear. A commander employs a penetration when there is no assailable flank, enemy forces are overextended and weak spots are detected in his defenses, or time pressures do not permit an envelopment.

Infiltration

8-17. An infiltration is a form of maneuver in which an attacking force conducts undetected movement through, or into, an area occupied by enemy forces to occupy an advantageous position in the enemy's rear while exposing only small elements to enemy defensive fires.

Frontal Attack

8-18. A frontal attack is a form of maneuver in which an attacking force seeks to destroy a weaker enemy force or fix a larger enemy force in place over a broad front.

ENGINEER PLANNING CONSIDERATIONS

8-19. To support planning, the ENCOORD develops an engineer estimate. The four key elements of the engineer estimate include developing an SOEO, refining an SOEO, identifying EMSTs, and the EBA. Planning for each form of offensive maneuver at the operational level requires the engineer commander to use the assured-mobility framework. Although the forms of offensive maneuver have different intentions, the planning phase must always begin with predicting the adversary's intent through a thorough understanding of the threat, his engineer capabilities, and how the terrain will effect operations. Geospatial products and information become the foundation and common reference for planning. Of all the forms of maneuver, knowledge of the threat's disposition is especially critical and required for an infiltration or penetration due to the requirements for stealth and surprise. A greater degree of planning is required for a penetration from the breach to the ultimate control of the decisive objective.

8-20. Planning includes how to best organize available engineer assets. The reverse planning procedure—identifying the combat engineering task required from the objective back to the start point—is essential for achieving the assured mobility fundamentals. As the ENCOORD resources each task, he identifies the shortages and develops a plan to ask for more resources, develops decision points to shift engineer assets, or receives a decision on where to accept risk. A frontal attack is generally a shaping operation that requires a broad front to fix a threat. Engineers may identify areas where no engineers are task-organized and elements of the maneuver force can provide necessary resources in critical areas. All of these plans require planning to protect the flanks of the maneuver from interdiction. Countermobility is the focus of flank protection while protection of the force focuses on the survivability effort and protection of priority assets.

8-21. Planning is the basis for developing a collection plan, which is the foundation for the fundamentals of prevention and detection. A thorough understanding of how the enemy fights allows the staff to develop a focused collection plan that resources ISR assets to prevent and detect potential impediments to force movement. During offensive operations, engineer units tend to have command relationships with maneuver commanders. OPCON is the most common command relationship for engineers during offensive operations because it allows engineers to be responsive and provides the maneuver commander the greatest flexibility.

8-22. Effective preparation for offensive operations requires engineer units to establish early linkups with the maneuver units and commanders they will support. As engineer units prepare for offensive operations, they should focus on inspections and combined arms rehearsals that address the avoidance of restrictive areas by creating combat trails or bypass routes. If the plan is to neutralize or breach through obstacles, the rehearsals should cover all the breaching tenets—intelligence, breach organization, breach fundamentals, mass, and synchronization. The rehearsal should also include the participants from each element of the breach organization. A current assessment of the ISR collection plan is required for final adjustments to the plan, and this may affect the fundamentals of assured mobility.

8-23. Executing the forms of maneuver translates into four types of offensive operations at the tactical level—MTC, attack, exploitation, and pursuit *(see FM 3-90).* Engineers may measure the successful execution of offensive operations when assured mobility facilitates orienting and massing combat power on the enemy. Successful execution includes integrating all of the BOS and other branches as they contribute toward their pieces of assured mobility. Engineers continue to update their running or continuous staff estimate and ensure it still meets the needs of the current situation. Changes that affect the functions of prevention and detection must be distributed to all forces, rapidly. As new impediments are detected, engineers should be organized to remain flexible and maintain momentum for the force so they can support avoiding or neutralizing threats to maneuver forces.

DEFENSIVE OPERATIONS

8-24. Defensive operations defeat an enemy attack, buy time, economize forces, or develop conditions favorable for offensive operations. Defensive operations alone normally cannot achieve a decision. Their purpose is to create conditions for a counteroffensive that allow for a regaining of the initiative. Other reasons for conducting defensive operations include retaining decisive terrain or denying a vital area to the enemy, attriting or fixing the enemy as a prelude to offensive operations, countering surprise action by the enemy, and increasing the enemy's vulnerability by forcing him to concentrate his force. While the offense is the most decisive type of combat operation, the defense is the stronger type. The inherent strengths of the defense include the defenders' ability to occupy his positions before the attacker and use the available time to prepare his defenses. Preparations end only when the defender retrogrades or begins to fight. The defender can study the ground and select defensive positions that mass the effects of his fires on likely approaches. He combines existing and reinforcing fires to disrupt, fix, turn, or block.

8-25. As a part of full-spectrum operations, the purpose of defensive operations is to impose US will on the enemy for decisive victory. Engineers play a significant role in defensive operations as a combat multiplier with a focus on combat engineering (M/CM/S), although geospatial engineering and general engineering are also major contributors. As such, engineers provide support throughout the depth of the battlefield, through continuous support to decisive, shaping, and sustaining operations.

8-26. Army forces conduct defensive operations simultaneously with offensive, stability, and support operations while performing joint, multinational, and interagency operations. Similarly, while engineers may be focused on one or more of the engineer functions, all five functions operate simultaneously in support of defensive operations. *Figure 8-2* shows that relationship.

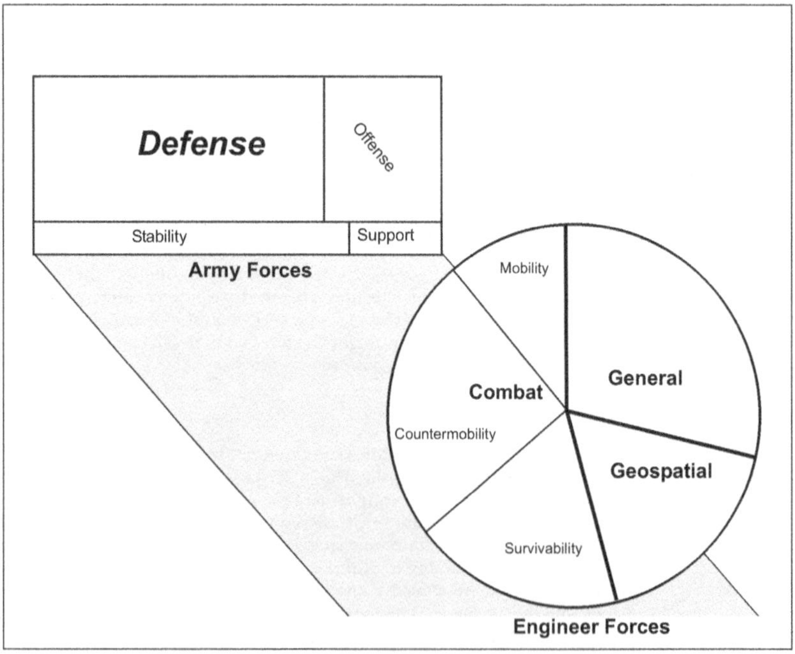

Figure 8-2. Generic Defensive Operations Relationships

8-27. Planning for defensive operations is inextricably linked to offensive operations and, for planning purposes, must consider the transition from offensive operations, and the follow-on offensive operations. During defensive operations, engineers use terrain products to best position the units within the defense. Engineers then work with intelligence staff to describe the threat functions to predict where the threat is likely to attack friendly forces. Engineers work in conjunction with intelligence personnel to determine which sensor capabilities to leverage and best predict and prevent the threat from maneuvering freely into the defended area. The consideration of counterattack planning or support for the mobile strike force is the same as in the offense. The ENCOORD works with other BOS's and uses the fundamentals of assured mobility to ensure that the counterattack force can mass its effects on the enemy for decisive operations. The type of defensive operation will define the amount of engineer effort required. An area defense will typically require a greater amount of effort. A mobile defense's effort will be to a lesser degree because it has greater flexibility and takes advantage of the terrain in depth. During defensive operations, engineer units tend to have support relationships to the maneuver commander.

8-28. During the preparation phase, engineer assets are postured with their task-organized C2 HQ and initiate engineer work effort. The focus is on resourcing the materials for creating obstacles and integrating them with friendly fire effects. Engineers are designated to support mobility operations and selected countermobility operations in support of the reserve or mobile strike force. The engineer commander and his staff coordinate for ISR assets to detect enemy engineer and breaching capabilities. Engineers provide protection from the threat systems through survivability operations to key support systems.

8-29. There are three types of defensive operations—area defense, mobile defense, and retrograde *(see FM 3-90)*. When executing defensive operations, engineers employ SCATMINEs and conventional mines and obstacles to develop EAs. They should also provide mobility to assist the reserve or mobile strike force with enemy destruction. Within the defense, engineers contribute to protecting the force by interdicting the threat's ability to maneuver, constructing life support areas, or digging in assets such as radars.

STABILITY OPERATIONS

8-30. Stability operations are part of full-spectrum operations intended to promote and protect US interests. This is done by influencing the threat, political, and information dimensions of the operational environment through a combination of peacetime developmental, cooperative activities and coercive actions. Stability operations are not necessarily sequential. They may occur independently, concurrently, or at the conclusion of combat operations. The focus is on sustaining the outcome achieved from combat operations to prevent the threat or the conditions for a threat to return and realize strategic results. Stability operations, unlike combat operations, may have an unambiguous threat. As discussed in *Chapter 2*, the continuum-of-relative interests model shows that groups may become more or less adversarial as the stability of the area changes. The engineer must be aware of the impact this will have on supporting stability operations.

8-31. Army forces may conduct stability operations simultaneously or in combination with offensive, defensive, and support operations while performing joint, multinational, and interagency operations. Similarly, while engineers may be focused on one or more of the engineer functions, all five functions operate simultaneously in support of stability operations. *Figure 8-3* shows that relationship.

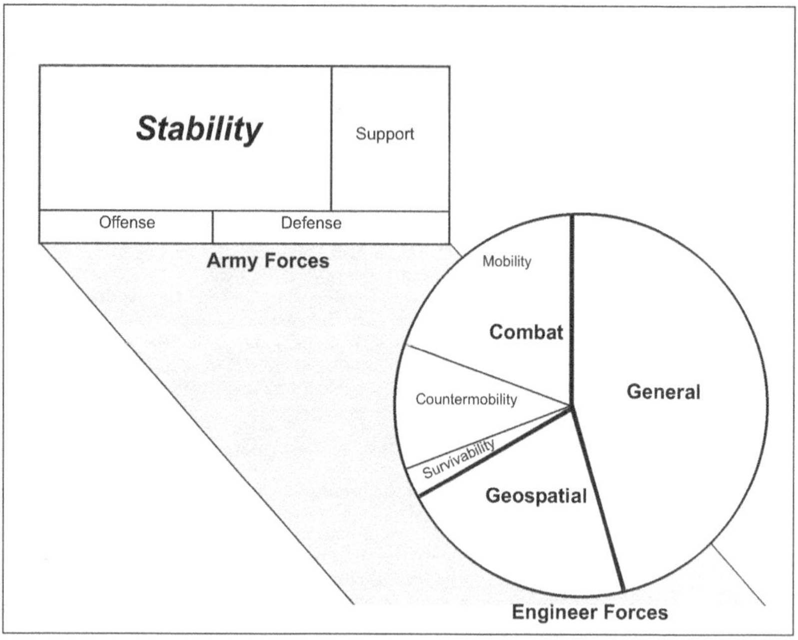

Figure 8-3. Generic Stability Operations Relationships

8-32. When planning for stability operations, engineers should consider a broad range of potential requirements. The EBA, although similar in format for combat operations, looks at different aspects of the terrain and friendly and threat capabilities. Terrain products continue to have a great deal of importance, but political and cultural considerations may be more important than strictly a combat terrain analysis. Terrain analysts work with the intelligence staff to develop a usable product for the commander to reflect this information if available. When analyzing the troops available, the ENCOORD should consider if there are HN, third-party NGOs, or other multinational forces involved. Interaction with these other parties requires engineers to address interoperability, common standards, and mutual agreements. Engineers should also plan for engineer units operating amongst civilians or in conjunction with NGOs and other international organizations.

8-33. The ENCOORD typically receives priorities from the joint staff on general engineering tasks. He then coordinates to secure the assets to perform these missions. As stability operations evolve, there may be periods of threat activity. If this occurs, the adversaries may use IEDs to target soldiers and affect the nation's will to continue the stability operation. Engineers are a major player, particularly in urban areas, as they plan general engineering tasks. Engineers are critical to predicting threat patterns in conjunction with the intelligence staff *(see FM 3-07)*. It is important for engineers to understand the goals and objectives of the higher HQ and address these issues with the maneuver commander to best assist him in preparing for expanding missions and considering the unique equipment and personnel capabilities.

8-34. Preparing for stability operations is more difficult than preparing for combat operations because of the broad range of potential missions engineers are expected to participate in during stability operations. An early on-the-ground assessment by engineers is absolutely critical to tailor the engineer force properly and to support the follow-on engineer contingency operations force logistically. Results of this assessment are quickly passed to deployment planners to ensure that an adequate engineer support force arrives in the AO in a timely manner. The failure to provide these engineers may cause inadequate troop bed down, sanitation, and force protection to the deployed force. This early on-the-ground engineer assessment identifies the—

- Status of the infrastructure in the AO, to include airfields, roads, ports, logistics bases, and troop bed-down facilities; real estate acquisition; environmental standards, conditions, and considerations; construction material supply; construction management; and line-haul requirements.
- Theater and situation-specific force protection requirements.
- Existing geospatial product availability and requirements for new terrain products.
- Specialized engineer requirements such as prime power, well drilling, and fire fighting support.
- Engineer C2 requirements, including HQ staffing, communications, and information systems support.
- Engineer liaison requirements, including linguists and civil affairs personnel.
- Requirements for officers with contracting officer's representative (COR) or USACE experience.
- Potential requirements for LOGCAP, contractor responsibilities, contract-construction procedures, and initial work areas.
- Potential requirement to establish a MICC to coordinate for HDOs.

8-35. There are ten types of stability operations and each is unique in its execution *(see FM 3-07)*. The ten stability operations include peace operations (PO), foreign internal defense (FID), security assistance (SA), humanitarian and civic assistance (HCA), support to insurgencies, support to counterdrug operations, combating terrorism (CBT), noncombatant evacuation operations (NEOs), arms control, and show of force. Engineers may or may not play a role in each type of stability operations. However, these areas are addressed for potential engineer considerations.

PEACE OPERATIONS

8-36. The broadest type of stability operation is POs, either as part of a UN, NATO, or multinational force or performed unilaterally. Peace operations inside peacekeeping operations (PKOs), peace enforcement operations (PEOs), and support to diplomatic efforts as its three basic forms.

Peacekeeping Operations

8-37. PKOs are military operations undertaken with the consent of all major parties in a dispute. They are designed to monitor and facilitate the implementation of an agreement (cease-fire, truce, or other such agreement) and support diplomatic efforts to reach a long-term political settlement *(see JP 1-02).* Before PKOs begin, a credible truce or cease-fire must be in effect, and the parties to the dispute must consent to the operation. PKOs take place following diplomatic negotiation and agreement among the parties to a dispute, the sponsoring organization, and the potential force-contributing nations. Engineers participate as part of a combined arms force and may construct and maintain roads, airfields, LZs, ports, pipelines, and other associated missions such as land mine detection and removal. Removal of mines by engineers during PKOs is based on tactical necessity. Humanitarian demining organizations provide the preponderance of mine removal.

Peace Enforcement Operations

8-38. A PEO is the application of military force or the threat of its use, normally pursuant to international authorization, to compel compliance with resolutions or sanctions designed to maintain or restore peace and order *(see JP 1-02).* PEOs are, by definition, coercive in nature and rely on the threat or use of force. However, the impartiality with which the peace force treats all parties and the nature of its objectives separates PEOs from war. The purpose of a PEO is not to destroy or defeat an enemy but to use force or threat of force to establish a safe and secure environment so that peace building can succeed. Engineers may participate in disarming to include seizing ammunition, collecting, and destroying weapons and supplies, closing weapons and ammunition factories, and preventing resupply.

8-39. HDOs are a part of the disarmament program *(see CJCSI 3207.1).* Because of the threat posed by land mines to peace and safety, HDOs have become a significant activity in POs. In many demining missions, military ordnance specialists and civilian organizations participate in partnership. Demining is ultimately an HN responsibility. US Army participation in HDOs focuses on mine-awareness education and, most importantly, training of HN personnel in the surveying, marking, and clearing or lifting of mines. US Army personnel do not remove mines (they are prohibited by DOD); however, they may assist and train others in demining techniques and procedures. Specialists, such as EOD personnel *(see Appendix L),* may be called on to destroy UXO, including munitions shells, fuses, grenades, and rockets. Special Forces, civil affairs, and PSYOP personnel are currently the primary means to execute the geographic combatant commanders' programs of HDOs. EOD and engineer personnel are also included in these programs and integrated into operational training missions.

Operations in Support Of Diplomatic Efforts

8-40. Military support of diplomatic efforts improves the chances for success in the peace process by lending credibility to diplomatic actions and demonstrating resolve to achieve viable political settlements. In addition to or as an integral part of PO, Army forces may conduct operations in support of diplomatic efforts to establish order before, during, and after a conflict. While these activities are primarily the responsibility of civilian agencies, the military can support these efforts within its capabilities. Army forces may support diplomatic initiatives such as preventive diplomacy, peacemaking, and peace building. Engineers may support executing PO with geospatial engineering support, lodgment, and theater infrastructure development, including the construct and repair of protective facilities, roads, airfields, parts, and troop life-support facilities.

FOREIGN INTERNAL DEFENSE

8-41. FID is a form of stability operation but is specifically a program to support friendly nations operating in or threatened with potential hostilities. FID promotes regional stability by supporting an HN program of internal defense and development (IDAD). These national programs free and protect a nation from lawlessness, subversion, and insurgency by emphasizing the building of viable institutions that respond to the needs of society. FID can include training; material, technical, and organizational assistance; advice; infrastructure development; and tactical operations. Military assistance is often necessary to provide a secure environment for these efforts to become effective *(see JP 3-07.1)*. Army engineer units and individuals can be tasked to provide this military assistance. Typically, engineers will be subject matter experts (SMEs) when tasked to support this program.

SECURITY ASSISTANCE

8-42. SA is the means through which the US provides defense articles, military training, and other defense-related services to eligible foreign governments or international organizations to further US national policies and objectives. These programs include foreign military sales, international military education and training, HCA, humanitarian assistance, humanitarian demining programs, international peacekeeping, the Warsaw Initiatives Program, and the Partnership for Peace Program. SA is a group of programs, not a mission assigned to Army units specifically. However, Army units and soldiers participate in SA programs through peacetime engagement activities and by training, advising, and assisting allied and friendly armed forces.

8-43. There are four primary methods of training:

- Mobile training teams are used when an HN element requires on-site training or needs surveys and assessments of training requirements. Engineers may deploy as part of a single-service, joint, or conventional force team. Each team is tailored to the training the HN requires.
- Extended training service specialists teams are employed on a permanent change of station to assist the HN in attaining readiness

on weapons or other equipment. These teams train the HN instructor cadre so the HN can assume responsibility for the training.

- Technical-assistance field teams are also deployed on a permanent change of station basis and train HN personnel in equipment-specific military skills.
- International military education and training and mobile education teams provide HN personnel with training opportunities in the continental US and in the host country. This training not only meets the immediate HN requirement of increased training, but also has a long-term impact of improving US and HN relations.

HUMANITARIAN AND CIVIC ASSISTANCE

8-44. HCA is a stability operation that provides assistance to the local populace by predominantly US forces with military operations and exercises. Such assistance must fulfill unit training requirements that incidentally create a humanitarian benefit to the local populace. The assistance that engineers may provide under HCA is limited to—

- Constructing rudimentary surface transportation systems.
- Well drilling and constructing basic sanitation facilities.
- Rudimentary construction and repair of public facilities.
- Detecting and clearing landmines, including activities relating to furnishing education, training, and technical assistance regarding detecting and clearing.

8-45. US forces, to include engineers, may be tasked to provide the C2 support necessary to plan and execute the ground portion of any humanitarian assistance operation. Engineers may also be tasked to provide the logistics support necessary to relieve human suffering or the forces to secure an area to allow the humanitarian relief efforts of other agencies to proceed. Engineer assistance may also include—

- Constructing and repairing rudimentary surface transportation systems, basic sanitation facilities, and rudimentary public facilities and utilities.
- Drilling water wells.
- Constructing feeding centers.
- Disposing of human and hazardous wastes.

SUPPORT TO INSURGENCIES

8-46. Engineers may provide limited support to insurgencies. On order, Army forces support insurgencies that oppose regimes that threaten US interests or regional stability. While any Army force can be tasked to support an insurgency, Army Special Operations Forces (ARSOF) usually receive these missions. Engineer support to insurgency forces is limited to providing geospatial products and construction of SOF operating bases located outside the AO. Engineer missions for counterinsurgency operations are similar to those for humanitarian and nation assistance water supply and sanitation improvements; road, airfield, and port construction; and multinational training.

SUPPORT TO COUNTERDRUG OPERATIONS

8-47. Two principles guide Army support to counterdrug operations. The first principle is using military capabilities to benefit the supported agency and to train our soldiers and units. The second is to ensure that the military members do not become directly involved in law enforcement activities. Engineers supporting domestic counterdrug operations perform missions focused on supporting local law enforcement agencies. Engineers are sensitive to the legal aspects of support to civilian authorities and abide by the Posse Comitatus Act. They are also aware of the capabilities of the threat, primarily heavily armed narcotics traffickers. Typical support tasks include—

- Constructing or rehabilitating law enforcement target ranges; helipads; and fuel-storage, billet, CP, and maintenance facilities.
- Producing geospatial products of likely counterdrug operations areas.
- Constructing or upgrading access roads for drug interdiction patrols.
- Clearing observation fields for counterdrug teams.

COMBATTING TERRORISM

8-48. After the events of September 11, 2001, the Army can expect to be involved with CBT. Terrorism is the calculated use of unlawful violence or threat of unlawful violence, intended to coerce or to intimidate governments or societies in the pursuit of goals that are generally political, religious, or ideological *(JP 1-02)*. CBT involves opposing terrorist actions across the threat spectrum. These actions include offensive and defensive components. The offensive form of action is counterterrorism (CT). The defensive form of action is antiterrorism. Army commanders at all echelons must protect their soldiers, equipment, and installations. Army personnel and units conduct antiterrorism anywhere in the world. Engineers may become targets for terrorists because of how and where they perform their missions, especially construction projects and other wide-area missions. Equipment parks and supply yards are large and difficult to defend. Soldiers operating equipment or hauling materials are vulnerable to ambush by fire, mines, and booby traps. In support of antiterrorism, engineer leaders develop force protection measures whenever they conduct engineer missions.

NONCOMBATANT EVACUATION OPERATIONS

8-49. NEOs are conducted to support evacuating noncombatants and nonessential military personnel from a foreign nation to an appropriate safe haven. These operations evacuate US citizens whose lives are endangered by war, civil unrest, or natural disaster. Such operations also may include evacuating selected HN citizens or third-country nationals. NEOs usually involve a swift insertion of a force, temporary occupation of an objective, and planned withdrawal upon completion of the mission *(see JP 3-07.5)*. Engineers that support a NEO generally operate as part of a joint force and may conduct a wide variety of tasks, such as—

- Constructing temporary facilities and protective structures in country or in another country for either US forces or evacuees.
- Providing needed geospatial products and data for the operation.

- Conducting route reconnaissance and mobility operations for land evacuation.
- Repairing airfields and clearing helicopter LZs for use in air-evacuation operations.

ARMS CONTROL

8-50. Another stability operation that requires limited engineer support is arms control. The overarching goal of arms control is to prevent or deter war, promote stability, reduce the potential damage of a conflict, and reduce defense expenditure. Arms control focuses on promoting strategic military stability. It encompasses any plan, arrangement, or process controlling the numbers, types, and performance characteristics of weapons, C2, logistics-support, and intelligence-gathering systems. Engineers may support executing arms control operations by providing geospatial products used to verify treaty compliance and by constructing logistics support facilities involved with the arms control process.

SHOW OF FORCE

8-51. A show of force is an operation designed to demonstrate US resolve by deploying or using military forces. The US conducts shows of force for three principal reasons. To bolster and reassure allies, to deter potential aggressors, and to gain or increase influence. Army units are not usually assigned the mission to conduct a show of force but rather to conduct other operations, such as those listed below, for the purpose of showing force. A show of force is normally executed as—

- A permanent forward deployment of military forces.
- Combined training exercises.
- The introduction or build-up of military forces in a region or area.
- An increase in the readiness status and level of activity of designated forces.

8-52. Engineer support to demonstrations and shows of force normally a joint and multinational effort. Engineer tasks are very similar to those described in POs. Overt use of engineers during shows of force may aid in the political intent of the operation.

SUPPORT OPERATIONS

8-53. Support operations provide essential services, assets, or specialized resources to help civil authorities deal with situations beyond their capabilities. Support operations usually involve actions that help civil authorities or NGOs provide the required DS to the affected population. Army forces may provide relief or assistance directly, when necessary; however, they normally support the overall effort controlled by another agency. When visualizing a support operation, commanders recognize that they will have to use a different definition of the enemy. In support operations, the adversary is often disease, hunger, or the consequences of disaster. These types of support operations are domestic support operations (DSO) and foreign humanitarian assistance (FHA) operations. They share four forms: relief operations; support to chemical, biological, radiological, nuclear, and high-yield explosive (CBRNE) consequence management; support to civil law enforcement; and community assistance. The types of support operations occur in varying degrees in DSO and FHA operations. Since domestic emergencies can require Army forces to respond with multiple capabilities and services, these operations may be conducted simultaneously during a given operation. *Figure 8-4*, shows the relationship of the five engineer functions to support operations. General engineering tasks provide the preponderance of support.

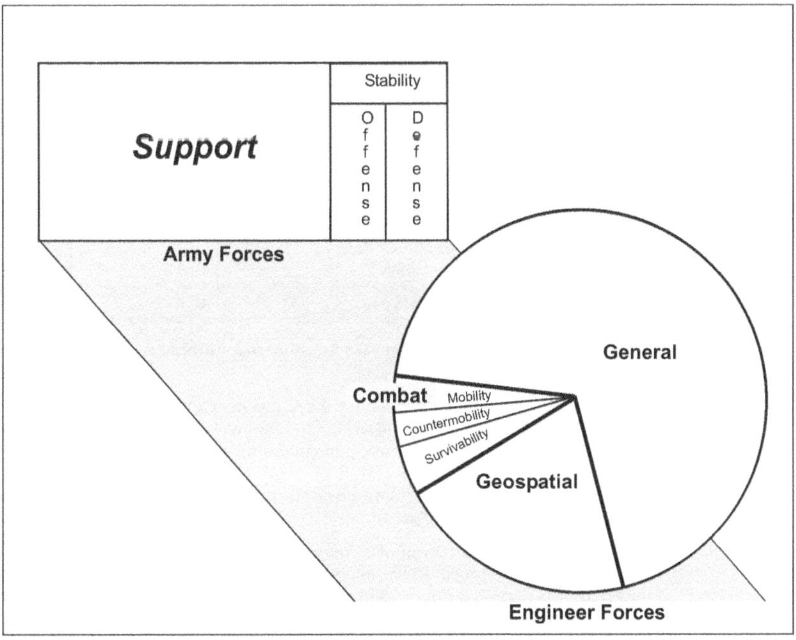

Figure 8-4. Generic Support Operations Relationships

8-54. Planning for support operations is significantly different from offense, defense, or stability operations—assessment of the threat is the reason. The threat may be a natural disaster with unpredictable behavior. Additionally, engineers must be aware of the number of statutes and regulations that restrict the Army's interaction with other government agencies and civilians for support operations. Different federal agencies have responsibility for the emergency support functions (ESFs) identified by Congress. *Table 8-1* contains a list of the ESFs and the federal and DOD proponents.

8-55. Army commanders will assume a support role to a designated agency. It is important for engineer commanders and staff to know who the proponent is

Table 8-1. ESFs and Proponents

ESF		
ESF	**Primary FED Agent**	**DOD POC**
Transportation	DOT	USTRANSCOM
Communications	NCS	OASD (C3I)
Public Works	DOD	USACE
Firefighting	USDA	USACOM
Info & Planning	FEMA	DOMS
Mass Care	ARC	DLA
Resource Support	GSA	DLA
Health/Medical Services	DHHS	OTSG (ARMY)
Urban Search & Rescue	FEMA	DOMS
Hazmat	EPA	DON
Food	USDA	DLA
Energy	DOE	USACE

and its responsibilities. Initial coordination and training with these agencies will facilitate the planning process.

8-56. There is usually little time for preparation of support operations. Relief operations and support to CBRNE consequence management require an immediate response. It is essential for commanders to have contingency plans developed and units identified so that they may prepare for such contingencies. Support to civilian law enforcement and community assistance allows greater leeway to plan and prepare.

8-57. During the execution of support operations, engineers may provide support to all four types. Relief operations respond to and mitigate the effects of natural or man-made disasters (including CBRNE incidents). They maintain or restore essential services and activities to mitigate damage, loss, hardship, or suffering. Relief is primarily a state, local, or HN responsibility. To support the efforts of local authorities or the lead agency, the President and SECDEF can employ Army forces before, during, or after an event to save lives; protect property, public health, and safety; or to lessen or avert the threat of catastrophe.

8-58. Engineers can expect to be involved in planning for support of relief operations with geospatial products and analysis of potential locations to establish LSAs. Engineers may be called on to provide manpower support or general engineering support to units with unique capabilities such as water purification operations, temporary shelter construction, power generation operations, and firefighting operations.

8-59. Support to domestic and foreign CBRNE consequence management is another major support operation that requires engineer support. It has, by far, the most extensive support requirements for military personnel. Other US government agencies have primary responsibility for responding to domestic terrorist incidents. Local authorities are the first to respond to a CBRNE incident. However, Army forces have a key supporting role and can quickly respond when authorized. In a permissive overseas environment, the President and SECDEF may make Army assets available to assist a foreign government after a CBRNE incident. Such assistance may be linked to concurrent relief operations. CBRNE incidents are deliberate or unintentional events that produce catastrophic loss of life or property.

8-60. Due to the requirement to respond to consequence management and to control a situation before it can get worse, engineers can expect to respond with mobility assets. Engineer equipment is best suited to remove rubble and debris associated with consequence management. Other contributors are public works and engineering support that includes technical advice and assessments, engineering services, construction management and inspection, emergency contracting, emergency repair of wastewater and solid waste facilities, and real estate support. Other engineering considerations include—

- The need for heavy equipment for camp construction and power generation.
- The emergency clearance of debris to allow for reconnaissance missions and the passage of emergency personnel.
- The temporary construction or repair of emergency access routes.
- The emergency restoration of critical public services and facilities.
- Technical assistance and damage assessment.
- Emergency demolition operations.
- An effective maintenance program which is vital to ensure that required support is provided.
- The consideration of all classes of supply, particularly Class IV.

8-61. Support to civil law enforcement includes support to CT and counterdrug operations, civil disturbances, and GS. The Army assists civil law enforcement by providing personnel, equipment, training, and expert advice within the limits of applicable laws. Units in state status (Title 32 USC) provide the primary source of military assistance to state and local law enforcement agencies. They may assist civil authorities in instances where federal units are precluded due to the restrictions of the Posse Comitatus Act. Engineers may have a role in supporting law enforcement. There support may include geospatial products, map support, and manpower support. There may be some general engineering requirements, particularly in the enforcement of counterdrug operations.

8-62. Community assistance applies the skills, capabilities, and resources of the Army to the needs and interests of the US and local communities. Supporting and participating in events and activities that benefit the Army and the civilian community build on a long tradition of America's Army helping American communities. Community assistance can have a large impact because AC, ARNC, and USAR units are located in thousands of towns and cities across the nation. What a command does, or fails to do, for the community will affect the attitudes of the American people on whom the Army depends for its support and existence. Every commander should identify opportunities to conduct initiatives that meet specific needs; have specific start points and end states; enhance readiness; and advance the interests of the nation, the Army, and local communities. Engineers may be called on to provide everything from hometown recruiters to support of holiday events. Installations may have memorandum of agreements with neighboring towns to provide emergency snow removal or other capabilities not readily available.

TRANSITIONS BETWEEN OFFENSIVE, DEFENSIVE, STABILITY OPERATIONS, AND SUPPORT OPERATIONS

8-63. The transition between full-spectrum operations is one of the most difficult tasks for engineer planners to prepare for. As forces transition between operations, there will be gaps or voids between combat forces and other support forces, resources, and/or agencies. The reason is because operations may transition more rapidly than expected or other forces or agencies are not readily available. As a result, engineers and other forces may expect to perform tasks not on their METL. *Figure 8-5, page 8-20* represents the broad scope of tasks that occur in each operation and the wide array of tasks that must occur to transition between them.

8-64. Engineer planners and commanders will be faced with the difficult challenge of planning for success and identifying what tasks engineers may be asked to perform during the void and having available resources, forces, or agencies to perform their requirements. Some tasks that engineers may expect to participate in while awaiting legitimate civilian authorities or agencies include—

- Coordinating with the civil-military operations center (CMOC) to identify the tasks NGOs will perform.
- Coordinating with the J4 to identify JCMEB requirements and priorities.
- Closing LSAs and other areas that will be handed over to the HN, NGOs, or other agencies.
- Reestablishing utilities, sanitation, infrastructure, and governmental responsibilities.
- Establishing a MICC to coordinate with the Humanitarian Operations Center and related demining operations.

8-65. In addition to these considerations, engineers will be affected by the decisions and priorities of higher HQ. These decisions and priorities are established at convened boards. Examples of joint boards that will affect transition operations include the—

- JCMEB. The JCMEB establishes policies, procedures, priorities, and overall direction for civil military construction and engineering requirements in the theater. A primary concern of the board is to deconflict requirements between the military and civilian portions of a joint operation. The JCMEB also arbitrates issues referred to it.
- JFUB. The JFUB is used to manage facilities. The JFUB evaluates and reconciles component requests for real estate, use of existing facilities, interservice support, and construction to ensure compliance with priorities established by the JCMEB.

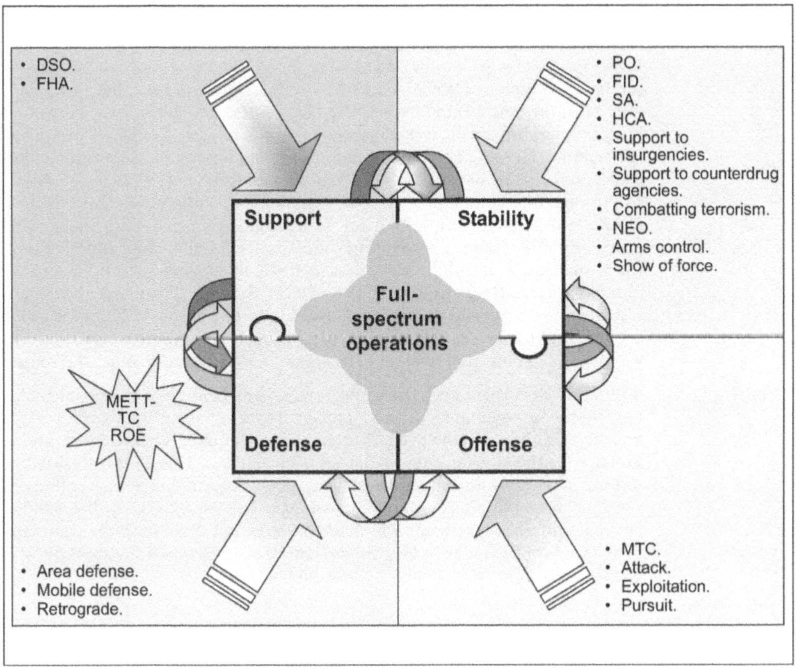

Figure 8-5. Full-Spectrum Operations

- JEMB. The JEMB board establishes policies, procedures, priorities, and the overall direction for environmental management requirements in the operational area.

8-66. The senior ENCOORDs must be involved with the J4 to coordinate priorities and synchronize the matching of capability requirements well in advance to ensure successful execution of the transition periods.

SPECIAL CONSIDERATIONS: URBAN OPERATIONS

8-67. Army commanders will likely determine that operations in the urban environment will be essential to mission accomplishment. They need to assess the relevance and impact of one or more urban areas as part of their mission. They will also need to determine whether full-spectrum UO may be the commander's sole focus or only one of several tasks nested in an even larger operation. Although UO can be conducted as a single battle, engagement, or operation, they will more often be conducted as a major ODSS operation requiring joint resources. *FM 3-06* provides a new framework—assess, shape, dominate, and transition—for UO. These are not phases or sequential operations but rather a means to visualize the fight.

8-68. Engineers provide critical support to any urban operation, whether foreign or domestic. *FMs 3-06* and *3-06.11* have more details, but commanders should understand that historically large numbers of engineers units have been task-organized for UO. Engineers provide unique geospatial products for the complex terrain of cities. Three-dimensional terrain products are available and continue to be developed. Assured mobility will be an important framework for commanders to use as maneuver commanders think of how to shape and dominate in urban terrain. General engineering tasks will be prevalent throughout all operations but will also be the major function during the transition to stability or support operations. Engineers will have to work closely with all branches of the M/CM/S BOS. They will have close coordination with EOD personnel to reduce IEDs and UXOs to minimize collateral damage with MP personnel to provide movement of civilians along routes, and chemical personnel to address chemical agents along the route.

8-69. Full-spectrum operations present a broad range of potential tasks to any engineer commander. It may appear daunting as he considers his METL and training plans. However, it is up to the commander to understand and prepare for the missions he can expect to encounter. There is no substitute for having a trained and disciplined unit in its core tasks. When called on to respond to a mission, commanders can expect assistance from the remainder of the Regiment to facilitate the unit's preparation. It is up to the commander to be aware of the potential considerations and understand the right questions to ask and explore to develop the best plans.

Chapter 9

Engineer Combat Service Support and Logistics Considerations

The troop engineer must have labor and transport with which to do his work, and be provided with a horse to inspect and speed its progress. He must serve many and accomplish much with little, yet, since he cannot do everything for all, he is beloved by none.

King Charles I (Articles of War 1625)

I don't know what the hell this 'logistics' is that Marshall is always talking about, but I want some of it!

Fleet Admiral E.J. King (to a staff officer in 1942)

Field Marshal Rommel once remarked that battles are won or lost on the back of logistics before the first rounds are fired. Admiral King's statement and Field Marshal Rommel's belief place a significance on logistics that underscores the relevance of logistic matters in doctrinal discussions. *JP 1* emphasizes that logistics sets the campaign's operational limits. The lead time needed to arrange logistics support and resolve concerns require continuous integration of logistic considerations into the operational planning process. This chapter will discuss special engineer logistic considerations and the internal implementation of CSS for engineer organizations.

FORCE SUSTAINMENT

9-1. Logistics is the science of planning and executing the movement and maintenance of forces. A force projection Army depends on making the right logistics decisions before the onset of operations. There is normally little time for last-minute logistics fixes when the decision to employ forces is made. Success is largely enabled through mental agility. Although the J4/G4/S4 is the lead for many aspects of support, engineers provide significant support to the logistician's overall plan. The logistics officer and staff planner should consider the—

- Force capabilities.
- Resources available in the AO.
- Construction standards.
- Environmental concerns.
- Contracting procedures.
- Force protection concerns.

9-2. Engineer force sustainment is critical to maintaining and multiplying combat power. Logistics operations must accurately anticipate engineer CSS needs. The ENCOORD staff assists the logistician in anticipating those unique engineer-related logistics. A mature, preplanned operational environment and extensive HN support cannot be assumed. The HN may not want or be able to deliver the necessary support in terms of quality, quantity, or timelines. Many engineer needs are unique requirements that demand the logistician to improvise, oftentimes causing the logistics system to become strained. Special engineer equipment is of low density, requiring intensive management to ensure that it is available for mission use. Engineer mission materials are normally bulky, heavy, and hard to transport. They must be requisitioned, transported, stockpiled, and issued in a streamlined manner. Engineer materials do strain the logistical system, but engineers balance the system with their contributions to theater logistics operations by constructing and upgrading logistics bases, troop bed-down facilities, airfields, ports, and MSRs. This chapter focuses on sustaining engineer units and engineer support to logistics operations. For more information, *see FMs 3-100-7, 3-0* and *4-0 (100-10).*

THE UNDERPINNINGS OF LOGISTICS

9-3. The objective of logistics is to ensure that operations succeed and facilitate the commander's ability to generate and mass combat power at a decisive time and place. Strategic logistics support wars, campaigns, and major operations. Operational logistics support encompasses those activities required to sustain campaigns and major operations and enable success at the tactical level of operations. Engineers closely support operational logistics operations in areas such as—

- Constructing base camps and bed-down facilities for force reception, onward movement, and sustainment.
- Opening and maintaining ports and airfields to support deployment, sustainment, and redeployment.
- Assisting in distribution management and movement of material and personnel by constructing MSRs and other logistics support facilities.
- Providing real estate management.
- Providing facilities operations and maintenance.
- Assisting in analyzing and integrating environmental considerations.
- Generating and distributing power.
- Providing enablers to facilitate JLOTS operations.

9-4. Operational and tactical logistics encompass all the CSS and engineer activities required to sustain the commander's ability to fight operations, battles, and engagements. Successful tactical logistics provide the right support at the right time and place to units. Engineers receive tactical logistics support from CSS agencies located at the maneuver brigade through the TSC level in the areas of manning, arming, fueling, fixing, moving, and sustaining operations. Each operational and tactical task is related or supportive of a specific BOS. It is important that engineer planners synchronize the implications of support for all types of forces and functions. Engineers support operational and tactical logistics operations in the following areas:

- Constructing FARPs.
- Digging ammunition supply points (ASPs) and corps logistics C2 nodes.
- Erecting fixed bridging along forward supply routes.
- Constructing the infrastructure for an ISB.

9-5. Regardless of the level of military operations, the engineer CSS structure and resource requirements depend on the METT-TC. The engineer logistics support structure fully supports the commander's intent and is integrated into the commander's concept of operation. Trade-offs between combat and general engineering capabilities directly affects this CSS capability.

COMBAT SERVICE SUPPORT CHARACTERISTICS

9-6. Successful CSS is balanced between being effective and efficient. CSS operations are characterized by being able to anticipate requirements, integrate joint and multinational CSS, improvise solutions, and be responsive and continuous. The characteristics of CSS (responsiveness, simplicity, flexibility, attainability, sustainability, survivability, economy and integration) are defined in *FM 4-0*. These characteristics facilitate effective, efficient CSS and enable operational success. They apply in war and contingency operations. These imperatives act as a guide for planners and operators to synchronize logistics on the battlefield. The engineer unit commander and staff understand and use these while planning engineer operations. The following paragraphs describe these eight characteristics and the engineer considerations for each.

9-7. Force commanders visualize and describe the concept of logistics support in concert with their CSS commanders. Commanders view CSS characteristics through an operational perspective to assist them in describing the considerations required to plan and execute successful operations. CSS characteristics are integrated throughout the operational framework, guide prudent planning, and assist the staff in developing the support plan. *FM 4-0* discusses the relationship between joint logistics and joint personnel principles and the US Army CSS characteristics). CSS characteristics seldom exert equal influence, and their importance varies by the situation. The commander identifies CSS characteristics having priority that, in turn, become the foundation for preparing the concept of logistics support.

RESPONSIVENESS

9-8. Responsiveness is the key characteristic of CSS. It means providing the right support in the right place at the right time. Responsiveness includes the ability to anticipate operational requirements and is the keystone of all logistics principles. Engineers ensure they identify all logistical requirements in advance, taking into consideration the support relationships of subordinate units. This information must be passed into CSS channels and tracked through delivery. It involves identifying, accumulating, and maintaining the minimum assets, capabilities, and information necessary to meet the support requirements. On the other hand, the force that accumulates enough material and personnel reserves to address every possible contingency usually cedes the initiative to the enemy.

9-9. Versatile CSS systems—

- Enhance the engineer unit's responsiveness.
- Adapt engineer change requirements without interrupting the flow of support.

9-10. In this respect, responsiveness is closely tied with improvisation. Theater logistics planners structure the logistics force to be versatile enough to complement engineer plans and operations yet be robust enough to ensure that engineer services are not interrupted. The structure is responsive enough to allow the engineer commander to seize and maintain the initiative.

9-11. Engineers plan to meet the changing requirements of the operation on short notice. The engineer sustainment system should be versatile enough to keep pace with rapid decision cycles and mission execution and react rapidly to crises or opportunities. Engineer planners are sensitive to engineer task-organization changes. Engineer units can normally respond to a change in task organization much quicker than theater CSS packages. Because of this, contingency engineer sustainment plans are normally developed.

9-12. The planner who anticipates is proactive—not reactive—while conducting before, during, and after operations. The ability of the force to seize and maintain the initiative, synchronize activities along the entire depth of the battlefield, and exploit success depends on the commanders', logisticians', and engineers' abilities to anticipate requirements. Engineers consider joint, multinational, contract civilian, and interagency assets when planning support for engineer operations. They—

- Use all available resources to the fullest, especially HN assets.
- Prioritize critical engineer activities based on the concept of operations.
- Anticipate engineer requirements based on wargaming by incorporating experience and historical knowledge.
- Do not think linearly or sequentially, but organize and resource for simultaneous and noncontiguous operations.
- Participate in and evaluate the engineer significance of each phase of the operation during the entire command estimate process, to include the mission analysis and COA development, analysis, and wargaming recommendation and execution.

SIMPLICITY

9-13. Simplicity is avoiding complexity and often fosters efficiency in planning and executing logistics operations. Mission-type orders and standardized procedures contribute to simplicity. Engineer commanders establish priorities and allocate classes of supply and services to simplify logistics support operations. Engineers use preconfigured loads of specialized classes of supply to simplify transport.

ECONOMY

9-14. Economy is the provision of support at the least cost. At some level and to some degree, resources are always limited. When prioritizing and allocating resources, the engineer commander and staff may not be able to provide a robust support package. Priority of effort will be established while balancing mitigation of risk to the operation. Engineer commanders may have to improvise to meet the higher intent and mitigate the risks. Commanders consider economy in prioritizing and allocating resources. Economy reflects the reality of resource shortfalls, while recognizing the inevitable friction and uncertainty of military operations.

FLEXIBILITY

9-15. Flexibility is the ability to adapt logistical availability based on changing situations, missions, and concepts of operation. Due to the inherently changing world environment, engineer missions will change. Engineers require flexibility and coordination with higher staffs to provide logistical support. As ENCOORDs work future branch or sequel plans, they ensure that logistical support is planned and resourced.

9-16. CSS plans and operations must be flexible enough to achieve responsiveness and economy. Flexibility may include improvisation, which is the ability to make, invent, or arrange what is needed from what is on hand. Improvised methods and support sources can maintain CSS continuity when the preferred method is undefined or not usable to complete the mission.

9-17. Extraordinary methods may be necessary to ensure success on the battlefield. Logistics planners attempt to push support to forward engineer units to ensure smooth combat operations. Sometimes this is not feasible or supportable. In such cases, engineers improvise by making, inventing, devising, or fabricating what is needed from what is on hand. One example is creating a demolition cratering charge using common fertilizer and diesel fuel.

9-18. Commanders must be aware of the environmental impacts of their actions. They must weigh the implications of holding out for logistical support against the environmental collateral damage that they will cause. They must ensure that a proper environmental risk assessment is done before beginning any action.

9-19. Specific damage assessment and repair procedures have been developed based on the need to improvise during the operation. Improvisation is not a substitute for good planning; requirements must be anticipated. However, improvisation can be a great strength; engineer personnel must recognize it as an advantage during emergencies.

ATTAINABILITY

9-20. Attainability is generating the minimum essential supplies and services necessary to begin operations. Commanders determine that minimum levels of support are acceptable to initiate operations. The engineer, in conjunction with the logistician, completes the logistic estimate and initiates resource identification based on the supported commander's requirements and priorities. An operation should not begin until minimum essential levels of support are on hand.

9-21. For engineers, attainability is at the very core of decisions that are made. Trade-offs may be necessary to attain a given goal or level and quality of product. While attainable, the cost may make other things unattainable. Since engineer materials must meet specific technical requirements, engineers work closely with the logistics staff to help them understand these requirements and obtain acceptable and suitable alternatives when trade-off decisions are required.

SUSTAINABILITY

9-22. Sustainability is a measure of the ability to maintain logistic support to engineers throughout the AO for the duration of the operation. Sustainability focuses on the engineer commander's attention for long-term objectives and capabilities of the engineer forces. Long-term support is a challenge for the engineer staff, which must not only attain the minimum essential materiel levels to initiate operations but sustain those operations through the end state. The ENCOORD must ensure that logistical requirements are known and are flowing based on the available transportation assets.

9-23. The engineer commander needs continuous logistics capability to gain and maintain the initiative. Pauses for rebuilding power impede momentum and rob the command of the initiative. Engineer planners synchronize all CSS assets to ensure that the support operation is not an impediment to the engineer commander. Continuity of operations is critical to success.

9-24. Engineers are either committed to the current operation or preparing for the next. The tempo of the battlefield requires a constant vigilance by the logistician and engineer commander to ensure a constant flow of support. Supplies are pushed (through unit distribution) forward whenever logistically feasible. This is especially crucial to engineer units because they do not usually have lulls in their operations that would allow them to resupply using the supply point method.

9-25. General engineering and contract support also support sustaining operations. General engineering involves constructing, repairing, operating, and maintaining infrastructure and facilities to enhance the provision of sustainment and services *(see JP 4-04 and FM 5-104)*. Contracting support provides supplies, services, and construction labor and materiel. Contracting often provides a responsive option or enhancement to support the force *(see FM 100-10-2)*.

SURVIVABILITY

9-26. The CSS characteristic of survivability is related but not exactly the same as the BOS of the same name. It is based on being able to protect support functions from destruction or degradation. Engineers contribute to ensuring logistics are survivable by constructing logistic bases and clearing LOCs.

INTEGRATION

9-27. Integration consists of synchronizing CSS operations with all aspects of Army, joint, interagency, and multinational operations. The concept of operations achieves integration through a thorough understanding of the commander's intent and synchronization of the CSS plan. Integration includes coordination with and mutual support among Army, joint, multinational, and interagency CSS organizations.

9-28. Operational and tactical plans integrate all logistics support so that it creates a synergy with the concept of operation. Engineer planners participate in and evaluate the logistics significance of each phase of the operation during the entire command estimate process. They create a clear and concise concept of support that integrates the commander's intent and concept of operation. This includes analyzing the mission; developing, analyzing, wargaming, and recommending a COA; and executing the plan.

9-29. Engineers will support joint and multinational operations. The theater commander integrates operations in their AOR, which often includes engineers from other services or countries and possibly civilian engineering contractors. FFE is one example of the integration of military and civilian engineers *(see Appendix H).*

LOGISTICS CONSIDERATIONS FOR ENGINEERS

PREDEPLOYMENT AND DEPLOYMENT PHASES

9-30. As soon as the commander begins the operational planning and development of COAs, the engineer planners should assess the potential support infrastructure and develop a CESP. The engineer should then apprise the logistics and operations planners of the infrastructure capabilities and options for engineer support. Since all military operations depend on a theater infrastructure system, base development should be placed high on the force projection theater commander's priority list. A foreign country's infrastructure cannot be developed overnight to support force projection operations. Normally, a country's infrastructure (to include roads, bridges, inland waterways, seaports, and airports) is built only to sustain indigenous population and industry, with minimal additional capacity available to support US and coalition forces.

9-31. The theater needs to have the capacity to move large numbers of units through its air and seaports. The theater also needs facilities to manage such functions as implementing C2, storing and transferring ammunition, maintaining equipment, storing and moving bulk petroleum, generating and distributing power, and staging and billeting rear area troops, which are necessary for a synchronized flow of support to occur.

9-32. Moving follow-on forces and supplies is critical to success. Engineer planners work closely with the logistics staff to develop a suitable transportation infrastructure (roads, bridges, and airfields). Anticipating engineer requirements is crucial to ensure that adequate time is available to complete a robust infrastructure. These facilities can also be improved with the foresight of using engineer assets before the operation and during contingency operations. HN or civilian contractor personnel can perform much of this work. Alternate plans should be developed for if HN support is not available, such as underdeveloped or uncooperative theaters.

9-33. Base development does not end once the operation begins. Base development increases, depending on the size of the force involved in the operation. Each time the force expands or contracts, planners review facilities and LOC requirements to ensure that they are adequate to accomplish the mission.

ENTRY AND OPERATIONS PHASES

9-34. The nature of engineer entry and operations phases places an extraordinary burden on the logistics structure. Rates of fuel consumption, repair parts, construction and obstacle materials, mines, and explosives dictate the commitment of a large amount of maintenance and transportation assets in support of engineers. Engineer construction of EPW and/or displaced civilian camps remains logistically intensive during this period. Engineer operations, by their nature, are dangerous. Engineer planners, therefore, anticipate and provide for replacing engineer losses.

POSTCONFLICT OPERATIONS

9-35. Upon the cessation of hostilities, engineer forces may be tasked to restore areas destroyed or damaged during combat operations or as a result of other factors. Depending on the political and social factors of an operation, devastation may require forces to do some restoration. This requires attention to detail in logistics anticipation planning and, most likely, in rotating follow-on engineer units working in concert with HN and civilian construction contractors. Access to ENCOM and USACE capabilities and expertise is vital as assessments are submitted to the JFCs staff for prioritization. Engineers can expect to support the requirement to enable the return to normalcy for the civilian population, including movement, commerce, and construction. Engineers must plan for success and be prepared to quickly transition to stability- or support-type missions.

ENGINEER AND THEATER COMBAT SERVICE SUPPORT RELATIONSHIPS

9-36. The relationship between the ENCOM and subordinate engineer elements and the senior CSS elements in the theater depends on both the echelon at which the elements are operating and the operational environment in which they are operating. The basic relationships are shown in *Figure 9-1.* The ENCOM and TSC (if it exists) command relationship with and between specialized commands is determined by the ASCC or ARFOR commander. The ENCOM routinely provides a liaison officer to the TSC support operations cell, which lacks organic engineer planning capability.

9-37. Specialized engineer detachments and teams (such as CRESTs or FEST–elements, MTOE utility detachments, or facilities engineering [FE] teams) are task-organized within engineer brigades and groups to provide area support services. These specialized detachments and teams may be temporarily under the ADCOM of supported ASGs, but typically provide support through DS or GS relationships.

LOGISTICS REQUIREMENTS

9-38. Engineer force sustainment maintains and multiplies combat power. Engineer commanders must understand their internal organizational CSS system and know where to obtain logistics support. Many engineer needs are

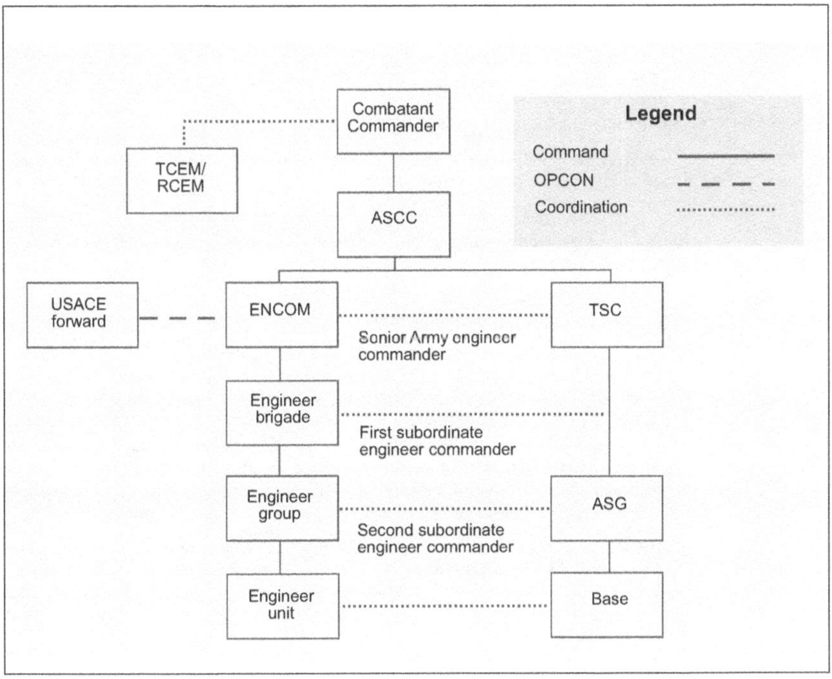

Figure 9-1. Engineer and Combat Service Support Relationships in Theater

unique requirements that stress the logistics system, such as massive requirements for Class IV barrier and construction materials. The key classes of supply are addressed in the following paragraphs:

9-39. Class III bulk and POL products are critical to the sustained operation of engineer equipment. Inadequate POL availability results in lost time that may affect the force commander's plan or result in a loss of momentum. While engineers are in a support relationship, the force commander should consider task-organizing fuel assets to the engineer force. Engineer equipment will operate for extended hours and may be out of synchronization with normal logistic resupply periods. Therefore, dedicated assets should be planned for and allocated. Engineer commanders will reflect on environmental considerations and take appropriate actions to manage risk when dealing with large amounts of Class III supplies.

9-40. Adequate Class IV supplies are central to the ability of engineer units to construct and maintain facilities in support of the ASCC's operations and the supporting sustainment base. For this reason, the senior Army engineer HQ (typically an ENCOM) plays a key role in establishing distribution protocols, stockage levels, and construction material allocation in theater. The joint force commanders, through TCEM/RCEM cells, often control critical Class IV supplies across the theater. The ENCOM may send a liaison to support the packaging of critical engineer Class IV supplies.

CONSTRUCTION MATERIALS AND SPECIAL CLASSES OF SUPPLY

9-41. In the TO, construction materials may be difficult to obtain. Required construction materials might be—

- Supplied by the CONUS through the supply system.
- Procured from neighboring countries within the region of operations.
- Obtained from local suppliers.
- Extracted from local natural sources.
- Produced by engineer units.

9-42. Ideally, construction materials should be procured from sources close to the construction site; however, this is often not possible. Because of the time and cost involved in moving construction materials, facility designs are often adapted to make maximum use of locally available materials. Quality control at the manufacture or delivery site is critical to ensuring that the materials meet any critical construction tolerances and standards. The quality of materials from HN suppliers may not meet these tolerances and standards or the necessary associated safety and environmental considerations.

9-43. Vertical materials, such as plywood and dimensional lumber, are Class IV items. These materials are universal commodities that are used by logisticians for overhead protection, flooring in high traffic areas, and shelving. However, equally important to the engineer are the materials that are instrumental in the success of horizontal (earthwork) construction.

9-44. Aggregate (crushed rock or river run and sand) are locally purchased or provided by engineer units. Other items, such as geotextiles or sand grid used to bolster poor soil conditions and improve trafficability, must be procured and shipped from CONUS. Soil stabilization and dust abatement require admixtures to the soil, such as lime, portland cement, asphalt emulsions, or resins. These products are bulky and may or may not be available within the region of current operations. Poor drainage can hinder operations. Therefore, ample supplies of culvert materials are a frequent concern of engineers.

9-45. Because of the multiple applications of many Class IV items, it is important that engineers remain involved in distribution decisions. This is particularly true regarding critical materials to ensure that they are not squandered on less critical missions. Engineers are not equipped to run Class IV yards, nor do they have the transportation assets to distribute Class IV supplies. However, the absence of engineer insight and critical assets management could lead to devastating theater shortages and mission impairment. Priority for transportation support is essential to supply operations, as Class IV items and engineer equipment are large and bulky and may require convoy credits or special haul assets.

9-46. Class V materials include ammunition, mines, and explosives. These materials are critical to the combat and CS operations that engineers conduct during combat operations. Additionally, these types of materials are usually not available for local procurement. Therefore, it is paramount for the ENCOORD to have an understanding of unit basic loads, anticipated missions, and availability of Class V resources. For instance, Class V Volcano mines may have a controlled supply rate (CSR). If this occurs, the ENCOORD should identify other types of mines available to provide a similar capability. Particular care must also be dedicated to planning for the transport of these HAZMATs with the transportation officer.

9-47. Class IX repair parts (less medical-peculiar repair parts)—all repair parts and components to include kits, assemblies, and subassemblies (repairable and nonrepairable)—required for maintenance support of all equipment. Engineers have a unique position in that their equipment tends to be low-density or civilian-leased equipment. Due to the low-density requirements for parts, the military logistic system may not be as responsive as local procurement of the commercially common parts. Engineers should be aware of all local vendors in the AO during the EDA.

9-48. Class X are materials to support nonmilitary programs, such as agricultural economic development, not included in Classes I through IX. These materials may be required by engineers during stability or support operations to assist local nationals and assist in economic engagement programs.

CONTRACTORS ON THE BATTLEFIELD

9-49. Contractors on the battlefield are addressed in *AR 715-9, FM 100-10-2,* and *FM 3-100.21.* Experience shows that there are never sufficient engineers to accomplish all missions for extended periods of time within the AO. The shortage of engineers may be partially overcome through the use of support contracts.

9-50. Engineers may be involved with a variety of contractors due to the large amount of general engineering requirements. Theater support contractors support deployed operational forces under prearranged contracts or contracts awarded from the mission area by contracting officers serving under the direct contracting authority of the theater principal assistant for contracting (PARC). They provide goods, services, and minor construction, usually through local vendors, to meet the immediate needs of operational commanders. Examples are organizations such as Brown and Root that work with engineer forces in many theaters.

9-51. External support contractors provide support to deployed operational forces that are separate and distinct from either theater support or support provided by system contractors. To support the mission, they may perform according to prearranged contracts or contracts awarded during the contingency itself. Engineers may receive logistical support from external support contractors while deployed on engineer specific deployments. Contracting officers who award and administer external support derive their contracting authority from PARC's outside the theater.

9-52. System contractors support deployed operational forces under prearranged contracts awarded by program executive officers, program managers, and the Army Materiel Command (AMC) to provide specific support to materiel systems throughout their life cycle, during both peacetime and contingency operations. These systems include, but are not limited to, vehicles, weapon systems, aircraft, C2 infrastructure, and communications equipment. Engineers have an increased likelihood of working with system contractors to assist with unique engineer equipment. When contractors support engineer operations, they must be managed, deployed, protected, and sustained.

9-53. The LOGCAP is a DA capstone program that employs contractor support to augment the Army's organic planning and CS and CSS capabilities. This program applies to CONUS and OCONUS. Before implementing LOGCAP, the combatant commander considers the use of AC and RC, other services, contract resources, and HN support. During a contingency operation, the combatant commander normally establishes an acquisition review board to determine the best means of fulfilling requirements. Board considerations include criticality, timeliness, quality, administration, effort, and cost. LOGCAP is used when contractor support is determined to be the most effective, expeditious, or cost effective method.

9-54. LOGCAP applies primarily in areas where multilateral or bilateral agreements or treaties exist. However, it can be employed in areas with formal HN agreements, where contractors are involved or peacetime support contracts exist. LOGCAP can also be used during mobilization to assist the CONUS support base and help units prepare for war or other contingencies. LOGCAP resolves shortfalls; it does not replace force structure. It includes all preplanned logistics-, engineering- and construction-oriented contingency contracts already awarded and peacetime contracts with contingency clauses. Preplanned weapon system sustainment contracts, ASCC contingency contracts, and the AMC support contracts are examples of contracts that fall under this program.

PLANNING CONSIDERATIONS

9-55. One of the primary responsibilities of the senior Army engineer is to forecast the types and the quantities of engineer materials required for the theater. *Chapter 3* described the CESP, which is generally used to establish the initial requirements during preconflict planning. *See CJCSM 3122.03A* for an example and further guidance for the CESP. Planning during the conflict requires good intelligence as to the damage caused by enemy action and good forecasting of the additional facilities that are needed. The Theater Construction Management System (TCMS) can be used as a guide in determining material requirements for necessary facilities. In many cases, existing facilities can be modified to meet military requirements, thus conserving Class IV materials and expediting construction completion.

9-56. For horizontal construction and nonstandard projects, material requirements must be estimated manually because TCMS provides standard Army designs. The senior Army engineer must also aid the logisticians in determining the adequacy and the most expeditious source of materials, either from local sources, regional sources, or a CONUS projection base. Materials from within the AOR may be from local manufacturers, commercial stockpiles, or HN government assets. Materials not locally available must either be procured out of theater (directly or through a service contract such as LOGCAP) or produced in theater by engineer units. Materials that fall in the latter category include aggregate, concrete, construction water, asphalt, and lumber. A local procurement system must be established to expedite procuring local materials. Such procurement may be restricted or centrally controlled in some theaters to avoid—

- Inflating the cost of construction materials in the HN.
- Creating a bidding war between all participants vying for materials or services within a specific area.
- Depleting specific resources.
- Overspending of operational funds.

DESIGN CONSIDERATIONS

9-57. When designing projects for the TO, designers consider the—

- Availability of construction materials.
- Local materials and property and their suitability.
- Construction practices to accommodate HN maintenance and repairs after US forces depart.
- HN environmental requirements.

9-58. A recent study by the Construction Engineering Research Laboratory (CERL) to update Class IV supply planning factors found that the requirement for local material, especially sand and gravel, could be substantial. Even in the most developed countries, problems exist in finding and hauling locally procured construction materials. Engineer planners develop realistic estimates of the available quantities of local materials as early as possible and assess the feasibility of the plans they have developed.

9-59. Many designs may not be practical because of logistics considerations. For example, although the TCMS's designs are adjusted for various climates (desert, tropic, and arctic), they may be difficult to construct because required construction materials are unavailable in the region. Suitable materials could be brought from the CONUS; however, the level or the length of the US commitment may not support this action. The engineers have developed theater-specific design books that consider regional requirements and standards. In European Command (EUCOM), the Red Book, and in CENTCOM, the Sand Book, are the theater standards.

9-60. Military designers must be knowledgeable of local construction standards and materials commonly used in the particular region. Designs must include the use of local materials or provide flexibility within the design for use of substitute materials. The construction standard for the TO is generally one of the following:

- Initial standard (up to six months of expected use).
- Temporary standard (up to twenty-four months of expected use).

9-61. Since the design life is short, only essential utilities, such as heating and cooling, are provided. This also reduces engineer material requirements. A heavy reliance on occupying existing facilities, either provided by the HN or leased, also minimizes construction material requirements. There are design requirements associated with existing facilities because modifications to structures are probable to meet the needs of the US force mission. As a contingency is drawn out and soldiers are deployed for greater lengths of time, the engineer should expect to upgrade facilities to enhance the QOL as the mission, time, and materials allow.

CONSTRUCTION AND OTHER SPECIAL CONSIDERATIONS

9-62. In the material estimation process, ten percent is usually added to the estimated quantity as an overage factor. The overage in the TCMS's bills of material (BOMs) provides for material loss, damage, and waste and minor field modifications. Although this overage may seem minor, the combined effect of material overage has a significant impact on the supply system. On the other hand, engineers should not succumb to pressures to eliminate this overage factor when ordering materials under the assumption that waste material can be eliminated through maximizing efficiency. Refuse material is unavoidable. However, all who are involved with actual construction can limit construction material waste by—

- Ordering and using optimum lengths.
- Providing clear designs (understandable to the construction unit).
- Inspecting the vendor's stocks to ensure satisfactory quality.
- Providing for proper storage, security, and handling of construction materials.
- Providing proper worker training to limit waste.
- Reusing materials.
- Constructing the facility right the first time.

9-63. Another construction consideration more fundamental than ensuring adequate quantities are ordered is securing transportation to haul materials to the construction site. For some materials, transportation is available within engineer resources. Aggregate, for example, can be hauled by engineer dump truck companies. Small quantities of dimensional lumber and plywood for small jobs can be hauled using organic assets within the combat heavy battalions. Larger quantities of Class IV supplies require coordination with the theater-level movement control center (MCC) and the Theater Army Area Command (TAACOM) to—

- Schedule convoys on MSRs.
- Allocate lift assets to move the material from the stockage points to the area of construction.
- Coordinate material drop-off requirements at engineer supply points (ESPs).

9-64. Beyond lumber and aggregate, other engineer items of bulk requiring coordinated lift assets include inland pipeline distribution systems (IPDSs), mines, and other special engineer Class IV barrier and Class V materials.

9-65. Unlike other supply classes, Class IV construction materials and Class II geospatial items are not provided based on documented consumption rates; there are no anticipated, preprogrammed surge rates for these items. With technological advances in material management, logistics planners can reduce the time required to get mission material to engineers. The management practices, however, cannot change the physical constants in the execution—

engineer mission materials are normally bulky, heavy, and require dedicated transportation. Prepositioning Class IV stocks reduces lift requirements during the initial stages of force projection.

9-66. Construction material is expensive and susceptible to weather damage; therefore, only limited quantities are on hand in the areas of major war plans. For this reason, the ASCC validates needs and initiates requisitions in advance of deployment or operations. Initial material forecasts are submitted by the senior Army engineer staff, using data from the TCMS and the JEPES and the base development plans in the theater-specific, mission-oriented CESP.

9-67. Successful execution of the theater's construction program depends on sufficient materials, logistics units to process the materials, and construction capability. Typically, during the early stages of a conflict, war damage repair and the construction of mission-essential facilities dominate engineer construction activities. The RSOI mission places heavy demands on engineers, as well as logisticians, to receive and launch the combatant forces into forward operating areas. As the theater matures, more substantial facilities are required and more construction forces are available. The ENCOM (or senior Army engineer) forecasts adequate construction materials that are flexible enough to meet the varied construction requirements throughout all operational phases. Working closely with the TSC's support operations cell, engineers ensure the control of Class IV materials. The ENCOM (or senior engineer commander) is a member of the RCEM and the TCEM teams. Given an understanding of both the combatant commander's and the ASCC's work priorities, the ENCOM (or the senior engineer commander) prioritizes the allocation of critical Class IV for the TSC Class IV manager, who implements the actual distribution of these materials.

9-68. Special logistics support considerations for topographic units are equally important. Equipment such as the multispectral imagery processor (MISP) and the Combat Terrain Information System (CTIS) require specialized critical low-density supplies. For example, during the Dayton Peace Accord negotiations, November 1995, the Defense Mapping Agency (DMA) (currently known as NIMA) distributed some 30,000 paper copies of maps covering the revised interentity boundary (IEB) on Bosnia-Herzegovina. As troops deployed to monitor the authoritative ceasefire line (the IEB and their respective buffer zones), thousands more copies of the 1:50,000 scale tactical line maps were needed. Engineer planners anticipated for demand surges on geospatial products before operations and recognized that topographic units were not equipped to run map warehouses, and did not have the transportation assets to distribute geospatial products.

9-69. Engineer bridging beyond the organic equipment available in the AO will either have to be purchased or transferred from another area of operations or from strategic and operational reserves stocks. It is the ENCOORD's responsibility to identify these materials and coordinate for their arrival.

ENGINEER ROLE IN PLANNING AND COORDINATION

9-70. The engineer's efforts to plan and coordinate engineer logistics efforts are essential to full integration of engineer units into the theater sustainment structure. All engineers and logisticians work closely to synchronize the logistics planning and coordination processes. They facilitate sound and timely plans or orders and necessary sustainment for engineer units.

9-71. After receiving a WO for a mission, the staff engineer immediately initiates an engineer logistics estimate as part of the EBA. This estimate is specifically focused on sustaining all subordinate engineer units. Class I, III, IV, and V supplies and personnel losses are the essential elements in the logistics estimate process. Close integration with the logistics staff can simplify and speed the estimate process by using their automatic data processing (ADP) systems. During continuous operations, the estimate process may need to be abbreviated due to time constraints. The staff engineer aggressively maintains an accurate logistics and combat status of all engineer units, which is critical to shortening the engineer logistics estimate process.

9-72. After developing an estimate to determine unit sustainment and mission supply requirements, the staff engineer compares the requirements with the reported status of subordinate units to determine the specific amount of supplies needed to support the operation. These requirements are then coordinated with the logistics staff to ensure that necessary supplies are identified and resourced. At the same time, the SES develops a required supply rate (RSR) to support engineer mission requirements. Based on how much of the required materiel is on hand and how much is needed, the staff engineer, in coordination with the logistics staff, assesses the availability of these supplies in stocks. They also analyze the capability to transport mission supplies to engineer units.

9-73. After identifying the requirements for unit sustainment and mission supplies and their availability, the staff engineer develops a projected engineer status, based on the current engineer sustainment status. The staff engineer then analyzes the requirements to support the plan and translates them into specific plans that are used to determine the supportability of COAs. After determining a COA, the specific engineer logistics input into the basic OPORD and paragraph 4 of the engineer annex are developed and incorporated. Current engineer sustainment operations may require redirection based on the new plan. If so, the OPORD and engineer annex is sent to subordinate engineer units for coordination and execution.

9-74. Engineer units provide unit and mission logistics status to the staff engineer so that the engineer can do a similar logistics staff planning process. Accurate and timely status reporting assists the staff engineer in providing accurate engineer unit status to the commander. It also energizes the staff engineer support to intercede in critical sustainment problems when necessary. The staff engineer ensures that mission-required supplies needed by engineer units to execute missions are integrated into logistics plans.

COMBAT SERVICE SUPPORT FOR ENGINEER ORGANIZATIONS

9-75. Engineer organizations are not organized with adequate internal and organic CSS to sustain the organization. They receive additional CSS from logistics elements at each echelon through their higher HQ support organization or, if in a command relationship, from the unit they are supporting.

9-76. Logistics elements at each echelon are organized to provide engineers with supply, maintenance, transportation, and field services. How the theater's logistics system supports a particular engineer unit depends on—

- That unit's organization.
- Its location in the TO.
- The command or support relationship under which it is operating.
- The maturity of the theater.

ARMY SERVICE COMPONENT COMMANDER

9-77. Operational-level engineer units are supported by subordinate ASGs of the ASCC's TSC. Other modular commands that may support engineers include personnel, transportation, finance, and medical commands. The ENCOM monitors logistics support to operational-level engineer forces.

JOINT TASK FORCE, JOINT FORCE LOGISTIC SUPPORT COMMAND, AND ARMY FORCES

9-78. Engineer units assigned to a JTF, Joint Force Logistic Support Command (JFLSC), or ARFOR are normally supported by an austere logistics support structure. A JFLSC may be established to support all forces in theater until an ASCC TSC structure is required. The senior JTF, JFLSC, or ARFOR engineer HQ monitors logistics support to deployed engineer forces.

CORPS

9-79. The COSCOM provides corps-level logistics support and health services support to corps engineer units and operational-level engineer units attached to the corps. The division support command (DISCOM) supports corps units attached to the divisions. All other corps units operating in division areas receive logistics support from COSCOM units operating in nearby areas. The division's medical support structure provides health services support for corps units operating in division areas on an area basis. Operational-level engineer units working in the corps's area generally receive support for common classes of supply and maintenance from the COSCOM. The corps engineer coordinates for required logistics support for all engineer units working in the area.

DIVISION

9-80. The DISCOM usually needs augmentation from the COSCOM to support corps engineer units attached to the division. In the heavy division, the forward support battalion (FSB) normally supports engineer equipment maintenance and Class IV and engineer Class V requirements. In light divisions, the main support battalion (MSB) supports engineer requirements.

Nondivisional engineer units working in the division area generally receive support for common classes of supply and common maintenance from the DISCOM. The DIVENG coordinates required logistics support for all engineer units working in the division area.

BRIGADE

9-81. The DISCOM forward support elements located in the BSA support the DIVENGs operating in the brigade's area. Nondivisional engineer units working in the brigade's area generally receive support for common classes of supply and common maintenance. The brigade engineer coordinates the required logistics support for all engineer units working in the brigade area.

MAINTENANCE SUPPORT

9-82. An effective maintenance program ensures that engineer units are capable of providing continuous, responsive support to combatant forces and the sustainment base. An effective supply system for repair parts and lubricants is also necessary. Engineer units consider the risks involved in mobility versus ample coverage for prescribed load list (PLL) repair parts. Logisticians anticipate and provide forward stockage of engineer-specific maintenance supply support requirements. The materiel management center (MMC) coordinates with tailored engineer forces for stockage selection and execution to maintain the dominant tempo of operations.

9-83. Military engineer units have the potential to use foreign commercial construction equipment. The source of this equipment is varied; for example, equipment may be—

- Provided by the HN.
- Leased from commercial venders.
- Obtained through a grant or loan.
- Bought by benefactor countries supporting the contingency operation.
- Purchased or leased through the LOGCAP.

9-84. Such was the case for the 34th Engineer Battalion (Combat) (Heavy) during the Gulf War. They deployed without their organic construction equipment and used the commercial equipment that the Government of Japan provided. Other battalions augmented their construction capabilities using additional leased equipment. These alternatives to increase construction capabilities pose challenges in maintenance to be considered, such as the—

- Lack of manuals and English labels on controls and switches.
- Lack of on-vehicle equipment (OVE), particularly safety equipment.
- Lack of service records or even a service schedule.
- Incompatibility or unavailability of repair parts within the Army's maintenance system.

9-85. The lack of manuals tends to make preventive maintenance haphazard. Nonstandard repair parts require mechanics to remove the part, take it to a vender to match up with a suitable replacement, and establish a PLL from scratch with no maintenance history. These challenges are not insurmountable; however, prior knowledge of the equipment condition and preventative maintenance status helps the commander establish a reputable maintenance program when assuming responsibility for foreign commercial construction equipment.

OTHER SUPPORT

9-86. The ENCOM and its operational units assist in managing critical Class IV and engineer Class V supplies, primarily in the establishment of theater policy or advisory capacity to the ASCC's logistics planner or the TSC. For other classes of supply and services, engineer units are consumers within the theater.

9-87. Engineer units obtain personnel service support (PSS), including finance services, from EAC PSS assets. Engineers need finance support to pay contractors and other local providers. The TSC's finance support center provides finance support. The nearest medical facility provides medical support, including medical supplies. The TSC's EOD detachments provide EOD support.

9-88. The ASG's supply and services battalion provides Class I supplies (rations). Rations are obtained based on the unit strength reports that the battalion Adjutant (U.S. Army) (S1) prepares. When working on an air base, engineer units obtain Class I support from the Air Force.

9-89. The property book officer (PBO), at the battalion level or in separate companies, requisitions Class II supplies from the ASG's supply company. Requisitions for regulated or command-controlled items are processed through the command channels. Engineer units request Class VII items through the material readiness officer at the MMC.

9-90. The ASG's supply and services battalion provides Class III supplies (POL). This is true for both bulk and packaged products. Requisitions from the PBO are required to obtain the needed supplies. When working on an air base, engineer units obtain Class III support from the Air Force.

9-91. Class V supplies (ammunition) are obtained from the nearest ASP that the TSC's ammunition battalion operates. Requisitions must be processed by the PBO. The TSC requires input on recommended stockage levels for engineer-specific Class V supplies, such as mines and demolition items, based on plans and anticipated usage rates.

9-92. Transportation and material transfer support are important when planning for Class IV and engineer Class V material. Supplies are often shipped by class. Transportation priorities for Class V supplies are often higher than those for Class IV supplies. Synchronizing mission Class IV and engineer Class V transportation and material transfer support is desirable to use mission loads effectively.

9-93. Logistics is an integral part of the engineer planning process. Engineer units that are not organic to a division are tailored for specific operations. Therefore, the staff engineer must articulate his CSS requirements from deployment planning—through the engineer estimate for mission loads—to the actual delivery of services and material. The coordination of units and resources facilitates the anticipation of engineer unit sustainment requirements and the execution of CSS. The staff engineer ensures that CSS planners understand engineer unit sustainment requirements and have a plan to meet those requirements.

ENGINEER COMBAT SERVICE SUPPORT COMMAND AND CONTROL

9-94. Each engineer CP has specific responsibilities in identifying unit and mission logistics requirements, estimating resources, integrating the logistics requirements into the operational planning cycle, and monitoring the execution of engineer missions supporting logistics operations.

REAR COMMAND POST ENGINEER SECTION

9-95. The rear CP engineer section—

- Is the engineer's primary integrator into rear CPs for executing logistics support for subordinate engineer units.
- Coordinates sustainment for current engineer operations and plans and prepares for implementing future operations.
- Maintains updated logistics status of engineer units.
- Provides the main CP engineer section with detailed logistics estimates to help formulate plans and orders.
- Ensures that engineer sustainment plans are synchronized with the logistics staff.

MAIN COMMAND POST ENGINEER SECTION

9-96. The main CP engineer section—

- Develops engineer sustainment plans.
- Writes the engineer logistics portions of the basic OPLAN or OPORD and paragraph 4 of the engineer annex.
- Integrates engineer sustainment through coordination with the main CP logistics cells.
- Ensures that the immediate engineer sustainment requests received from the tactical command post (TACCP) or assault CP are forwarded to the rear CP.

TACTICAL OR ASSAULT COMMAND POST ENGINEER SECTION

9-97. The TACCP or assault CP engineer section has limited capability to impact engineer logistics support from this location. Its primary logistics duties are receiving and forwarding reports and influencing the redirection of sustainment priorities for forward operating engineer units.

ENGINEER BRIGADE AND GROUP COMBAT SERVICE SUPPORT CELLS

9-98. The engineer brigade and group CSS cells—

- Support the brigade and group S1 and S4 officers in developing the engineer sustainment plans and writing paragraph 4 for brigade and group OPORDs and OPLANs.
- Monitor current engineer logistics status through periodic personnel and logistics status reports from subordinate units.

- Recommend logistics priorities to the brigade or group commander.
- Identify critical personnel and supply shortages, along with maintenance or transportation problems, that affect engineer unit and mission sustainment.
- Redirect logistics support.

9-99. The engineer brigade CSS cell maintains constant communication with subordinate engineer unit logistics sections, the rear CP engineer section, the COSCOM CP, and the G4. The engineer group CSS cell maintains constant communication with the brigade CSS cell, the subordinate engineer units, and the supporting maneuver unit logistics support units, if required.

ENGINEER BRIGADE AND GROUP HEADQUARTERS AND HEADQUARTERS COMPANY COMMAND POSTS

9-100. The engineer brigade and group headquarters and headquarters company (HHC) establish CPs at or near the brigade or group TOC. The HHC CP is responsible for sustaining the brigade and group CP. This includes establishing accounts with designated logistics support units, setting up life support areas for CP personnel, and locating vehicle maintenance areas. The HHC CP may also be designated as a Base Defense Operations Center (BDOC) or Base Cluster Operations Center (BCOC).

Appendix A

Engineer Field Manuals and Related Joint Publications

This appendix shows the comprehensive hierarchy of Army and joint engineer field manuals and the relationships between them.

A-1. *Table A-1, page A-2,* is a hierarchy chart of the Engineer Branch proponent manuals and the related joint engineer manuals. It is current as of the publication of this manual, but is subject to change as manuals are updated, renumbered, and published. Always refer to the latest official publications.

A-2. Army and engineer proponent manuals are in the process of being renumbered to align them with the joint manuals numbering system. *Table A-1* lists manuals by the current number, followed by the old number in parenthesis. In the new numbering system—

- The prefix 3 represents alignment with the category of joint publications operations manuals.
- The suffix 34 represents an engineer publication.
- The extensions designate the mission content (such as combined arms or engineer operations) or a reference focus for the manual.

A-3. All Army manuals can be downloaded and viewed through the Reimer Digital Library (RDL). Manuals that require passwords can be accessed by applying for a password on the web site. The RDL is located at http:// 155.217.58.58/atdls.htm.

The JPs can be located at http://www.dtic.mil/doctrine. Those of specific interest to engineers include—

- JP 2-03.
- JP 3-15.
- JP 3-34.
- JP 4-04.

Table A-1. Engineer FMs and Related Joint Publications

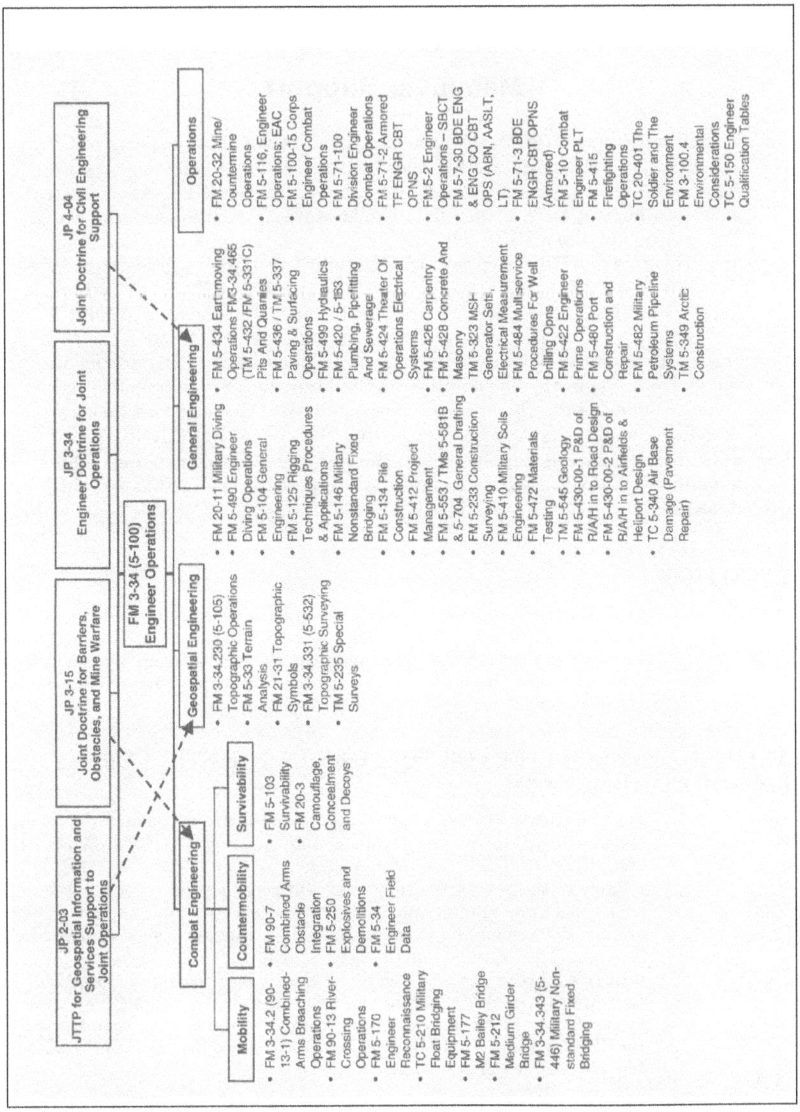

Appendix B

Maneuver Support

Maneuver Support (MANSPT) and the grouping of those aspects of CS activities and elements are not recent ideas, although the operational reality of this is only recently being worked with in the SBCT as the Army looks at other ways to functionally organize staffs to enhance performance. The creation of a MANSPT cell, just like the creation of a fire effects cell and others, is an initial step toward what staff structures will look like in the future. As such, this appendix will look at MANSPT, its definition, today's realities, and its likely future. The primary premise for a MANSPT cell is built around the M/CM/S BOS. At the core of MANSPT are the engineer, chemical, MP, and other branches such as the EOD portion of the Ordnance Branch. By establishing the Maneuver Support Battle Lab and consolidating the chemical, engineer, and MP schools at Fort Leonard Wood, Missouri, the Army is furthering the concept of MANSPT and developing relationships between the three primary branches. This appendix will discuss current and future opportunities for this evolving concept.

DEFINITION

B-1. MANSPT is defined as staff integration of the M/CM/S BOS, with the remaining BOS focused on enabling assured mobility for the friendly force. Staff integration is focused on enhancing tactical freedom of maneuver and force protection using the assured mobility imperatives and fundamentals as the framework.

TACTICAL LEVEL: MANEUVER SUPPORT IN THE STRYKER BRIGADE COMBAT TEAM

B-2. The SBCT MANSPT cell is organic to the SBCT HQ and is responsible for planning, integrating, and synchronizing MANSPT for all SBCT operations. It is composed of an engineer and an MP section. Although not organic to the MANSPT cell, significant coordination is required with the chemical section to integrate MANSPT considerations into the cell. While no organic EOD capability exists in the SBCT, this too is an area of MANSPT that must be integrated into this cell. If the brigade engineer is also the senior officer with the MANSPT cell, he may also function as the MANSPT coordinator. In this case, he must rely heavily on the MP and chemical sections for their expertise to ensure integrated and synchronized MANSPT to the SBCT. For more information related to the specific duties of the MANSPT coordinator, *see FM 3-21.31*.

OPERATIONAL-LEVEL CONSIDERATIONS

B-3. The key aspects or imperatives of MANSPT include the following:

- Understand the environment—consider the space, air, water, land, subterranean, weather, light, infrastructure, hazards, and populations.
- Enable theater access—access installations through air, land, and seaports of embarkation (SPOEs) and move between them; mitigate the adverse effects on the environment, and protect and facilitate multiple entry points.
- Provide assured mobility—guarantee the force commander the ability to deploy, move, and maneuver at all times without interruption.
- Deny enemy freedom of action—leverage the environment to isolate, deny, impede, or canalize the enemy.
- Enable force protection and security—provide cueing, early warning information, protected movement of forces between operating areas, and layered protection.
- Engage and control populations—control EPWs, refugees, detainees, internees, mass transiting civilian populations, and disaster evacuees.
- Neutralize hazards and restore the environment—reduce or eliminate hazards through avoidance, mitigation, neutralization, and restoration of the environment.

THE ROAD AHEAD

B-4. MANSPT as a future concept and framework for the future force is much broader than the SBCT and encompasses the means to enable, enhance, and protect the strategic, operational, and tactical freedom of action of the force, as well as the denial of comparable freedom of action to the adversary. MANSPT accomplishes this by shaping, leveraging, or mitigating the effects of the operational environment. For an offensively oriented, maneuver-based force, the focus is on enabling force movement and maneuver at all levels of war and echelons and across the full spectrum of military operations. This is critical with the increased emphasis on operational and strategic maneuver of ground forces for the future force.

B-5. MANSPT concentrates on two interrelated components: freedom of maneuver and force protection. For a maneuver-based force, there is significant overlap and synergy between these two components, with MANSPT assets, systems, and soldiers able to readily support both components simultaneously. This translates into an effective and efficient application of forces. MANSPT takes on added dimensions and significance due to the increasing emphasis on battlefield frameworks that encompass extended, dispersed, distributed, noncontiguous, and 3-D battlefield and nonlinear operations, to include forcible and early entry operations and operational maneuverability. Additionally, the Army's vision of the operational environment recognizes that adversaries will apply a wide array of conventional and unconventional or asymmetric means to deny the access and movement of friendly forces, starting with their strategic deployment platforms and continuing within the TO and in individual tactical AO.

Finally, the physical environment, to include terrain, infrastructure, weather, hazards, and the presence of the local populace, will affect any ground force, sometimes more than enemy actions will.

B-6. For further thoughts on the future of MANSPT and its application as a mission area see the soon to be published *TRADOC Pamphlet 525-3-25*.

Appendix C

Field Force Engineering

FFE is a focused area for the Engineer Regiment that specifically targets improving general (and to a lesser degree geospatial) engineering support to contingency operations. The focused area reduces the ad hoc solutions previously applied to meet contingency requirements. These requirements have grown, and the Engineer Regiment has responded by doctrinally accepting the best ad hoc solutions, using lessons learned and actively seeking to add technological advances as they impact the engineer mission. Another reason for the development of FFE as a focus area is to provide specialized engineer support to help to quickly establish and maintain the necessary infrastructure to support forces in contingency AOs. The nature of contingencies has made it necessary to provide installation-type engineer support to tactically deployed combined arms forces through the creation and use of base camps. In Bosnia alone, over 2,500 engineer soldiers from a host of countries established and maintained base camps. During the Bosnia and Kosovo operations, a modular expert personnel package called the Base Camp Coordination Agency (BCCA) was developed and deployed to assist these efforts. The original package, partially staffed by members of the ENCOMs, consisted of a cell with a tailored capability of engineer skill sets. After 1997, the package consisted largely of civilian and military personnel from elements within the USACE, using military individual mobilization augmentation personnel, mobilized RC personnel, civilian personnel from elements within USACE, and selected individuals within CONUS (largely pickup teams, not units). These teams quickly develop and implement techniques, tactics and procedures to accomplish their mission. Typical missions included base camp design and closure, force protection analysis, ASP upgrade, MSR maintenance and basic DPW missions in support of base camps. A book of standards, commonly referred to as the "Red Book," was developed by United States Army, Europe (USAREUR), Deputy Chief of Staff, Engineers (DCSENGR), and provided a blueprint for planning and requirements relating to base camp construction and maintenance. A similar book of standards know as the "Sand Book" was developed for the CENTCOM AOR. Additionally, a critical piece of FFE and the reach capability includes the associated technology of the TeleEngineering communication suite.

THE CHALLENGE

C-1. The challenge for engineers is to be prepared to support contingency operations and all missions within the full spectrum of conflict. *Figure C-1, page C-2* graphically demonstrates the changing requirements for engineers over time in any given contingency operation. Different portions of the Engineer Regiment will play varying roles in supporting these requirements at different times. Added to this are the factors of METT-TC that may limit or require the use of one or more portions of the Engineer Regiment. At a particular part of a contingency operation, it may be desirable to limit either the footprint of engineers in the AO, or at least the number of green suit engineers. In this case, extensive use of reach capability and/or the use of USACE civilians may be preferable to military engineers out of the ENCOMs. In another contingency operation, the threat level may be so high that the decision is made to keep as many civilians as possible out of the AOR, leaving a pure military engineer solution. In any case, FFE allows for a doctrinal solution to meet each challenge in an organized fashion.

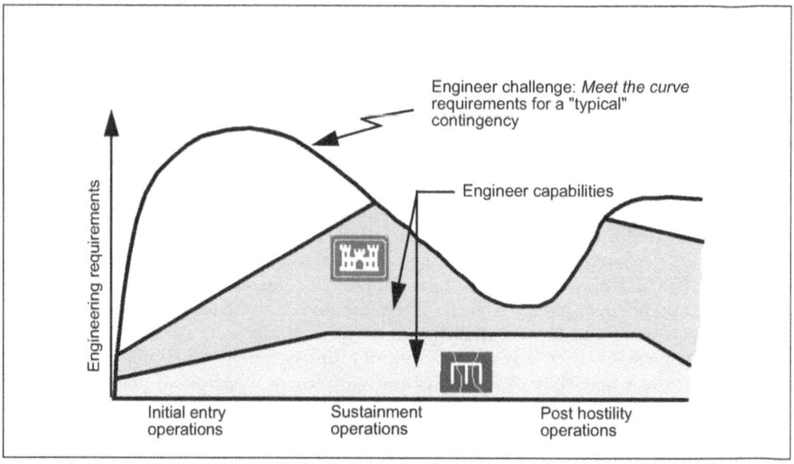

Figure C-1. Engineer Support Shortfall

OVERVIEW

C-2. The application of FFE is Army-wide, and, due to the joint nature of operations, applicable to the support of not only Army forces, but other Services and governmental agencies outside of DOD. *Figure C-2* highlights the majority of capabilities the Engineer Regiment offers to support contingency operations. FFE works to provide seamless general (and associated geospatial) engineering support for the war fighter by fusing the capabilities resident in USACE, USAES, ENCOMs, public works, and civilian contractors. Modular teams are available to meet the needs of C2, augmentation of ASCC, or other headquarters for engineering staff, theater planning, facilities engineering, troop construction, topographic engineering, contract construction, technical engineering, real estate acquisition, and environmental engineering. The mission of FFE is to support the senior engineer in theater by providing focused planning, coordination, and execution of nontactical engineer missions, including the construction, maintenance, technical engineer support, and final disposition of infrastructure.

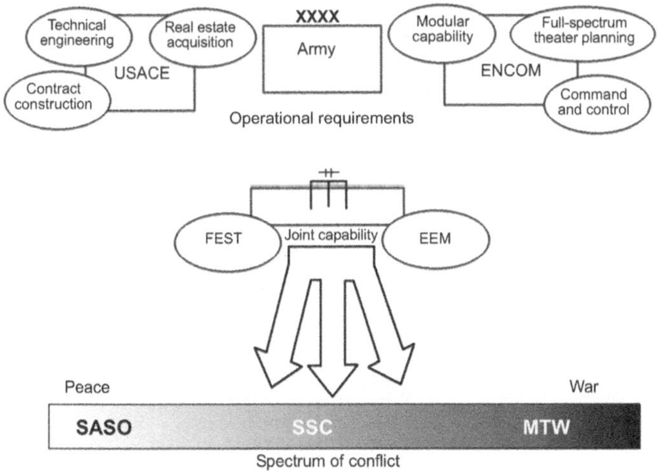

Figure C-2. Field Force Engineering

PLANNING STRUCTURE

C-3. The USACE and the ENCOMs provide the primary elements to create the integrated planning structure for FFE support. Operational missions addressed include: LOC integrity, AOR infrastructure stability, construction management, contractor oversight, and oversight of facilities maintenance missions. The forward elements are responsible for daily interaction with the MACOMs and joint staff as required. They provide the senior engineer with a level of expertise and linkage with other engineer capabilities that will allow comprehensive construction and facilities maintenance planning. *Figure C-3, page C-4* demonstrates the links of the USACE and the ENCOMs with the various combatant commanders, JFCs and ASCC commanders in both peace and war. The current established relationship with particular combatant commanders and other HQs is identified in Appendix I under the USACE and later in this appendix. The ENCOMs currently have permanent forward planning cells with all of the ASCCs except the United States Army Forces Southern Command (USARSO). The ultimate goal of these linkages is to provide proactive and appropriate support for any type of operation. It establishes a more unified approach to using all assets of the Engineer

Regiment across the spectrum of conflict and acknowledges the reality of needed tailored engineer support assets.

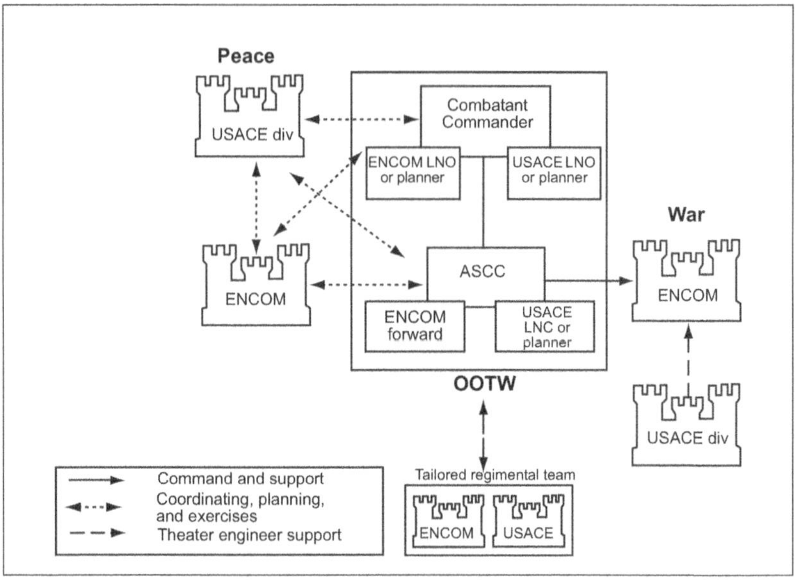

Figure C-3. Field Force Engineering Planning Structure

MODULAR DEVELOPMENT

C-4. The Army's first major challenge in a contingency operation is mobilizing and deploying forces into theater. The space availability on military strategic sealifts and airlifts will always constrain the size and content of the deploying force. Early deployment of one or more engineer modular packages provides specific and necessary expertise from the beginning of operations. Modular deployment minimizes the footprint of units and personnel. As the theater matures and sealifts and airlifts become available, more robust and complete capabilities may be brought into theater if necessary and desirable. There is recognition that most of the focused expertise in the FFE is nontactical in nature and may be supported by civilian contractors. Reach may be used to obtain much of the specialized support required. One method for this is the use of the USACE TeleEngineering Emergency Operations Center (TEOC) and a forward deployed TeleEngineering communication suite. Examples of modular capabilities are listed below:

C-5. USACE personnel are organized for employment as FESTs. The FEST planning cell provides direct augmentation to the combatant commander, JFC, and ASCC staff engineers; or the ENCOM if they are the senior engineer element. The FEST execution team provides C2 for missions for which the USACE has the lead.

- A forward engineer support team-main (FEST-M) is a deployable USACE organization that executes the USACE mission in the AO. It is usually subordinate to the senior engineer commander in the AO.
- A forward engineer support team-augmentation (FEST-A) is a deployable planning augmentation cell that augments the engineer staff of other organizations including the combatant commander, JFC, ASCC, and ENCOMs. If a FEST-M is also deployed, the FEST-A will come under the command of the FEST-M commander.
- A CREST is a deployable 5- or 6-person team and typically a component of a FEST. A CREST can quickly execute real property (land and facilities) leases at forward locations.
- An environmental support team (ENVST) is a deployable 4-person environmental team, typically a component of a FEST. An ENVST provides baseline and other environmental studies at forward locations. It is staffed with environmental specialists that can provide technical expertise to staff engineers and ultimately the commander of the combined arms force.
- A base development team (BDT) is a nondeployable team that can quickly provide base development engineering and planning and facilities design for staging bases, base camps, forward operating bases, displaced persons camps, and any similar requirement.
- An infrastructure assessment team (IAT) is a nondeployable team that provides engineering infrastructure assessments for military deployments and civil military operations in forward areas. Focus areas for the IAT are infrastructure related to USACE missions and aspects of the AO impacting contract construction, including roads, utilities, water resources, and HNS.

C-6. ENCOMs typically support a combatant commander through the ASCC and plans for flexible modular entry into theater via early entry modules (EEMs). EEMs may range from several soldiers to a functional ENCOM C2 element. They have the ability to put soldiers into theater within days of mobilization through the use of three primary means. This includes the use of AC and Title 10 active USARNG and USAR soldiers, hasty mobilization teams (a new process recently endorsed by the Chief of USAR) and soldiers on temporary tours of active duty and active duty for special work. For situations requiring military organizations, the ENCOMs contain the following types of organizations *(refer to Appendix D as well):*

- A facility engineer group (FEG) is a 405-person TDA or TOE organization responsible for design and technical engineering support, including facility and environmental management, BASOPS, force protection, and other areas of expertise. Only one FEG exists in the force structure, and it is under the C2 of the 416th ENCOM during peacetime operations. An FEG is comprised of six 15-person TO&E FED and thirty-person TDA facility engineer teams (FETs) aligned under four geographic facility engineer centers (FECs) that provide additional peacetime command, control, and administrative support.

Each organizational element of the FEG is assigned a unique Unit Identification Code (UIC) for rapid and efficient deployment.

- Three FEDs are war-traced to each of the two major contingency operations (MCOs) ENCOMs and are further aligned as one per ASG. During peacetime, the detachments provide environmental and facility assessments, engineer design, and other support to the USAR engineer, with the other half of their effort dedicated to exercise participation, engineering planning, and design review. The FEDs are deployable elements structured to provide the full range of DPW facilities management for troop concentrations in a theater of operations. Their composition includes all the necessary disciplines to manage, operate, design, and assess bare base facilities and their component systems.
- FETs deploy during war to conduct DPW, BASOPS, environmental, and facility assessment missions. When deployed, they come under the C&S of the senior engineer in their particular AO. During peacetime, they provide environmental and facility assessments, engineer design, and support to the USAR engineer. FETs are deployable organizations that can support contingency operations.

C-7. A partial list of potential missions performed by deployed FFE units and organizations includes—

- Project management, to include supervising work done by military engineers from all branch services, LOGCAP contractors, and local contractors.
- Expertise in force protection engineering and engineering services, including areas such as wastewater treatment, nontactical power generation, water production, snow and ice removal, and entomology.
- Expertise in theater operational (such as in the IPDS) and infrastructure pipeline construction.
- Expertise in mining and demining operations.
- The development and maintenance of base camp standards according to ASCC guidance.
- The development of prioritized facilities requirements with supporting construction engineers.
- The management of the installation and regional and AOR master plan development process, including the—
  - Development of base camp master plan(s).
  - Review and approval of base camp master plans developed by others.
  - Development of regional master plans in support of the operational support structure.
- Base camp maintenance, including—
  - Structures.
  - Utilities.
  - Roads and airfields.
  - Provisions for water.
  - Wastewater treatment.

  - Solid waste disposal.
  - Training ranges.
- Management of base camp facilities use, including—
  - Environmental compliance.
  - Maintenance of real property records.
  - Plans for future expansions or contractions of base camps.
  - Quality control for construction projects and facilities upgrades.
  - The establishment of a fire prevention program and management of fire fighting teams.
  - Facility engineering supplies management.

LEVERAGING TECHNOLOGY

C-8. Reach and TeleEngineering are terms used to describe the ability to conduct engineering analysis and support using expertise (professional engineers, scientists, and technicians), as well as databases that are not resident in the theater or the AO. The source of this expertise and data is real time via secure voice and video information technology to the forward mission site. This results in faster response times and the application of greater levels of expertise examining and solving problems and providing solution options. An additional set of benefits includes a reduced deployment footprint, increased associated support and force protection requirements, and a general reduction in cost.

Appendix D

Army Engineer Organizations and Capabilities

Army engineers operate as an integral member of the joint force to provide a complete range of engineering capabilities to the JFC across the full spectrum of military operations. Whether performing combat engineering missions, providing general engineering support throughout the operational area, or providing geospatial engineering support to commanders, engineers do not fight alone. They are an important part of the combined arms team, increasing the team's combat effectiveness.

The engineer force structure has been developed to support various missions from front to rear, with more survivable and mobile forces in the forward CZ. Any engineer unit may be pushed forward, if the mission dictates.

A wide variety of engineer units provide specialized technical capabilities required to accomplish essential tasks throughout the theater. The engineer architecture depicts an organization that is responsive to commanders at all echelons. This appendix addresses all Army engineer organizations and select architectures. For information on USACE organizations and capabilities, *see Appendix I.*

The following information identifies the types of engineer units available by their base TOE or TDA. For further information, consult the USAES, Directorate of Combat Developments.

SECTION I – OPERATIONAL-LEVEL UNITS

HEADQUARTERS, ENGINEER COMMAND

D-1. The ASCC commander tailors the engineer structure to the theater requirements based on staff advice from the DCSENGR. The operational-level engineer commander provides C2 operations and a central organizational framework for the engineer effort. Engineer forces outside the corps focus on reinforcing and augmenting corps engineer efforts, developing the theater support base, and maintaining an infrastructure for sustainment. This focus involves—

- Planning.
- Ensuring combat, geospatial, and general engineering support at the operational level.
- Coordinating all theater engineer assets.

- Providing construction guidance, RPMAs, LOC sustainment, engineer logistics management, and base development.

D-2. Engineers must be closely tied into current and future operations because they provide versatility to the operational commander. All engineer units focus on operations in the CZ. Additionally, they support the theater by providing general engineering support at the operational level. The engineer's operational-level topography unit and a variety of specialized engineer teams and detachments support or augment engineer forces throughout the theater. Combat heavy engineers weight the main effort and provide sea, air, and land operational and strategic mobility. A typical operational-level engineer organization is shown in *Figure D-1*. *FM 5-116* discusses the operational-level engineer function.

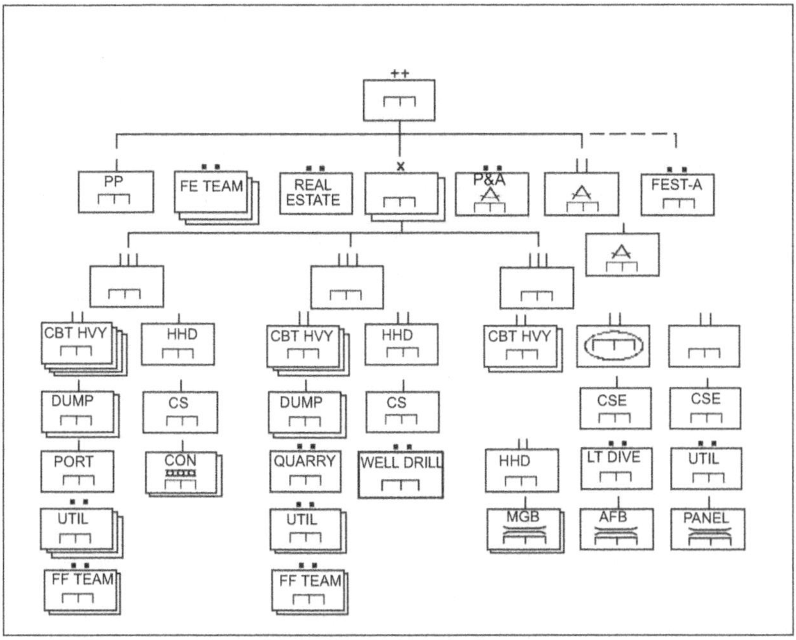

Figure D-1. Generic ENCOM

D-3. Theater construction management often accommodates multiservice requirements. The combatant commander may direct establishing a RCEM to control all theater level engineering. The Army ENCOM commander can perform this role if the combatant commander designates the ASCC commander as the RCEM and the ASCC designates the ENCOM commander as the agent. To support force projection requirements, an engineer element may deploy in sections to meet highly variable workloads and situations.

D-4. Theater infrastructure development is a primary concern of the ENCOM. The ENCOM is responsible for developing plans, policies, procedures, and programs for engineer support for the ASCC. This support includes requirements determination, M/CM/S (at the operational level), other general engineering (ADC, military construction, engineering design, construction materiel and maintenance, and repair of real property), and geospatial engineering (production of military geographic intelligence). Engineer units are responsible for infrastructure planning, development, construction, and maintenance *(see Figure D-2).*

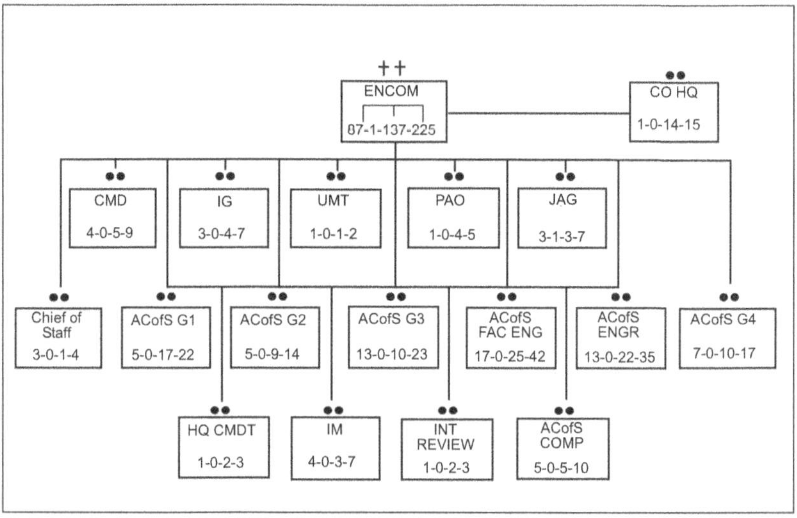

Figure D-2. Headquarters and Headquarters Company, Engineer Commands

D-5. Typical operational-level engineer missions include the following:

- Planning; designing; supervising, coordinating and providing guidance and technical assistance; constructing or rehabilitating general troops; and providing contractual support and support to the Army, other services, and allies throughout the theater to include the construction of—
  - Airfields, ports, pipelines, bridges, roads, railroads, and inland waterways.
  - Hospitals, base camps, EPW and civilian internee compounds, bulk petroleum storage and distribution systems, dry cargo and ammunition storage areas, and equipment maintenance facilities.

- Missile sites, air defense emplacements, protective shelters, and local security measures.

- Performing emergency runway repairs that exceed Air Force capabilities.
- Conducting rear operations missions, to include ADC.
- Augmenting ASG RPMA capabilities.
- Providing and coordinating topographic, geospatial, and military geographic intelligence support.
- Maintaining critical LOCs.
- Managing HN and contractual construction and labor engineering efforts, to include US, indigenous, and third-country personnel.
- Providing fire fighting support, as specified by the field commander.
- Providing environmental support.

D-6. The HQ element of the ENCOM provides staff supervision over all engineer operations in the COMMZ and directs engineer support to all Army, joint, and multinational commands and other theater elements, as directed by the JFC and ASCC commander. It provides policy and technical guidance to all Army engineer units in the theater. The HQ element maintains a technical relationship with the ASCC staff engineer to help establish engineer policy for the theater and maintains required coordination links with JFC and multinational command engineering staffs.

D-7. Several critical specific functions of the ENCOM HQ include—

- Monitoring engineer support for real-property management and coordinating reports from engineer organizations and additional support elements supporting TSC operations, HNS, and contracting directorates.
- Monitoring requirements and priorities for Class IV and engineer Class V materials and coordinating the flow of these materials with the TSC supply and maintenance directorate, the distribution management center (DMC), and MMC.
- Performing the EBA and coordinating with the ASCC engineer, G2, and TSC support operations section to communicate critical enemy capabilities, affecting support facilities and operations.

D-8. *FM 5-116* provides more details on engineer operations in a theater.

HEADQUARTERS, ENGINEER BRIGADE, THEATER ARMY

D-9. The HQ, Engineer Brigade *(see Figure D-3)* is normally assigned as an ASCC and may be attached to an ENCOM. This HQ—

- Commands, controls, conducts planning for, and supervises attached engineer units that are engaged in CS operations and the construction and rehabilitation of facilities in support of a TO.
- Commands, controls, plans, and supervises the attached units' activities.
- Allocates resources in support of engineer operations.

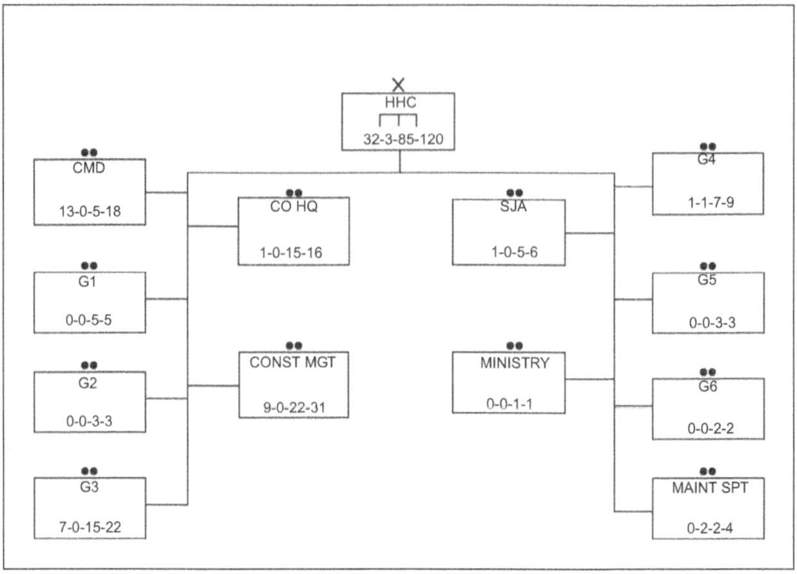

Figure D-3. Headquarters and Headquarters Company, Engineer Brigade (Theater Army)

- Supervises engineer units that are constructing and rehabilitating roads, structures, air-landing facilities, and petroleum storage and distribution facilities.
- Supervises HN personnel and administers contractual construction and labor.

D-10. Utility teams (TOE 05530LH00) normally augment the capabilities of this unit when facilities engineering support is required.

HEADQUARTERS, ENGINEER GROUP, FACILITY

D-11. The United States Army Facility Engineer Group (USAFEG) is a composite TDA and TOE organization headquartered in Chicago, Illinois. This USAR unit contains the Army's highest concentration of 21Ds and facility engineer personnel. The USAFEG is the Army's only FEG. (There are no AC or ARNG equivalent units.) The USAFEG consists of four FECs and a mobilization planning support cell (MPSC).

D-12. The FECs are O6 directorates and are located throughout the country in the northeast, southeast, northwest, and southwest. Each of the three FECs consists of 10 seven-person FETs that are dispersed geographically from their FEC HQ. The fourth FEC is comprised of six 15-person FEDs, which are TOE organizations. These FEDs are war-traced with three detachments going to each of the two Army ENCOMs.

D-13. The MPSC provides mobilization and training support to the FECs and their teams. They also support FORSCOM during activation in times of need.

D-14. The USAFEG has a diverse membership that, with civilian skills, include licensed professional engineers, environmental engineers, electrical engineers, mechanical engineers, civil engineers, architects, construction contractors, facility managers, and project managers. The civilian experience of these engineers uniquely qualifies them to perform their peacetime and wartime missions.

D-15. In peacetime, the USAFEG supports the United States Army Reserve Command (USARC). Teams are responsible for assessing over 1,200 USAR facilities comprising 25 million square feet of workspace. Assessments focus on facility maintenance requirements, environmental compliance, port maintenance, energy conservation, and safety. As force protection becomes increasingly relevant to the safety of soldiers, the teams could also assess and recommend security improvements to meet these new requirements.

D-16. The USAFEG wartime mission is to provide facility engineer services, including base camp design and limited master planning. In times of national emergency or contingency operations to Army components worldwide, this mission is accomplished using the small footprint TDA component— a highly modular skill structure engineer unit—the FE team. In these situations, a seven-person team is mobilized and deployed into the theater of operations for each base camp that is or will be operational. Properly constructed and maintained operations and logistics facilities are instrumental in sustaining military operations in theater.

D-17. In contingency operations, the FET is normally assigned to the ASG and has a coordination relationship with the ENCOM. The missions of these teams range from the construction design and management of building base camps, as soldiers are initially deployed into an area, to managing the maintenance and repair of established bases. In all cases, the FET operates as a DPW. As a TDA unit, the FET has no organic communication, transportation, or defensive equipment and must rely on the ASG for this support.

D-18. During major combat operations, the FEDs aligned with the appropriate theater can be activated to provide those same services. These 15-man units bring their civilian engineering experience and knowledge to assist in all facility work, including base camp planning, design, and layout. These soldiers utilize their civilian skills as master planners; civil, mechanical, structural, and electrical engineers; contracting specialists; and project managers to provide a force multiplier in these essential task areas. These units are aligned with the ENCOM responsible for the theater.

D-19. As the DPW for contingency base camps, the FET and/or FED, in coordination with the base camp commander,—

- Controls, supervises, and coordinates all engineer and construction assets, including troop labor, contracted HN construction assets, and LOGCAP construction.
- Plans and designs projects for infrastructure and operations and living facilities.

- Allocates engineer troops, materials, and equipment for construction projects.

HEADQUARTERS, ENGINEER GROUP, CONSTRUCTION

D-20. The HQ of this engineer group is assigned to a divisional engineer brigade, ASCC, or corps engineer brigade. This engineer group commands and controls assigned or attached units to include engineer combat heavy battalions, corps combat battalions, and separate engineer companies and detachments *(see Figure D-4)*. It possesses a design and management section for specific engineering tasks at EAC and can command, control, plan, and supervise engineer units with construction capabilities throughout the depth of the AO. Normally, the HQ commands and controls three to seven engineer battalions. When task-organized to a division, the group may assume a command or support responsibility for divisional engineer assets. This HQ—

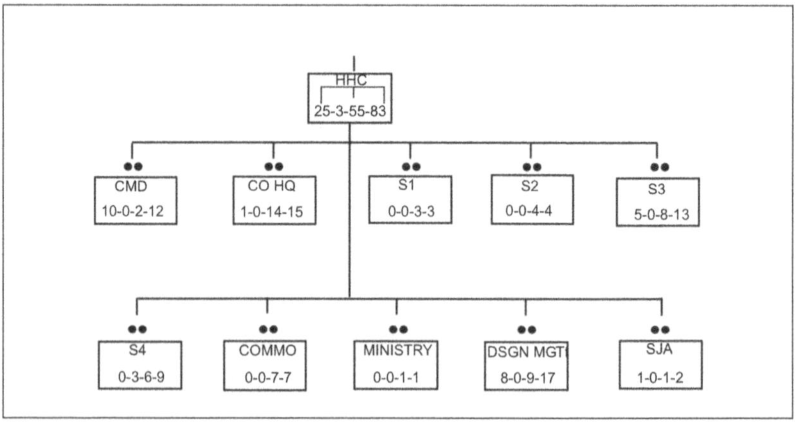

Figure D-4. Headquarters and Headquarters Company, Engineer Group, Construction

- Plans, supervises, and coordinates the activities of the assigned and attached engineer units engaged in survivability, general engineering, and limited mobility and countermobility operations at the operational level.
- Plans and supervises units performing general engineering and selected combat engineering tasks such as constructing and maintaining combat routes, MSRs, bridges, fords and river crossing sites, landing strips, heliports, port facilities, railroads, and battle positions. This includes the engineer technical reconnaissance associated with these missions.
- Plans and supervises the construction of base camps, logistical bases, EPW and displaced civilian camps, pipelines, railroads, and so forth.
- Plans and supervises ADC operations.

ENGINEER BATTALION, PRIME POWER

D-21. The engineer battalion, prime power is assigned to an ASCC and attached to either an ENCOM or a TA engineer brigade. This battalion consists of one HHC and two line companies *(see Figure D-5)*. Each line company contains an HQ section and six prime power platoons. This battalion—

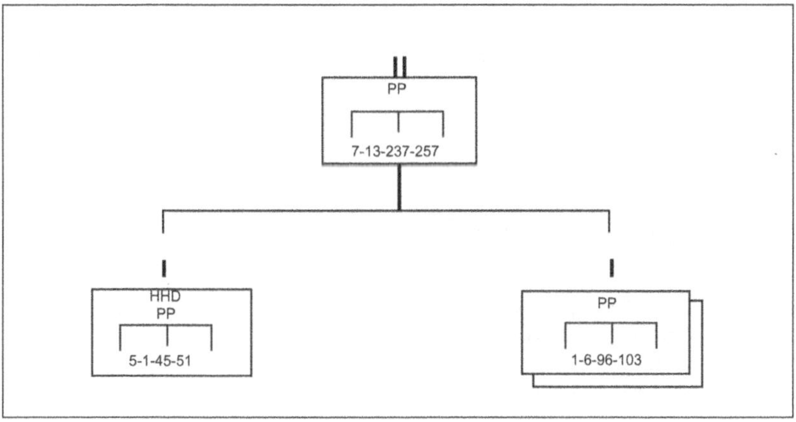

Figure D-5. Engineer Battalion (Prime Power)

- Generates electrical power and provides advice and technical assistance on all aspects of electrical power and distribution systems in support of military operations.
- Produces electrical power, up to 36 megawatts, in support of C2 sites, hospitals, weapons systems, and logistics support areas and provides relief for tactical generators at fixed sites and critical facilities.
- Repairs and maintains organic power production and distribution equipment, distributing power produced with organic systems.
- Operates, maintains, and performs minor repairs to other electrical power production equipment, to include HN fixed plants.
- Provides electrical engineering support, such as limited design and analysis capabilities, to commanders and senior engineers for all aspects of electrical power systems.
- Performs electrical surveys and electrical-related COR assistance.
- Manages and coordinates worldwide prime power requirements.
- Supports RPMA and power reliability enhancement programs.
- Designs and constructs power plants and distribution systems to the panel box for base camps.

- Provides quality control (QC) for contractor designs and the construction of electrical power plants and power distribution systems.

D-22. *FM 5-442* provides more details on engineer prime power operations.

HEADQUARTERS AND HEADQUARTERS DETACHMENT, ENGINEER BATTALION

D-23. Assigned to an engineer brigade, an ASCC, or an engineer group, the number of battalion headquarters and headquarters detachments (HHDs) authorized is based on the number of separate engineer companies and teams. This unit or detachment provides—

- Command, control, and administrative support for separate engineer companies and teams, to include construction support companies, pipeline companies, port construction companies, dump truck companies, and utilities teams.
- C2 for three to seven companies and teams performing engineer-related tasks.

HEADQUARTERS AND HEADQUARTERS COMPANY, ENGINEER TOPOGRAPHIC BATTALION

D-24. The headquarters and headquarters company, engineer topographic battalion is normally assigned to an ENCOM. It—

- Provides C2 for operating, planning, and supervising topographic units in a TO.
- Provides DS and GS maintenance for topographic, reproduction, air-conditioning, and power generation equipment to support engineer topographic units.
- Provides technical supplies for and performs maintenance on organic equipment, as well as unit maintenance on communications and electronics equipment.
- Maintains liaison with allied topographic units, NIMA, and appropriate support unit staff elements.

SECTION II – SEPARATE OPERATIONAL-LEVEL COMPANIES

ENGINEER COMPANY ECHELONS ABOVE CORPS TOPOGRAPHIC BATTALION

D-25. Normally assigned to a TA, this engineer company (EAC) topographic battalion—

- Provides geospatial engineering support to EAC.
- Compiles controlled, semicontrolled, and uncontrolled image maps and mosaics.

- Revises existing maps and other geospatial data within its capabilities.
- Drafts special maps, overprints, overlays, and other geospatial products.
- Reproduces monochrome and multicolor maps, map substitutes, overlays, overprints, and other topographic products by offset lithography and photocopy.
- Provides terrain intelligence and terrain analysis products.
- Performs topographic surveys and provides survey information to the EAC.
- Interprets and measures remote-sensed imagery.
- Extends horizontal and vertical controls into corps and division areas.
- Stores and distributes special topographic products that the company produces.
- Provides a survey information system and maintains digital point-positioning databases (DPPDBs).

ENGINEER COMPANY (CONSTRUCTION SUPPORT)

D-26. Normally assigned to an engineer battalion (heavy) to provide additional horizontal construction capability at the operational level, the engineer company (construction support)—

- Provides construction support, equipment, and personnel for rock crushing, bituminous mixing, paving and storage facility construction, and airfield operations.
- Provides support to one engineer brigade or group engaged in construction projects that—
  - Require up to 150 to 225 tons per hour (tph) of crushed rock and sand from rock quarries and gravel pits for a two-shift operation.
  - Require up to 75 tph of washed and sized precrushed rock for a two-shift operation.
  - Require up to 150 to 225 tph of bituminous mixes and blends for paving projects for a one-shift operation.
  - Require equipment and operators to support selected construction for a two-shift operation and personnel with the capability of supervising contractual labor and indigenous personnel and assisting in the supervising contractual construction.

ENGINEER COMPANY (DUMP TRUCK)

D-27. Normally assigned to an engineer brigade or group, the engineer company (dump truck) is often linked to an engineer construction company. It—

- Operates dump trucks for moving bulk materials in support of other engineer units.
- Provides a haul capacity of up to 600 tons of bulk material (such as gravel, earth fill, and crushed stone) per cycle.

ENGINEER COMPANY (PIPELINE CONSTRUCTION SUPPORT)

D-28. The engineer company (pipeline construction support) is normally assigned to an engineer brigade, ASCC, or engineer group. It typically augments a combat heavy battalion to give it the capability to perform pipeline operations. The engineer company (pipeline construction support)—

- Provides technical personnel and specialized equipment to assist combat heavy battalions or other construction units in constructing, rehabilitating, and maintaining pipeline systems.
- Provides advisory personnel to support up to three engineer companies engaged in pipeline construction, pipe coupling, storage tank erection, and pump station and dispensing facility construction.
- Provides specialized tools, equipment, and personnel to operate on a two-shift basis.
- Is capable of transporting 21,000 linear feet of 6-inch pipe, 16,200 linear feet of 8-inch pipe, or 9,000 linear feet of 12-inch pipe over unimproved roads in two lifts.

ENGINEER COMPANY (PORT CONSTRUCTION)

D-29. The engineer company (port construction) is normally assigned to an engineer brigade, ASCC, or engineer group. It typically augments a combat heavy battalion with specialized equipment required for port construction and repair operations. The engineer company (port construction)—

- Provides specialized engineer support in developing, rehabilitating, and maintaining port facilities, including LOTS operations.
- Constructs, rehabilitates, and maintains offshore facilities, including mooring systems, jetties, breakwaters, and other structures required to provide safe anchorage for ocean-going vessels.
- Constructs, rehabilitates, and maintains piers, wharves, ramps, and related structures required to load and off-load cargo.
- Constructs facilities for roll-on/roll-off (RO/RO), break-bulk, and containerized cargo handling.
- Maintains tanker discharge facilities, including repairing or replacing existing POL jetties and submarine pipelines.
- Provides limited dredging and removes underwater obstructions.
- Installs the offshore petroleum discharge system (OPDS) in support of Army LOTS operations when no naval units are assigned.
- Provides operators for selected items of equipment for a two-shift operation.
- Depends on the pipeline construction engineer company for the radiographic inspection of pipeline welds, when required, and the lightweight diving team for underwater construction, survey reconnaissance, and recovery.

SECTION III – SEPARATE ENGINEER TEAMS (OPERATIONAL-LEVEL)

ENGINEER TEAM, FIRE FIGHTING HEADQUARTERS

D-30. The engineer team, fire fighting HQ provides C2 and administrative support. One HQ team (AA) can control three to seven fire fighting teams (AB). A team commander serves as the fire marshal of an installation/facility or within the AOR. An AA team—

- Plans for fire defense on an installation.
- Conducts fire-prevention inspections.
- Conducts fire investigations.
- Establishes a fire department communications network between the HQ, the MP, the airfield, and the fire fighting teams.
- Commands the fire fighting teams.
- Maintains and refills fire extinguishers.
- Makes minor repairs to fire hoses.
- Inspects and maintains fixed fire-protection systems on an installation/in an AO.
- Coordinates for the resupply of fire fighting assets, agents, self-contained breathing apparatus (SCBA); air, and fuel.
- Coordinates mutual aid with other services and HN fire-protection assets.

D-31. *FM 5-415,* provides more details on engineer fire fighting operations.

ENGINEER TEAM, FIRE TRUCK (AB)

D-32. The team provides fire protection, administers first aid, provides an initial response to HAZMAT incidents, and implements a fire prevention program. A commander's primary task list determines the team's assignments. An AB team—

- Provides crash and rescue support for medical evacuation (MEDEVAC) and normal flight or maintenance standbys.
- Conducts fire prevention inspections on an installation or airfield.
- Provides C2 of the nonfire fighting assets used to support natural cover fire fighting operations (such as heavy equipment and personnel).
- Conducts fire fighting operations (such as structural, crash/rescue, and natural cover) on an installation and in an AO.
- Provides emergency medical assistance to victims.
- Conducts an initial response to HAZMAT incidents.
- Conducts the training of unit level fire brigades.
- Assists with medical resources during mass casualty incidents.
- Assists in HNS, as a commander requires.

ENGINEER TEAM, QUARRY

D-33. A quarry team is normally assigned or attached to an engineer battalion (heavy) and does the following:

- Performs rock-crushing operations, which increases the capabilities of the construction group in major horizontal construction projects such as roads, storage facilities, and airfields.
- Provides personnel and equipment for a 24-hour period to operate the 75 (68.25 metric tons) or 150-tph (136.5 metric tons) crushing plant and issue its product to users.
- Provides personnel and equipment for drilling and blasting operations that are required to produce raw stone for operating the 75 (68.25 metric tons) or 150-tph (136.5 metric tons) crushing plant.
- Is capable of hauling 30 tons (27.3 metric tons) of rock per trip from the quarry to the processing plant.

ENGINEER TEAM, WELL-DRILLING HEADQUARTERS

D-34. Assigned or attached to an engineer brigade or group, this team can be deployed to support units from divisional to operational level. For more information *see FM 5-484.* The team HQ—

- Provides personnel and equipment for drilling and developing water wells.
- Is capable of drilling and casting two complete water well holes of 5 7/8 inches (14.92 centimeters) in diameter to a depth of 2,500 feet (758.33 meters).
- Installs casings, screens, and pumps and develops the well to provide water at the wellhead.
- Can sustain two-shift operations.

ENGINEER TEAM, WELL-DRILLING

D-35. The engineer team, well-drilling, is assigned or attached to an engineer brigade or group, and can be deployed to support other units. This team—

- Provides personnel and equipment for drilling and developing water wells.
- Is capable of drilling and casting two complete water well holes of 5 7/8 inches (14.92 centimeters) in diameter to a depth of 2,500 feet (758.33 meters).
- Installs casings, screens, and pumps and develops the well to provide water at the wellhead.
- Can sustain two-shift operations.

ENGINEER HEAVY DIVING TEAM

D-36. The engineer heavy diving team is normally assigned to a TA and attached to an ENCOM to support commanders in ports, harbors, and coastal zones. This team—

- Assists in constructing port facilities, LOTS structures, and floating barriers.
- Repairs damaged piers, docks, wharves, seawalls, and breakwaters.
- Clears underwater obstructions and marks navigational waterways.
- Reduces and emplaces underwater obstacles and mines.
- Reduces structures with underwater demolitions.
- Recovers sunken materiel and vessels.
- Installs and maintains vessel moorings.
- Repairs Army lighters and vessels.
- Installs and maintains the underwater portion of off-shore petroleum and water distribution systems.
- Protects land forces, vessels, and underwater structures from underwater threats, reducing the probability of underwater structural damage.

ENGINEER LIGHT DIVING TEAM

D-37. An engineer light diving team is assigned to a port construction company, an ASCC, or a corps HQ. This team—

- Performs scuba, lightweight, or deep-sea surface diving to a maximum depth of 190 feet (57 meters). Diving is done to support light-salvage, harbor clearance, underwater pipeline, fixed bridge, and port construction repair and rehabilitation.
- Performs ship husbandry, LOTS, underwater demolition, cutting, welding, and multiple diving operations.
- Depends on the control and support team for specialized supplies, DS and GS maintenance of life-support systems, and augmentation of personnel and equipment for deep sea and heavy salvage operations.
- Performs scuba diving in forward combat areas to perform river crossing site reconnaissance (near shore and far-shore); assists in preparing landing sites, riverbank, and exit routes for crossing forces; clears underwater obstacles as part of combat operations; and supports assault float bridge operations.
- Repairs damaged bridges, locks, dams, pipelines, canals, and levees.
- Constructs underwater bridge structures, obstacles, and floating barriers.
- Recovers submerged weapons systems.
- Protects land forces, river crossing equipment, and underwater structures from underwater threats and deceives enemy forces of friendly underwater and waterborne intentions.

ENGINEER TEAM, REAL ESTATE

D-38. An engineer team, real estate, is assigned to an ASCC. This team performs functions related to acquiring, using and disposing of real property for military purposes. A real estate team—

- Manages real estate.

- Investigates and processes real estate claims.
- Records, documents, and prepares reports on the real estate in any area that the Army uses, occupies, or holds.
- Coordinates with other agencies or friendly HNs to execute joint real estate functions.

ENGINEER TEAM, UTILITIES

D-39. An engineer team, utilities, is assigned to an ASCC, TAACOM, ASG, or a corps HQ. This unit provides limited facility engineering support in the areas of carpentry, masonry, electrical, plumbing, and road maintenance and repair.

ENGINEER TEAM, TOPOGRAPHIC PLANNING AND CONTROL

D-40. An engineer team, topographical planning and control, is assigned to an HHC, engineer topographic battalion. This team—

- Performs geospatial operational planning for units and agencies engaged in producing and supplying military geographic information and topographic products.
- Determines the requirements and provides programs for and coordination of engineer topographic units assigned or attached to the theater.
- Coordinates with the NIMA, the HN or allied nations' topographic support activities, and the higher HQ to accomplish the mission.

SECTION IV – CORPS ENGINEER UNITS

HEADQUARTERS, ENGINEER BRIGADE, CORPS

D-41. Refer to *Figure D-6, page D-16* for a diagram of the HQ engineer brigade corps. This brigade—

- Provides C2 operations to assigned and attached units and coordinates engineer activities.
- Plans and coordinates the operations of the engineer units that are engaged in CS and constructing and rehabilitating facilities in support of the corps or the airborne corps.
- Plans and supervises the activities relating to river crossing, barrier placement, and counterobstacle and countermine operations.
- Supervises the engineer units that are constructing, rehabilitating, and maintaining trails, structures, air landing facilities, and petroleum storage facilities.
- Supervises indigenous and HN personnel and administers contractual construction and labor.
- Provides an engineer staff element to the corps or the airborne corps HQ.
- Plans and supervises terrain intelligence and topographic operations.

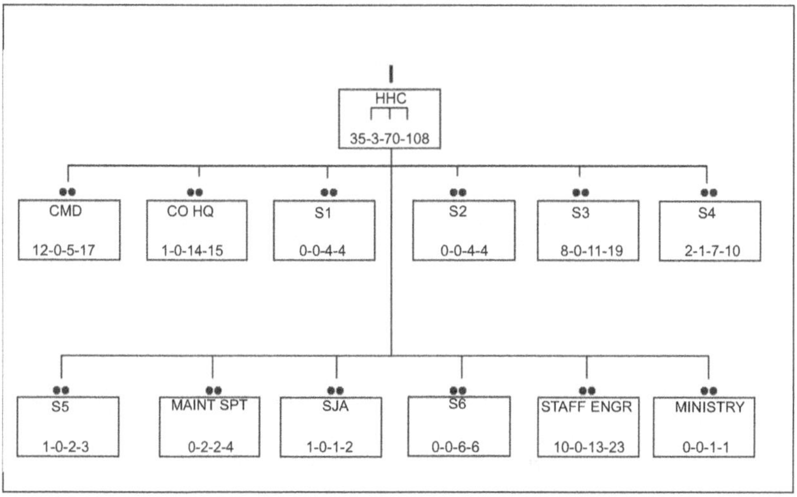

Figure D-6. Headquarters and Headquarters Company, Engineer Brigade, Corps

ENGINEER COMBAT BATTALION, CORPS, MECHANIZED

D-42. Engineer battalion, corps mechanized, is normally assigned to a combat group HQ. This battalion consists of one HHC and three line companies *(see Figure D-7)*. The number of allocated mechanized battalions is based on the number of heavy divisions allocated to the corps. This battalion—

- Increases the combat effectiveness of supported units by accomplishing combat (M/CM/S) and limited general engineering tasks.
- Reinforces heavy divisional engineer units, heavy brigades, and ADC engineer units.
- Reorganizes and fights as infantry, when required.
- Participates in joint military operations, when required.
- Constructs tactical obstacles, defensive positions, and fixed and floating bridges.
- Supports combined arms breaching and river crossing operations.
- Emplaces and maintains assault bridges to span twelve 60-foot gaps, simultaneously.
- Constructs, repairs, and maintains landing strips, heliports, CPs, LOCs, and tactical routes, culverts, fords, and other general engineering horizontal-related tasks.
- Provides technical advice, assistance, and training in mine warfare, field fortification, and camouflage, demolition, and engineer reconnaissance techniques.

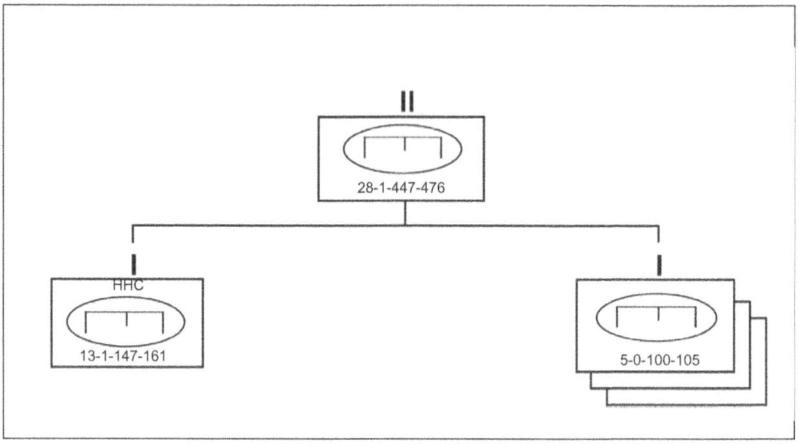

Figure D-7. Engineer Battalion, Corps Mechanized

- Emplaces conventional and SCATMINEs.

ENGINEER COMBAT BATTALION, CORPS, WHEELED

D-43. The engineer combat battalion, corps, wheeled, is normally assigned to a combat group HQ. The number of wheeled battalions allocated is based on the number and type of divisions in a corps. An engineer combat battalion consists of one HHC and three line companies *(see Figure D-8, page D-18)*. This battalion—

- Increases the combat effectiveness of the corps by accomplishing combat (M/CM/S) and general engineering tasks.
- Reinforces divisional engineer units, when required.
- Reorganizes and fights as infantry, when required.
- Participates in joint military operations, when required.
- Commands, controls, and supervises assigned and attached units.
- Provides engineer support in constructing obstacles and defensive positions.
- Emplaces conventional and SCATMINEs.
- Employs fixed and floating bridges.
- Performs general engineering tasks (such as constructing, repairing, and maintaining landing strips, heliports, CPs, LOC, tactical routes, culverts, fords, supply installations, buildings, and structures) and other horizontal related construction tasks, as required.
- Supports limited combined arms breaching operations.
- Provides engineer support in river crossing operations.

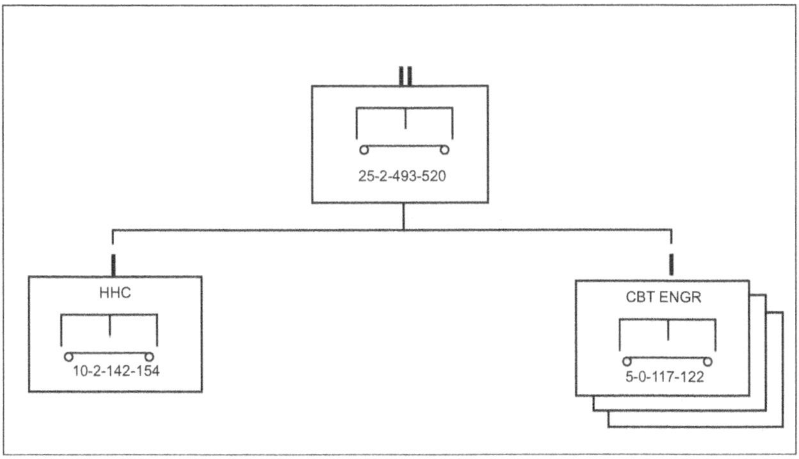

Figure D-8. Engineer Battalion, Corps Wheeled

- Provides technical advice, assistance, and training in mine warfare, field fortification, and camouflage, demolition, and engineer reconnaissance techniques.
- Supports ADC missions.

ENGINEER COMBAT BATTALION, CORPS, AIRBORNE

D-44. The engineer combat battalion, corps airborne, is normally assigned to a combat group HQ supporting an airborne corps. This battalion consists of one HHC and three line companies *(see Figure D-9)*. The battalion—

- Increases the combat effectiveness of supported units by accomplishing combat (M/CM/S) and limited general engineering tasks.
- Augments engineer operations and capabilities with engineer equipment to support light force operations.
- Can parachute or be air-delivered to work sites.
- Reinforces divisional engineer units, when required.
- Reorganizes and fights as infantry, when required.
- Participates in joint military operations, when required.
- Provides C2 operations and supervises assigned and attached units.
- Provides engineer support in constructing obstacles and defensive positions.
- Employs fixed and floating bridges.
- Provides engineer support in river crossing operations.
- Performs general engineering tasks (such as constructing, repairing, and maintaining landing strips, heliports, CPs, LOCs, tactical routes,

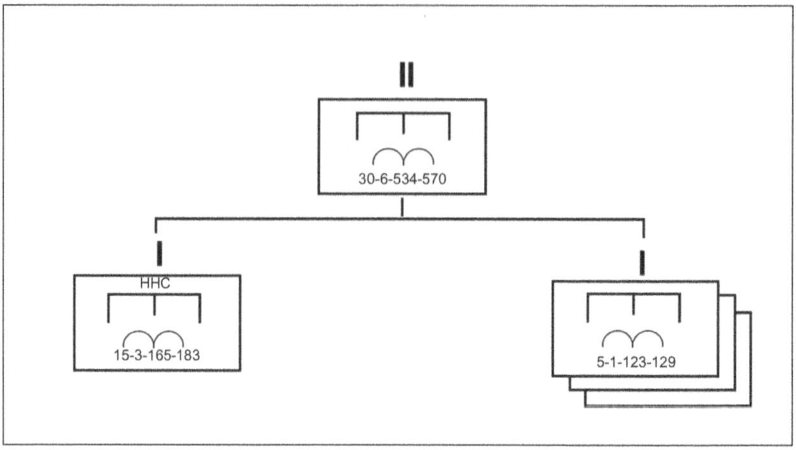

Figure D-9. Engineer Combat Battalion, Corps, Airborne

culverts, fords, supply installations, buildings, and structures) and other horizontal related construction tasks, as required.

- Conducts limited breaching operations and assists in assaulting fortified positions.
- Provides technical advice, assistance and training in mine warfare, field fortification, and camouflage, demolition, and engineer reconnaissance techniques.

ENGINEER COMBAT BATTALION, CORPS, LIGHT

D-45. The engineer combat battalion, corps, light, is normally assigned to a combat group HQ. This battalion consists of one HHC and three line companies and normally supports a light division *(see Figure D-10, page D-20).* This battalion—

- Increases the combat effectiveness of supported units by accomplishing combat (M/CM/S) and limited general engineering tasks.
- Augments engineer operations and capabilities with engineer equipment to support light force operations.
- Reinforces divisional engineer units, when required.
- Reorganizes and fights as infantry, when required.
- Participates in joint military operations, when required.
- Commands, controls, and supervises assigned and attached units.
- Provides engineer support when constructing obstacles and defensive positions.
- Provides engineer support in river crossing operations.

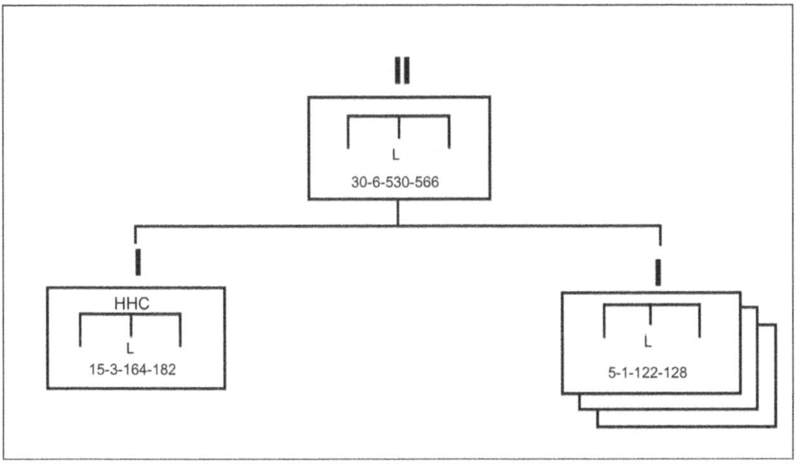

Figure D-10. Engineer Combat Battalion, Corps, Light

- Performs general engineering tasks (such as constructing, repairing, and maintaining landing strips, heliports, CPs, LOCs, tactical routes, culverts, fords, supply installations, buildings, and structures) and other horizontal related construction tasks, as required.
- Conducts limited breaching operations and assists in assaulting fortified positions.
- Provides technical advice, assistance, and training in mine warfare, field fortification, and camouflage, demolition, and engineer reconnaissance techniques.

ENGINEER COMBAT BATTALION, HEAVY

D-46. Engineer combat battalion, heavy is assigned to a corps, based on one battalion per division, and at the operational level, based on the number of battalions per the workload driven requirements. This battalion consists of one headquarters, headquarters and headquarters company, and three line companies *(see Figure D-11).* This battalion—

- Increases the combat effectiveness of the division, corps, and TAs forces by accomplishing combat (M/CM/S) and limited general engineering tasks.
- Constructs, rehabilitates, repairs, maintains, and modifies landing strips, airfields, CPs, MSRs, supply installations, building structures, bridges, and other related structures as required, normally to the rear of the division.
- Repairs, and on a limited basis, reconstructs railroads and sewage and water facilities.

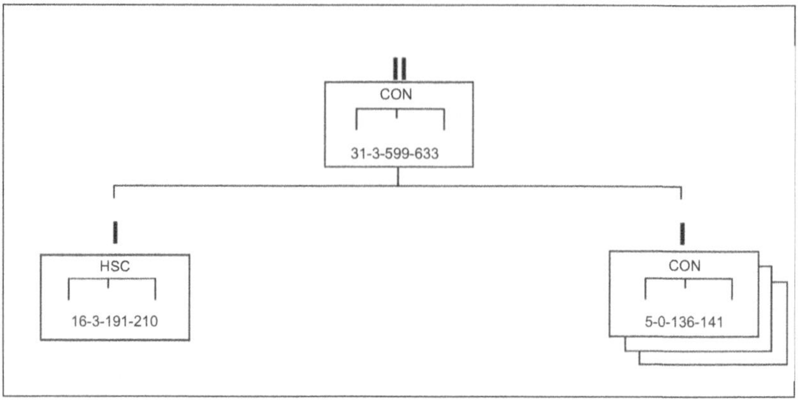

Figure D-11. Engineer Combat Battalion, Heavy

- Supervises contractual construction, skilled construction labor, and unskilled indigenous personnel.
- Constructs protective obstacles to degrade enemy mobility in rear areas.
- Clears obstacles as part of an area clearance operation, not as part of an assault-breaching operation.
- Provides bituminous paving operations and quarrying and crushing operations, rehabilitates ports, constructs petroleum pipelines and storage facilities, distributes power, and restores and constructs major airfields (when specialized personnel and equipment are attached).
- Performs rear area security operations, when required.
- Provides field engineering assistance and support to divisional engineers preparing protective positions.
- Performs rear area operations, including infantry combat missions, within the limitations of organic weapons and equipment.
- Conducts area damage clearance and restoration operations.

SECTION V – SEPARATE CORPS COMPANIES

ENGINEER COMPANY, COMBAT SUPPORT EQUIPMENT

D-47. This engineer company is normally assigned to a corps and attached to an engineer brigade or group *(see Figure D-12, page D-22)*. The number of companies allocated is based on the number of combat wheeled and mechanized battalions assigned to a corps AO. This company—

- Supports engineer combat operations within corps and division areas by conducting general engineering tasks.

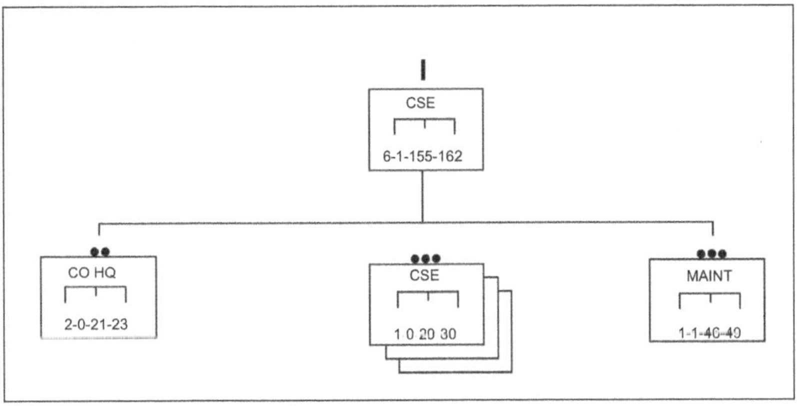

Figure D-12. Combat Support Equipment Company

- Performs survivability and countermobility tasks, general engineering along MSRs and combat trails in other corps close operation areas, and countermobility operations in a corps area.
- Provides manned engineer construction equipment to construct, rehabilitate, repair, maintain, and modify landing strips, airfields, base camps, CPs, MSRs, and LOCs.

ENGINEER COMPANY, LIGHT EQUIPMENT, AIRBORNE

D-48. This engineer company, light equipment, airborne, is normally assigned to a corps and is usually aligned with the engineer combat battalion, corps, air assault/light. This company—

- Augments engineer operations with the capabilities to support airborne operations with engineer equipment that is downsized and rapidly deployable. Can parachute or be air-delivered to work sites.
- Provides earthmoving equipment to support survivability and general engineering missions.
- Provides a cross-country dump truck capability of about 50 cubic yards or 45 tons (40.95 metric tons) per lift.
- Works with airborne battalions in early deployment with forces to establish forward logistics bases until heavier corps and theater engineer assets arrive.

ENGINEER COMPANY, LIGHT EQUIPMENT, AIR ASSAULT

D-49. This engineer company, light equipment, air assault, is normally assigned to a corps and is usually aligned with the engineer combat battalion, corps, air assault and light. This company—

- Augments engineer operations with the capabilities to support air assault and light operations with engineer equipment that is downsized and rapidly deployable.
- Provides earthmoving equipment to support survivability and general engineering missions.
- Provides a cross-country dump truck capability of about 50 cubic yards or 45 tons (40.95 metric tons) per lift.
- Works with air assault, light battalions in early deployment with force protection forces to establish forward logistics bases until heavier corps and theater engineer assets arrive.

ENGINEER COMPANY, MULTIROLE BRIDGE

D-50. The engineer MRBC, is normally task-organized to a corps or a divisional engineer battalion or a combat engineer group to support bridging operations *(see Figure D-13)*. This company—.

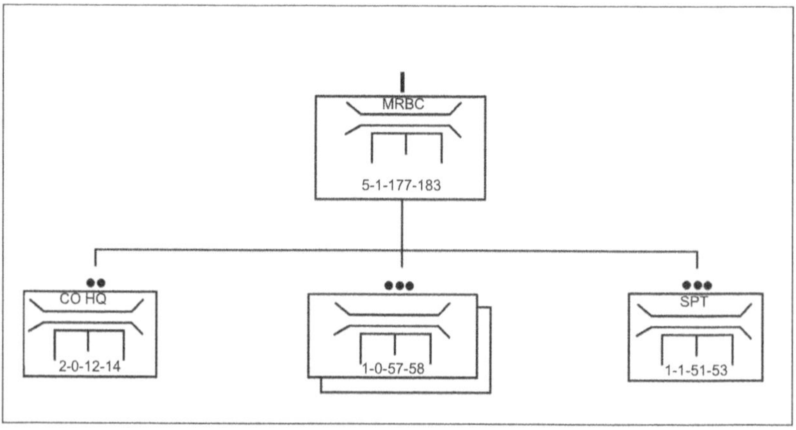

Figure D-13. Engineer Company, Multirole Bridge

- Provides personnel and equipment to transport, assemble, disassemble, retrieve, and maintain all standard US Army bridging systems:
  - Performs fixed bridging with the medium girder bridge (MGB). Maintains four MGB sets with sufficient components for the assembly of bridges with various spans and classes. Under normal conditions, the sets provide four 31.1 meters (102 feet) of Class 16 bridges (reinforceable to Class 70 with reinforcement kits) or two 49.4 meters (162 feet) of Class 60 bridge (reinforceable to Class 70 with reinforcement kits).

  - Performs float bridging using the ribbon bridge. Maintains about 213 meter (705 feet) of bridge with load classes up to Class 140 depending on current speed, or maintains four rafts with load classes up to class 140 depending on current speed.
  - Performs float bridging using the ribbon bridge. Maintains about 213 meters (705 feet) of Class 75 (tank) and Class 96 (wheel) float bridge or 6 rafts of Class 75 (tank) and Class 96 (wheel) bridge, based on a 0 to 1 meter per second (0 to 3 feet per second (fps)) water velocity.
- Provides transportation for palletized load system (PLS) configured cargo.
- Provides one panel bridge set (Bailey bridge) with sufficient components and a cable reinforcement set for erecting bridges of various spans and load classes. The set includes two 24.4-meter (80 feet) double-truss, single-story (Class 50 wheeled/Class 60 tracked) bridges without a cable reinforcement set and one 58.5-meter (180 feet) triple-truss, single-story (Class 50 wheeled/Class 60 tracked) bridge with a cable reinforcement set. Additional bridge components for spans over 100 feet are available.
- Provides technical supervision to assist other engineer units in bridge construction.
- Constructs bridges using organic personnel (in emergencies).

D-51. *FM 90-13* discusses the use of the Bailey bridge

ENGINEER COMPANY (CORPS), TOPOGRAPHIC

D-52. This engineer company (corps), topographic is normally assigned to a TA engineer battalion (topographic). This company—

- Provides geospatial engineer support to the corps.
- Compiles controlled and uncontrolled photomaps and mosaics.
- Revises existing maps and other topographic data within its capabilities.
- Drafts special maps, overprints, overlays, and other topographic products.
- Reproduces monochrome and multicolor maps, map substitutes, overlays, overprints, and other topographic products by offset lithography and photocopy.
- Provides terrain intelligence and analysis products.
- Performs topographic surveys and provides survey information.
- Interprets and measures remotely sensed imagery.
- Extends horizontal and vertical controls into corps and division areas.
- Stores and distributes special topographic products.
- Provides a survey information system and maintains the DPPDB.

SECTION VI – DIVISION ENGINEER UNITS

HEADQUARTERS, HEADQUARTERS DETACHMENT, ENGINEER BRIGADE, HEAVY DIVISION

D-53. Headquarters, headquarters detachment, engineer brigade, heavy division—

- Provides staff planning for C2 operations and supervision of engineer units that are assigned, attached, or supporting units engaged in combat (M/CM/S), geospatial, and general engineering tasks.
- Advises the division commander and staff on engineer operations and their impact on division operations.
- Plans and coordinates engineer operations for the units constructing tactical obstacles, defensive positions, and fixed and floating bridges; breaching or clearing obstacles; and conducting river crossing operations.

ENGINEER COMBAT BATTALION, HEAVY DIVISION

D-54. Three engineer combat battalions are organic to an engineer brigade, heavy division. Each battalion consists of one HHC and three line companies *(see Figure D-14)*. The engineer combat battalion, heavy division, increases the combat effectiveness of a division by accomplishing combat (M/CM/S) and limited general engineering tasks; it may fight as infantry, when required. Special capabilities of this battalion include, but are not limited to—

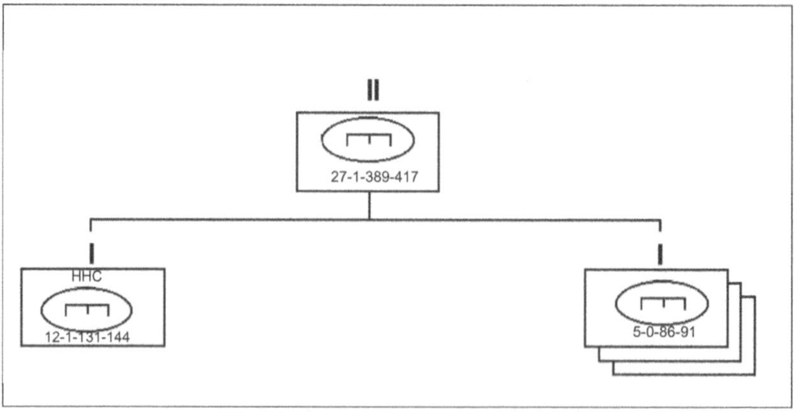

Figure D-14. Engineer Combat Battalion, Heavy Division

- Emplacing and maintaining assault bridges to span twelve 20 meter (60-foot) gaps simultaneously.

- Supporting combined arms breaching operations.
- Constructing tactical obstacles and defensive positions.
- Performing expedient repair of essential combat trails, bridges, fords, and roads in the brigade area.
- Performing engineer technical reconnaissance in support of maneuver units and other elements of the division.

ENGINEER COMBAT BATTALION, FORCE XXI DIVISION

D-55. There are three engineer combat battalions per FXXI division—one per maneuver brigade. Each battalion consists of one HHC and three line companies. The engineer combat battalion, FXXI division, increases the combat effectiveness of a division by accomplishing combat (M/CM/S) and limited general engineering tasks; it may fight as infantry, when required. Special capabilities of this battalion include, but are not limited to—

- Emplacing and maintaining assault bridges to span twelve 20 meter (60-foot) gaps simultaneously.
- Supporting combined arms breaching operations.
- Constructing tactical obstacles and defensive positions.
- Performing expedient repair of essential combat trails, bridges, fords, and roads in the brigade area.
- Performing engineer technical reconnaissance in support of maneuver units and other elements of the division.

ENGINEER COMBAT BATTALION, ENHANCED HEAVY SEPARATE BRIGADE

D-56. One engineer combat battalion is organic to a maneuver-enhanced (armor/mechanized), heavy separate brigade. The engineer combat battalion consists of one HHC and three line companies *(see Figure D-15).* This battalion increases the combat effectiveness of an enhanced heavy separate brigade by accomplishing combat (M/CM/S) and limited general engineering tasks; it may fight as infantry, when required. Special capabilities of this battalion include, but are not limited to—

- Emplacing and maintaining assault bridges to span twelve 20 meter (60-foot) gaps simultaneously.
- Supporting combined arms breaching operations.
- Constructing tactical obstacles and defensive positions.
- Performing expedient repair of essential combat trails, bridges, fords, and roads in the brigade area.

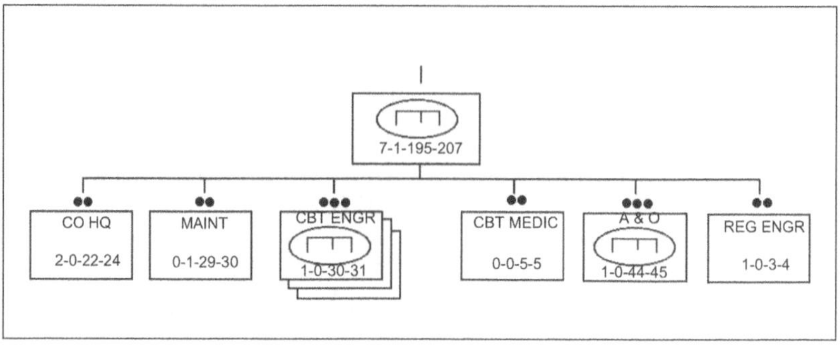

Figure D-15. Engineer Combat Battalion, Enhanced Separate Heavy Brigade

- Performing engineer technical reconnaissance in support of maneuver units and other elements of the brigade.

SECTION VII – LIGHT DIVISION ENGINEER UNITS

ENGINEER BATTALION, AIRBORNE DIVISION

D-57. Organic to an airborne division, an engineer battalion, airborne division, consists of one HHC and three line companies *(see Figure D-16, page D-28* It increases the combat effectiveness of an airborne division by accomplishing combat (M/CM/S) and limited general engineering tasks; it may fight as infantry, when required. This battalion—

- Prepares and maintains expedient combat routes in the forward battle area, to include ingressing and egressing to block positions and river crossing sites and conducting expedient repair of essential bridges, fords, and culverts.
- Assists in assaulting fortified positions.
- Constructs tactical obstacles.
- Supports combined arms breaching operations.
- Performs engineer technical reconnaissance in support of maneuver units and other elements of the division.

D-58. Organic to an airborne division, an engineer battalion, airborne division, consists of one HHC and three line companies *(see Figure D-16).* It increases the combat effectiveness of an airborne division by accomplishing combat (M/CM/S) and limited general engineering tasks; it may fight as infantry, when required. This battalion—

- Prepares and maintains expedient combat routes in the forward battle area, to include ingressing and egressing to block positions and river crossing sites and conducting expedient repair of essential bridges, fords, and culverts.
- Assists in assaulting fortified positions.

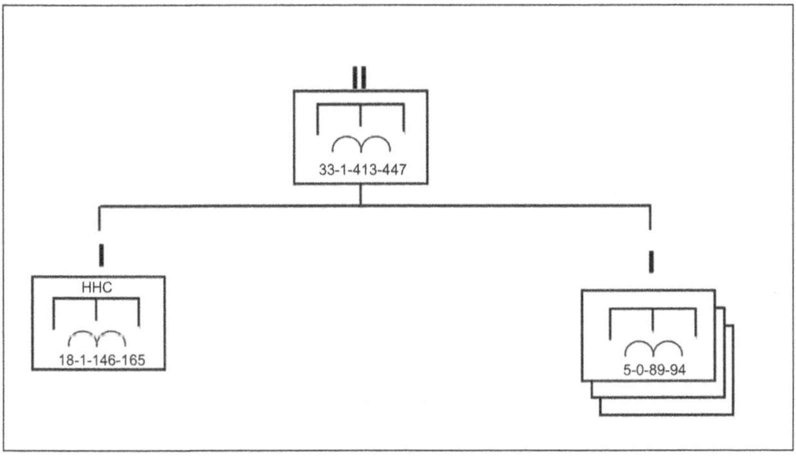

Figure D-16. Engineer Battalion, Airborne Division

- Constructs tactical obstacles.
- Supports combined arms breaching operations.
- Performs engineer technical reconnaissance in support of maneuver units and other elements of the division.

ENGINEER BATTALION, AIR ASSAULT DIVISION

D-59. Organic to an air assault division, this battalion consists of one HHC and three line companies *(see Figure D-17)*. It increases the combat effectiveness of an air assault division by accomplishing combat (M/CM/S) and limited general engineering tasks; it may fight as infantry when required. This battalion—

- Prepares and maintains expedient combat routes in the forward battle area, to include ingressing and egressing to block positions and river crossing sites and conducting expedient repair of essential bridges, fords and culverts.
- Assists in assaulting fortified positions.
- Constructs tactical obstacles.
- Supports combined arms breaching operations.
- Performs engineer technical reconnaissance in support of maneuver units and other elements of the division.

ENGINEER BATTALION, LIGHT INFANTRY DIVISION

D-60. Organic to a light infantry division, this battalion consists of one HHC and three line companies *(see Figure D-18)*. It increases the combat effectiveness of a light infantry division by accomplishing combat (M/CM/S)

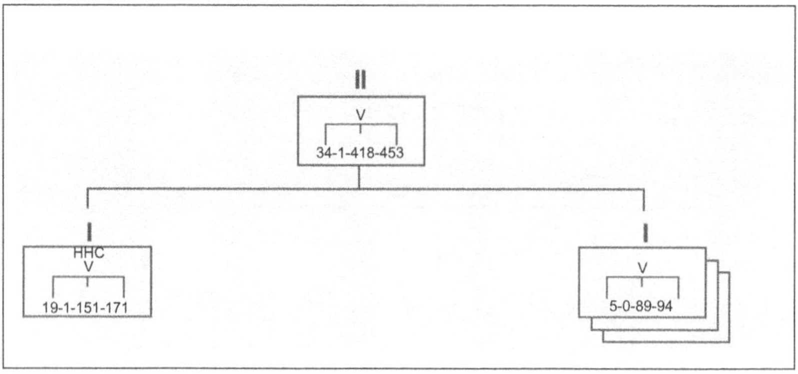

Figure D-17. Engineer Battalion, Air Assault Division

and limited general engineering tasks; it may fight as infantry, when required. This battalion—

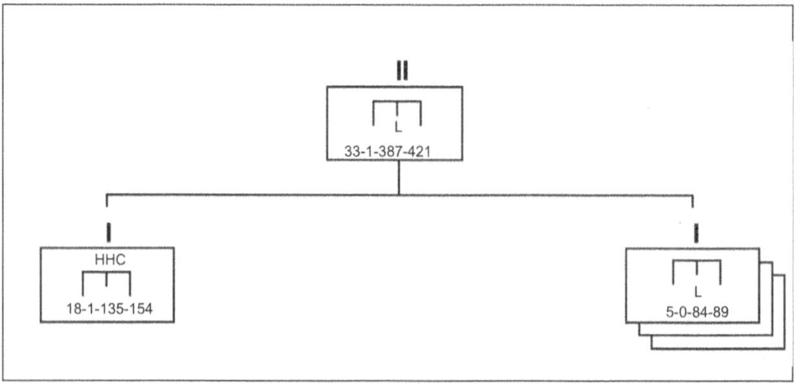

Figure D-18. Engineer Battalion, Light Infantry Division

- Prepares and maintains expedient combat routes in the forward battle area, to include ingressing and egressing to block positions and river crossing sites and conducting expedient repair of essential bridges, fords, and culverts.
- Assists in assaulting fortified positions.
- Assists maneuver forces in the assault breach of obstacles and minefields.
- Constructs tactical obstacles.
- Supports combined arms breaching operations.

- Performs engineer technical reconnaissance in support of maneuver nits and other elements of the division.

SECTION VIII – SEPARATE COMPANIES

ENGINEER COMPANY, ARMORED CAVALRY REGIMENT

D-61. Organic to an ACR, this company *(see Figure D-19)* increases the combat effectiveness of an ACR by accomplishing combat (M/CM/S) and limited general engineering tasks; it may fight as infantry, when required. This engineer company—

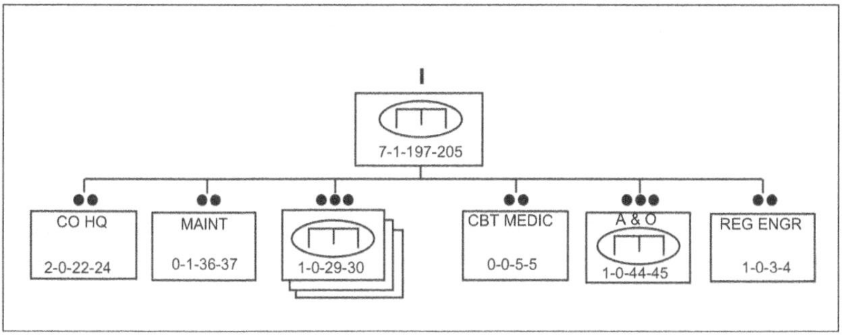

Figure D-19. Engineer Company (Armored-Cavalry Regiment)

- Advises the maneuver commander on engineer unit capabilities and the impact on the maneuver plan.
- Prepares and maintains essential combat trails in forward areas.
- Conducts expedient repair of essential bridges, fords, and roads.
- Provides, emplaces, and maintains assault bridges to span six 20 meter (60-foot) gaps simultaneously.
- Constructs tactical obstacles, defensive positions, and fixed and floating bridges.
- Supports combined arms breaching operations and assault river crossing operations.
- Constructs, repairs, and maintains CPs, combat trails, site damage, chemical decontamination sites, and logistics field sites.
- Performs engineer technical reconnaissance in support of maneuver units and other elements of the regiment.

ENGINEER COMPANY, LIGHT ARMORED CAVALRY REGIMENT

D-62. Organic to a light ACR, this company increases the combat effectiveness of a light ACR by accomplishing combat (M/CM/S) and limited general engineering tasks; it may fight as infantry, when required. This company—

- Provides limited combat (M/CM/S) engineering capability to support one cavalry squadron.
- Advises the maneuver commander on engineer unit capabilities and the impact on the maneuver plan.
- Prepares and maintains combat trails in forward areas; repairs essential bridges, fords, and roads expediently; conducts breaching operations; and constructs tactical obstacles and defensive positions.
- Performs engineer technical reconnaissance in support of maneuver units and other elements of the regiment.

ENGINEER COMPANY, HEAVY SEPARATE BRIGADE

D-63. Organic to a heavy separate brigade, this company increases the combat effectiveness of a heavy separate brigade by accomplishing combat (M/CM/S) and limited general engineering tasks; it may fight as infantry, when required. This company—

- Advises the maneuver commander on engineer unit capabilities and the impact on the maneuver plan.
- Prepares and maintains essential combat trails in forward areas.
- Conducts expedient repair of essential bridges, fords, and roads.
- Provides, emplaces, and maintains assault bridges to span six 20 meter (60-foot) gaps simultaneously.
- Constructs tactical obstacles, defensive positions and fixed and floating bridges.
- Supports combined arms breaching operations and assault river crossing operations.
- Constructs, repairs, and maintains CPs, combat trails, damaged sites, chemical decontamination sites, and logistics field sites.
- Performs engineer technical reconnaissance in support of maneuver units and other elements of the brigade.

ENGINEER COMPANY, STRYKER BRIGADE COMBAT TEAM

D-64. Organic to a SBCT, this company *(see Figure D-20, page D-32)* increases the combat effectiveness of an SBCT by accomplishing combat (M/CM/S) and limited general engineering tasks; it may fight as infantry, when required. This company—

- Advises the brigade commander, the staff, and other maneuver commanders on allocating engineer resources and capabilities available to assist in accomplishing the unit's missions.
- Prepares and maintains essential combat routes in the brigade area, to include ingressing and egressing to block positions and river crossing sites and conducting expedient repair of essential bridges, fords, and culverts.
- Assists in assaulting fortified positions.

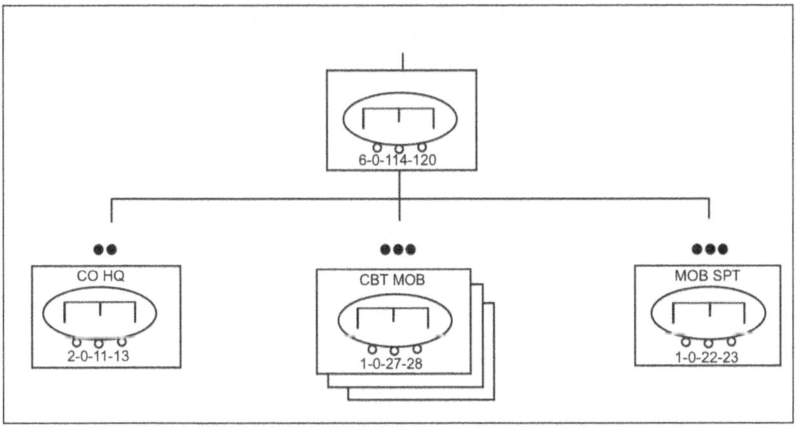

Figure D-20. Engineer Company (Stryker Brigade Combat Team)

- Supports combined arms breaching operations.
- Performs engineer technical reconnaissance in support of maneuver units and other elements of the brigade.

ENGINEER COMPANY, SEPARATE INFANTRY BRIGADE

D-65. Organic to a separate infantry brigade, this company increases the combat effectiveness of a separate infantry brigade by accomplishing combat (M/CM/S) and general engineering tasks; it may fight as infantry, when required. This company—

- Advises the brigade commander, the staff, and other maneuver commanders on allocating engineer resources and capabilities available to assist in accomplishing the unit missions.
- Prepares and maintains essential combat routes in the brigade area to include ingressing and egressing to block positions and river crossing sites and conducting expedient repair of essential bridges, fords, and culverts.
- Assists in assaulting fortified positions.
- Supports combined arms breaching operations.
- Performs engineer technical reconnaissance in support of maneuver units and other elements of the brigade

SECTION IX – SEPARATE TEAMS

TOPOGRAPHIC TERRAIN DIRECT SUPPORT TEAM, HEAVY DIVISION

D-66. Organic to a heavy division, the topographic terrain DS team is normally collocated with the G3/G2 planning and operations staff. It provides staff advice and assistance to a supported division and control of terrain analysis teams (heavy) in DS of geospatial missions.

TOPOGRAPHIC TERRAIN ANALYSIS TEAM, HEAVY DIVISION

D-67. Organic to a heavy division, the topographic terrain analysis team—

- Produces terrain intelligence for a heavy division.
- Provides qualified personnel who collect, evaluate, and disseminate terrain data.
- Analyzes the effects of terrain on military operations.
- Advises the supported commander on all terrain-related matters.

TOPOGRAPHIC DETACHMENT, FORCE XXI DIVISION

D-68. Organic to a FXXI division, the topographic detachment—

- Produces terrain intelligence for a FXXI division.
- Provides qualified personnel who collect, evaluate, and disseminate terrain data.
- Analyzes the effects of terrain on military operations.
- Advises the supported commander on all terrain-related matters.

TOPOGRAPHIC TERRAIN ANALYSIS TEAM, LIGHT DIVISION

D-69. Organic to a light division, topographic terrain analysis team—

- Produces terrain intelligence for a light division.
- Provides qualified personnel who collect, evaluate, and disseminate terrain data.
- Analyzes the effects of terrain on military operations.
- Advises the supported commander on all terrain-related matters.

FACILITY ENGINEER TEAM

D-70. The FE team is a TOE (SRC 05530L100) and a reserve TDA unit with specific installation responsibilities. TOE teams are generally assigned to an ENCOM and/or a TSC or ASG. In times of national emergencies and contingency operations, it may be mobilized and attached to an ASCC, TAACOM, ASG, corps HQ or a staff element of a TF HQ. When three or more FE teams are operating in a theater, an engineer support group is usually formed. It provides an additional depth of FE capability and has C2 over its assigned teams. The engineer support group has the capability to perform the same missions as the FE teams. The group—

- Manages engineer resources in support of facility and civil engineering; performs master planning and allocates resources, including inspecting facilities and identifying, prioritizing, and conducting work; conducts planning boards; and develops facility and civil engineering projects.
- Manages real property and controls real estate engineer teams and coordinates their activities.
- Manages housing and space use.
- Manages the theater environmental compliance and prevention programs, to include environmental compliance assessments, recommendations for corrective actions, and proper reporting.
- Performs limited design capability, including preparing drawings, specifications, and cost estimates in support of facility and civil engineering projects.
- Reviews the designs of contract architects and engineers to ensure that they conform to the user requirements, missions, and codes.
- Manages utilities services, and maintenance and repair efforts and can control engineer utility teams and coordinate their activities.
- Inspects and ensures that the quality standards of construction projects by contract or troop labor are met.
- Manages BASOPS, including sanitation and landfill operations and can control and coordinate the activities of fire fighting and utility teams.
- Performs limited supervision of troop labor and indigenous personnel.
- Manages facility engineering supplies through assigned unit.

Appendix E

Navy Engineer Organizations and Capabilities

Naval engineering units are composed of the subordinate elements of the naval construction force (NCF), better known as Seabees. They are deployable naval engineering units with the primary mission of providing responsive contingency construction support for US military forces in the operational area. The common thread that is woven throughout all SEABEE units is responsiveness and flexibility.

NAVY CAPABILITIES

NAVY ENGINEER UNITS

E-1. The NCF performs generalized and specialized construction missions in support of the joint force. Naval mobile construction battalions (NMCBs) possess robust vertical construction, bridging, and heavy earthmoving capabilities. NMCBs construct roads for supply routes, extend or construct airfields, construct ammunition supply points, install expeditionary airfields, and provide all types of force bed-down and logistic facilities. NCF units are necessary to reinforce and augment the Marine Corps's limited general engineering capability. The normal employment of the NCF is as a major subordinate element within the Marine air-ground task force (MAGTF) to maximize engineering capabilities available to the MAGTF commander.

SPECIALIZED CAPABILITIES

E-2. Navy engineer units have specialized capabilities for performing engineering work at the water and shore interface in support of amphibious operations or other beach operations.

NAVAL CONSTRUCTION FORCE

E-3. NCF SEABEE units are organized under two naval construction brigades that include both AC and RC units. They are under the administrative command (ADCOM) of the naval component commander of their home port geographic combatant commander. When deployed during peacetime, their OPCON chain normally falls under the theater naval component commander. During contingencies when NCF SEABEE units deploy to support Marine forces, their OPCON chain falls under the theater Marine forces component commander.

E-4. Amphibious construction battalions (ACBs) are organized under two naval beach groups (NBGs) that fall under the Atlantic/Pacific surface forces. The NBGs are OPCON and ADCON to the Atlantic and Pacific fleets. Like the NCF SEABEEs, OPCON is normally chopped to the theater combatant commander and delegated to the naval component command in which the units are deployed.

E-5. The NCF is organized, trained, and equipped to construct, maintain, operate, and repair advanced bases and their associated logistic pipelines. The force also provides disaster control and relief assistance and performs civic action projects to complement military or other national programs. NCF units are organized for quick and effective response, as required by planning assumptions and guidance. When directed, SEABEE assets are transferred to the COCOM (command authority) of the combatant commander.

E-6. In executing assigned projects, NCF units maintain a significant self-defense capability for their construction sites and can be employed as part of a perimeter defense force. All SEABEE units are equipped with small arms. The larger units (NMCBs) have organic, indirect-fire weapons systems (mortars). Their weapons are identical to those in the United States Marine Corps (USMC) inventory. A USMC advisor is resident to the NMCB staff, and NMCB personnel receive semiannual training on military skills and tactics.

E-7. SEABEE resources are continually forward deployed to provide a quick response to any location where a contingency may occur. The forward-deployed presence of SEABEE resources minimizes lift requirements and ensures prompt engineering support to the commanders. The following SEABEE units make up the NCF:

- Naval construction brigade (NCB).
- Naval construction regiment (NCR).
- NMCB.
- Construction battalion maintenance unit (CBMU).
- Construction battalion unit (CBU).
- Underwater construction team (UCT).
- Naval construction force support unit (NCFSU).
- ACB.

NAVAL CONSTRUCTION BRIGADE

E-8. The NCBs provide forces to fulfill the operational requirements of a combatant commander. There are two active NCBs. The NCB commander normally exercises C2 over two or more construction regiments. The NCBs are deployable but rely on subordinate units for tactical and logistic support. When deployed, NCB personnel can perform limited work in an NBC environment, possessing individual personal protective gear.

NAVAL CONSTRUCTION REGIMENT

E-9. The NCRs, both AC and RC, act as C2 elements for subordinate NCF units and report to the NCBs. There are two line NCRs and two home port training and logistical support NCRs. The regiment primarily functions as a planning, training, and oversight group and is subordinate to the command exercising OPCON of NCF units. The home port NCR performs administrative, training, project-tasking, and logistical support functions for

NCF units. The logistical support provided by the home port NCR includes planning for and movement of personnel.

NAVAL MOBILE CONSTRUCTION BATTALION

E-10. The NMCBs' primary mission is the expeditionary construction of advanced base facilities in support of the Navy, USMC, other services, and allied and coalition partners engaged in military operations. NMCBs construct base and troop bed down facilities; install, repackage, and redeploy Bailey bridges and MGBs; and conduct defensive combat operations, as required by the operational commander. They may also conduct rapid runaway repair (RRR) and BDR for airfields. Additional functions include the repair, the maintenance, and the construction of shore facilities and LOC during contingency, emergency, or disaster recovery operations.

E-11. NMCBs are operationally self-sustainable for up to 60 days, requiring only replenishment of consumables (Classes I, II, and V limited to 5, 3, and 15 days, respectively). They are capable of limited operations in an NBC environment, possessing individual personal protective gear, limited chemical and radiological detection equipment, and decontamination apparatus capable of decontaminating facilities and equipment organic to the unit. Each battalion is organized into one HQ and four line companies. Other configurations of the battalion's personnel and equipment, such as those pre-positioned on maritime pre-positioning force ships (MPSs), have been tailored to meet various missions. An NMCB can task-organize and deploy away from its main body a number of detachments, details, and teams, depending on the operational tasking and theater requirements. Units may be reconfigured into modular packages. An NMCB can organize and deploy the following standard organic detachments and teams:

AIR DETACHMENT

E-12. The air detachment is a readily deployable (within 48 hours), self-sustainable, company-sized detachment, completely air-transportable via a C-130 lift, with a minimum of 89 personnel and 300 short tons of equipment. The unit is capable of sustained operations in a contingency or wartime environment for 30 days without resupply (except for consumables). The air detachment's primary mission is to repair war damage and construct urgent projects required by OPLANs and/or OPORDs during the early stages of a contingency operation.

REINFORCED AIR DETACHMENT

E-13. The reinforced air detachment includes two additional platoons and is capable of providing increased construction and engineer support tailored to the operational scenario.

OPERATIONS DETACHMENT

E-14. The operations detachment is a reinforced, company-sized detail deployed from the main body of the NMCB, to perform light to medium construction and engineer support.

REINFORCED OPERATIONS DETACHMENT

E-15. The reinforced operations detachment is expanded to two companies in strength and is capable of providing increased construction and engineering support tailored to the operational scenario.

LIMITED OPERATIONS DETACHMENT

E-16. The limited-operations detachment is smaller, typically platoon-sized, deployed from the NMCB, and able to perform light construction and engineering support.

ENGAGEMENT TEAM

E-17. The engagement team is a squad-sized detachment deployed from the NMCB for task-specific light construction projects that are accomplished within 30 to 120 days.

TRAINING TEAM

E-18. The training team is a fire team-sized detachment deployed from the NMCB to provide training and oversight of task-specific construction and engineering projects performed by others.

MAIN BODY

E-19. The main body is the core of the deployed NMCB. The size of the main body ranges from 812 personnel at full wartime end strength with no detachments to as few as 250 personnel with all other NMCB personnel deployed with detachments.

MARITIME PREPOSITIONING

E-20. Three NMCB tables of allowances (TOAs) (P25M) are configured for maritime pre-positioning within the three existing strategically deployed maritime pre-positioning force squadrons (MPSPRON), in support of the Marines. Each TOA P25M is configured and spread-loaded on three ships within a squadron in support modules that can be deployed in various sizes. The modules are configured as: core, basic, and heavy. The core module contains enough tools and equipment to sustain 250 SEABEEs ashore in support of a Marine expeditionary unit (MEU). The basic module contains additional vertical capability and can be added to the core module, should the mission require it. The heavy module contains additional horizontal capability and can be added to any of the other core modules, with or without a basic module as the situational tasking dictates. All of the modules together, plus the fly in echelon, which contains NMBC personnel, weapons and communications equipment not on the MPSRONs, constitute a full NMCB. The MPS survey, liaison, and reconnaissance party (SLRP); the MPF advance party; and the MPF offload preparation party (OPP) are not organic elements of the NMCB; they are temporary, subordinate elements of the MPF. Each element includes a commander, an MPF, naval coastal warfare, Navy support element (NSE), Navy cargo handling and port group, an MPF MAGTF, and NCF and fleet hospital personnel. The SLRP, advance party and OPP only exist temporarily to facilitate establishing the MPF MAGTF ashore.

- The MPF SLRP is a detachment of personnel that deploys as the lead element in support of an MPF operation. The size of the SLRP is situation-dependent and typically consists of three to eight NCF personnel.
- The MPF advance party is the advance element of an NMCB which assists with offload operations and receives NMCB equipment and supplies in support of the MAGTF in the MPF.
- The MPF OPP deploys to the MPF squadron four days before offload operations to prepare the equipment for debarkation.

CONSTRUCTION BATTALION MAINTENANCE UNIT

E-21. The CBMU provides follow-on public works (PW) operations to maintain and repair existing advanced base shore facilities or facilities constructed by NMCBs during contingency operations. As a secondary mission, a CBMU can backfill a PW organization at existing bases during emergency operations or due to increased operational requirements. The unit is capable of equipping, manning, and maintaining water production, as well as steam and electrical power generation and distribution systems for advanced base facilities of up to 5,000 personnel. They perform war damage repairs to central base camps, power, sewage, POL, and water systems. A CBMU also operates and maintains automotive and construction equipment, including materials handling equipment (MHE). Possessing personal protective gear, limited chemical detection equipment, and decontamination apparatus capable of limited decontamination of equipment organic to the unit, the unit is capable of limited operations in an NBC environment. CBMUs are operationally self-sustainable for up to 60 days, requiring only replenishment of consumables (Classes I, II, and V limited to 5, 3, and 15 days, respectively).

CONSTRUCTION BATTALION UNIT

E-22. CBUs support Navy fleet hospitals. Normally, two CBUs deploy to provide the engineering, constructing, maintaining and repairing services necessary to support a fleet hospital. Following the erection of fleet hospital facilities, the CBU provides operation, maintenance and repairs and augments security for the facility. The unit is capable of deploying within 48 hours and erecting a CZ fleet hospital within 10 days. The unit can perform limited operations in an NBC environment when attached to the fleet hospital, which provides CBU personnel with individual protective gear and limited chemical-detection equipment. CBUs have a secondary mission to conduct disaster recovery.

UNDERWATER CONSTRUCTION TEAM

E-23. UCTs are specially trained and equipped units that provide underwater engineering, construction, repair, and inspection capabilities to meet the Navy and Marine Corps or joint force operational requirements.

E-24. UCTs perform complex inshore and deep ocean underwater construction tasks in any climate, including the arctic. They provide ocean bottom surveys for the appropriate site selection of underwater facilities. They are capable of diving and working to a depth of 60 meters (190 feet) and rely on scuba gear and/or surface-supplied air. Typical projects include measurement conversions, underwater repair of wharves, piers, pipelines, moorings, boat ramps, and underwater cable systems. The unit can also support offshore petroleum discharge equipment by installing single-anchor leg moorings. An UCT possesses individual protective gear and limited chemical detection capabilities and can perform limited operations in an NBC environment. UCTs have a secondary mission to conduct disaster recovery, humanitarian assistance, and civil-action operations.

E-25. A UCT is divided into three identical squad-sized air detachments and a platoon-sized sea echelon. The unit can deploy as one unit or separately. Each air detachment carries a surface decompression chamber as part of its TOA. The sea echelon accompanies larger unit equipment for sustained operations. The shore component of a UCT is used for equipment and follow-on logistics coordination.

NAVAL CONSTRUCTION FORCE SUPPORT UNIT

E-26. NCFSUs provide logistics-oriented construction and engineering support for the NCR as well as other NCF units, including specialized civilian engineer support equipment (CESE), materials, repair parts, and technical expertise. NCFSU echelons are attached to other units to manage, maintain, and inventory material, transportation, and construction equipment to augment the NMCB TOA. NCFSUs have secondary missions to provide limited construction and defensive capabilities during a contingency. NCFSUs are operationally self-sustainable for up to 60 days, requiring only replenishment of consumables (Classes I, II, and V limited to 5, 3, and 15 days, respectively). The unit provides support in the following areas:

- The operation, maintenance, and repair of local- and long-haul transportation equipment.
- The operation and support of quarry and rock-crushing operations, asphalt and concrete production and placement, and soil analysis and stabilization equipment.
- The production and storage of potable water using nine reverse osmosis purification units 600 gallons (2500 liters per hour) and nine 3,000 gallon (12000 liter) water bladders.
- The construction material management, including requisitioning, expediting, receiving, controlling, storing, issuing, and delivering material.
- The advanced facility planning, designing, estimating, and engineer support (as required) to execute construction projects.
- The maintenance, custody, inventory control, and issuing of special SEABEE support equipment.

AMPHIBIOUS CONSTRUCTION BATTALION

E-27. There are two ACBs in the NCF. The ACBs provide over-the-shore logistics movement and construction support to amphibious forces during the initial assault and early phases of an operation. ACBs are part of the NSE and report to the NBG, which is responsible for in-stream off loading of maritime ships in support of amphibious military operations. They are not placed OPCON or ADCON to an NCB. The primary tasks of the ACBs are to—

- Provide, assemble, and operate causeway barge ferries.
- Provide side-loadable warping tugs and powered causeway sections for installing piers, elevated causeways systems (ELCAS), ship-to-shore bulk fuel, and water systems and the OPDS.
- Provide, assemble, maintain, and operate the ELCAS and amphibious assault bulk fuel and bulk water systems.
- Install the OPDS, with support from two UCT air detachments.
- Establish and operate a support camp and provide limited construction support for the NBG.

E-28. The unit can operate in an NBC environment and has limited detection and decontamination capability.

ARMY-NAVY ENGINEER CONSIDERATIONS

E-29. During force projection operations, the initial US naval engineering capability available in theater is most likely NMCB air detachments and MAGTF amphibious forces. NMCBs are also deployed quickly to force projection theater locations to construct necessary naval facilities. The Army engineer staff must consider the following when coordinating joint engineer plans and operations with the Navy:

- Requesting the latest engineer intelligence data from deployed or deploying NMCB air detachment elements to assist in identifying force projection theater Army engineer requirements and enemy engineer capabilities. (These requirements include soils data, the availability of construction materials, and HN construction support.)
- Establishing engineer staff links between the Navy forces (NAVFOR) and ARFOR engineer staff sections through the JTF or theater engineer staff and HQ.
- Providing necessary Army engineer LNO support.
- Developing the joint task-organization relationships that enhance NCR capabilities, following the deployment of Army engineer units.
- Assessing the need for NMCB support following the arrival of Army combat and construction units in theater.
- Determining if NMCB units need augmentation from Army combat and construction units.
- Developing procedures for Army engineer units to be able to acquire Class IV construction materials from NCFSUs.

E-30. For additional information on NCF units, *refer to Operational Naval Instruction (OPNAVINST) 5450.46K.* For additional information on ACB units, *refer to OPNAVINST 3501.93D.*

Appendix F

Air Force Engineer Organizations and Capabilities

Air Force mobility engineering forces, organized as Prime base engineer emergency force (BEEF) or RED HORSE units, provide support ranging from expeditionary civil to general engineering services across the range of military operations. Both units have limited combat engineering capabilities focused on the defense of deployed forces and base denial. Capabilities include airfield construction, maintenance, repair, and sustainment operations, as well as other specialized functions, such as EOD, fire, and limited NBC detection and response capability. Air Force engineering units can deploy either as a part of an air expeditionary force (AEF) or as detached units operating in support of specific missions and operational taskings.

AIR FORCE CAPABILITIES

AIR FORCE ENGINEER UNITS

F-1. Prime BEEF and RED HORSE units are deployed in unit type code (UTC) sets. Depending on the mission and engineering requirements, UTCs can be added or subtracted as needed. Air Force engineering units can deploy either as part of an AEF or as detached units operating in support of specific missions and operational taskings. Prime BEEF lead UTC teams to deploy with the necessary command, control, and communications (C3) to conduct independent engineer; fire, crash, and rescue, NBC defense; and EOD operations to establish, sustain, defend and recover a bed-down location. Follow teams combine with lead teams to satisfy support at larger locations. Prime BEEF teams deploy with their individual protective clothing, hand tools, and selective team equipment; however, most construction equipment and supplies and all vehicles (except EOD) must be provided at the deployment location. Prime BEEF forces can conduct construction; maintenance; repair; fire protection, NBC, including disaster preparedness planning, training, detection, and decontamination within service limitations; force protection; and EOD operations. RED HORSE squadrons and their associated UTC configurations provide highly mobile, largely self-sufficient, rapidly deployable echelons to support major force bed-down requirements and repair heavy war damage. RED HORSE units are stand-alone squadrons, not tied to peacetime base support. They provide a dedicated, flexible airfield and base heavy construction and repair capability that allows JFCs to move and support missions as requirements dictate. RED HORSE squadrons are organized and deployed for austere, independent operation to execute heavy horizontal and vertical construction projects. The projects include site development; construction and repair of runways, taxiways and ramps, roads, and revetments; heavy earthwork; and construction and repair of facilities and infrastructure. General capabilities shared by Prime BEEF and RED HORSE include site surveys, bare base construction using mobility assets (Harvest Falcon and Harvest Eagle), concrete and asphalt paving, and utility system installation (water, waste and electrical). RED HORSE expands on this capability by providing automatic building machine (K-span) support, large-scale concrete and paving operations, heavy horizontal and vertical construction, water well drilling, water purification, and quarry operations.

SPECIALIZED CAPABILITIES

F-2. Air Force engineer units have specialized capabilities to support all aspects of airfield operations, including RRR, the installation of aircraft arresting systems, war damage repair, and force bed down. By deploying civil engineering teams, Air Force engineer units also perform other engineer functions, such as fire protection (structural, aircraft crash fire and rescue, and HAZMAT response), EOD (munitions support, antiterrorism, and civil assistance), and NBC support (including disaster preparedness planning, training, detection, and decontamination within service limitations). Specialized support includes installing and operating emergency airfield lighting systems, mobile aircraft arresting systems, reverse osmosis water purification, munitions storage site construction, and HAZMAT response.

CIVIL ENGINEERING TECHNICAL AND CONTRACT SUPPORT

F-3. A variety of other supporting organizations separate from Prime BEEF and RED HORSE units provide technical engineering, environmental, and contract support for the joint force. The Air Force Civil Engineering Support Agency (AFCESA) provides technical support and training and administers the Air Force Contract Augmentation Program (AFCAP). Additionally, AFCESA provides specialized civil engineer maintenance, inspection, and repair teams to assess pavement and runway conditions, as well as teams to conduct field maintenance and repair of specialized equipment. The Air Force Center for Environmental Excellence also provides a wide range of environmental program support and contract services.

PRIME BASE ENGINEER EMERGENCY FORCE

F-4. Prime BEEF is the primary organizational structure for supporting mobility and in-place contingency requirements. The principal objective of deploying Prime BEEF teams is to bed down and support an AEF. Force bed down is generally divided into three categories: aircraft, personnel, and infrastructure support. Aircraft support provides the maintenance shops, hangars, squadron operations centers, munitions storage, fuel storage, and other facilities directly supporting the flying mission. Personnel support provides the housing, dining facilities, latrines, showers, administrative offices, and other indirect-support facilities. Infrastructure support provides the utility systems, solid and hazardous waste disposal, roads, and communications that serve the bed-down site. Bed-down locations range from main operating bases with adequate existing facilities to bare bases (BBs) with no facilities other than runways, taxiways, and aircraft parking aprons with a source of water.

F-5. Prime BEEF tasks executed during the deployment include the following:

- Airfield support operations—pavements, lighting, and mobile aircraft arresting systems.
- Fire protection—aircraft crash, fire, and/or rescue; structural fire; and HAZMAT response operations.
- Fuel systems set up and support.
- EOD operations and planning.
- NBC training, reconnaissance, and operations.
- Force protection and base defense.
- RRR.
- Expedient facility repair and rapid utility repair.

F-6. Prime BEEF UTCs include—

- 4F9EA—Prime BEEF AEF team A.
- 4F9EB—Prime BEEF AEF team B.
- 4F9EP—Prime BEEF AEF team C.
- 4F9SA—staff augmentation team.
- 4F9FL—engineer sustainment team.
- 4F9AP—power production team.
- 4F9DA—NBC full-spectrum threat response heavy team.
- 4F9DB—NBC full-spectrum threat response light team.
- 4F9DC—NBC full-spectrum threat response augmentation team.
- 4F9DD—NBC full-spectrum threat response theater and/or joint task force planning and support staff.
- 4F9DE—NBC full-spectrum threat response contamination control team.
- 4F9DF—lightweight NBC reconnaissance system.
- 4F9DG—NBC collective protection system.
- 4F9DH—open air contamination control area set.
- 4F9DJ—full-spectrum threat response personnel augmentation team.
- 4F9DK—lightweight NBC reconnaissance team.
- 4F9DL—NBC collective protective equipment set.
- 4F9FJ—fire protection incident command team.
- 4F9FN—fire protection management augmentation.
- 4F9FP—fire protection operations team.
- 4F9XA—EOD C2.
- 4F9XB—EOD en route.
- 4F9X1—EOD lead team.
- 4F9X2—EOD follow team.
- 4F9X3—EOD base support team.
- 4F9X4—EOD augmentation.
- Special teams include—
  - 4F9AC—civil engineering maintenance, inspection, and repair team—aircraft arresting systems, power systems, heating,

ventilation, and air conditioning (HVAC) systems and bare base systems.

- 4F9AC—pavement evaluation team.
- Prime BEEF self-sustainability package.
- Snow removal team.
- En route airlift support team.
- EOD contingency team.
- Disaster preparedness high-threat teams.
- Armored base recovery vehicle team.
- Armored munitions clearance vehicle team.

RAPID ENGINEERS DEPLOYABLE HEAVY OPERATIONS REPAIR SQUADRON, ENGINEER

F-7. RED HORSE squadrons accomplish major construction (both vertical and horizontal) in forward locations, often in advance of the main deploying force. With their organic capabilities and stand-alone operations, RED HORSE squadrons are essential assets for early entry operations and joint reception, staging, onward movement, and integration (JRSOI) preparation. In addition to providing heavy horizontal (earthmoving and pavements) and vertical (facility and utility skills including POL, structural, electrical, mechanical, and power generation) capabilities, RED HORSE units also have the following special capabilities:

- Concrete and asphalt batch plant and paving operations.
- Mobile aircraft arresting system and navigational aid (NAVAID) installation.
- Field dispensary and air transportable clinic operations.
- Automated building machine (K-span) operations.
- Material testing.
- Quarry and rock-crushing operations.
- Revetment erection and facility hardening.
- Water well drilling and sanitation system construction.
- Road, airfield, and ramp construction.
- Expedient site and airfield repair.
- Base denial operations.

F-8. RED HORSE mobility UTCs deploy with self-sustainment capabilities not found in Prime BEEF units, including organic logistics, services, medical, transportation and financial management, and contracting capabilities. Team equipment can be airlifted, shipped via sealift, drawn from prepositioned stockpiles, or obtained from HN sources (contracting). Each RED HORSE squadron has the resources to conduct site development and heavy repair work at four different locations. The mobility echelons include a—

- 4F9R-1 advanced team, which is deployable in 16 hours (12 hours with prepositioned assets). Priority tasks include advanced airfield surveys, site layout, and planning for the establishment and future development of an operational location during contingencies. The team is self-sustaining at the deployed location, when water and fuel is available, for five days.

- 4F9R-2 team, which forms the hub of the construction capability. It is largely self-sufficient at the deployed location for 30 days. This team includes the C2 function of the entire squadron and contains personnel, skills, vehicles, and equipment to execute heavy horizontal and vertical construction. It can deploy its full mission capability (MISCAP) in 96 hours and deploy personnel in only 24 hours.
- 4F9R-3 UTC, which contains the specialized skills necessary to complete site development, construct and repair runways, taxiways and ramps, and complete other horizontal construction. This UTC provides heavy horizontal and light vertical construction capabilities in austere environments. It can deploy its full MISCAP in six days and deploy personnel only in twenty-four hours.
- 4F9R-4 UTC, which contains the specialized skills necessary to construct and repair facilities and infrastructure. It has limited capability to do earthwork, road, and airfield operations. This UTC provides the heavy vertical and light horizontal construction capabilities in austere environments. It can deploy its full MISCAP in eight days and deploy personnel in only 24 hours.
- 4F9H1, heavy equipment package, designed to supplement 4F9R-2 and 4F9R-3 teams and provide full heavy earthwork operations. The package should come by sealift, but many items could come by C-5 or C-17 aircraft if necessary. This package includes T-9 dozers, belly scrapers, 4-cubic yard front-end loaders, size 7 graders, and a sheepsfoot roller.
- 4F9H2, specialized building construction, including K-span, a large crane, forms for footings and stem walls, and concrete placement tools.
- 4F9H3, well drilling, including a well drilling rig, tow vehicle, trencher, casing, and drill bits.
- 4F9H4, asphalt batch plant, including asphalt batch plant and conveyor belts.
- 4F9H5, concrete batch plant, including a concrete batch plant, and a concrete transit, mixer, and conveyor.
- 4F9H6, quarry operations, including rock drills, a crusher, and rock dumps.

PRIME READINESS IN BASE SUPPORT

F-9. Prime Readiness in Base Support (RIBS) units are worldwide combat morale, welfare, recreation, and services (MWRS) forces, organized and trained for wartime support. The Prime RIBS program organizes forces that can deploy with 22 to 28 hours notice to support global or major regional conflict operations on military operating bases (MOBs), collocated operating bases (COBs), forward operating locations (FOLs), APODs, aerial ports of embarkation (APOEs), and BBs or to support essential MWRS missions at critical CONUS bases. Prime RIBS units can—

- Provide initial food service, billeting, recreation programs, and mortuary operations support for up to 1,200 people.

- Support an independent or dependent combat aviation squadron of 16 to 24 fighter aircraft or a significant aviation deployment less than squadron size in a major deterrent force posture.
- Support, when augmented, organizational field laundry operations, personnel fitness programs, and tactical field exchange resale operations.

ARMY AND AIR FORCE ENGINEER CONSIDERATIONS

F-10. During force projection operations, the initial available Air Force engineering capabilities in theater will probably be RED HORSE elements that establish APODs. Prime BEEF and RIBS units are also deployed quickly to force projection theater locations to operate at major air bases. The Army staff engineer should consider the following when coordinating joint engineer plans and operations with the Air Force:

- Requesting the latest engineer intelligence data from deployed or deploying RED HORSE elements to assist in identifying force projection theater Army engineer requirements and capabilities. These requirements include soils data and the availability of construction materials and HN construction support.
- Establishing engineer staff links between the Air Force forces (AFFOR) and ARFOR engineer staff sections through the JTF or theater engineer staff and HQ.
- Providing the necessary Army engineer LNO support.
- Developing the joint task organization relationships that enhance RED HORSE and Prime BEEF capabilities, following the deployment of Army engineer units.
- Assessing the need for RED HORSE airfield maintenance and repair support following the arrival of Army construction units in theater.
- Determining if Prime BEEF units need augmentation from Army construction units, especially in the area of RRR.

F-11. For additional information on Air Force engineers and their doctrine and associated TTPs, refer to Air Force Doctrine Document (AFDD) 2-4, and AFDD 2-4.4.

Appendix G

Marine Corps Engineer Organizations and Capabilities

The Marine Corps is organized into three Marine expeditionary forces (MEFs). Each MEF contains a command, ground combat, aviation combat, and CSS element. Each element, except the command element, contains organic engineer support. The MEF can form smaller task-organized MAGTFs to conduct expeditionary operations in littoral environments or other places as directed. A MAGTF ranges in size from a full MEF to a Marine expeditionary brigade (MEB), down to a MEU or smaller unit. MAGTFs normally contain task-organized elements from each of the primary engineer units.

MARINE CORPS CAPABILITIES

EXPEDITIONARY FORCE

G-1. The engineer force is organized to accomplish specific tasks of limited duration. Marine Corps civil engineering is temporary in nature due to its organization and mission, which focus on providing expeditionary engineering support to the MAGTF. Civil (general) engineering capabilities include but are not limited to construction of expeditionary airfields, FOBs, LZs, roads for supply routes, and small-scale construction operations. The Navy provides civil engineering support to the Marine Corps through NCF units under the OPCON of a MAGTF commander. NCF units are necessary to reinforce and augment the Marine Corps's civil engineering capability.

SPECIALIZED CAPABILITIES

G-2. The specialized capabilities of Marine engineers include erecting standard bridging and rafting, constructing nonstandard bridging, providing mobile electric power, producing and storing potable water supplies, and providing EOD. Marine engineers have the capability to construct and maintain expeditionary airfields up to 600 meters (2,000 feet) in length with arresting gear and perform RRR. Marine engineers also maintain and operate bulk fuel systems capable of storing up to 600,000 gallons. Marine engineers are integrated with their combat elements under the MAGTF concept of operations.

G-3. Geospatial intelligence is not a combat engineer function in the USMC. A topographic platoon resides in the production and analysis (P&A) company of the intelligence battalion within the MEF. It provides tailored terrain, littoral and infrastructure studies, and factor overlays. It is a MEF-level asset. It disseminates products such as graphic decision aids to support IPB and COA development. These topographic platoons use the same ArcView software as Army topographic units.

COMMAND STRUCTURES

COMBAT ENGINEER BATTALION

G-4. The Combat Engineer Battalion (CEB) *(Figure G-1)* enhances the M/CM/S of the Marine division and provides close combat engineer support and limited general engineer support to the Marine division. It enhances the momentum and tempo of maneuver units and helps shape the battlespace for the Marine division. The CEB contains a headquarters and service (H&S) company, engineer support company, and four combat engineer companies. An engineer support company contains heavy engineer equipment (bulldozers, cranes, and forklifts), motor transport equipment (including dump trucks), and utilities equipment (shower units, mobile electric power generators, water purification equipment).

G-5. Of note, the CEB does not have AVLBs. Their AVLBs are organic to their tank battalions. Also the MK-1545 mine clearing line charges ([MICLICs], triple shot) are organic to the assault amphibian vehicle (AAV) battalion. Normally, a combat engineer company, reinforced with detachments of equipment and personnel from the support company and H&S company, provides the close combat engineer support to an infantry regiment. A combat engineer platoon, reinforced with detachments of equipment and personnel from a support company and an H&S company, provides close combat engineer support to an infantry battalion. A combat engineer squad provides close combat engineer support to an infantry company. While these are typical relationships, the actual level of combat engineer support depends on the situation and the mission analysis. The capabilities of a CEB include field fortifications, planning, and installation.

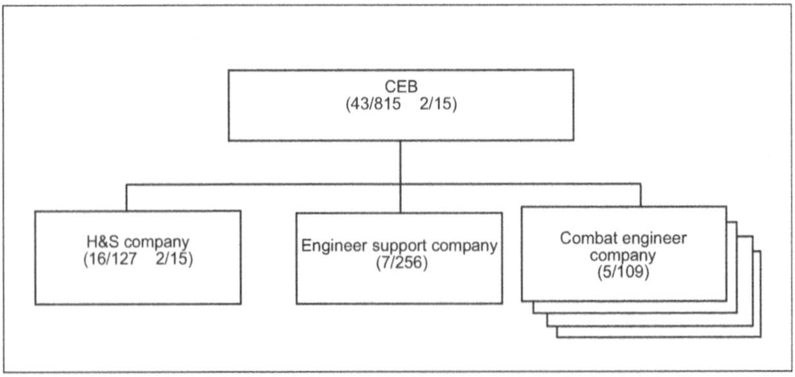

Figure G-1. Combat Engineer Battalion

G-6. The CEB provides the following capabilities and support:

- Conducts engineer reconnaissance to support the division's intelligence collection plan.

- Plans, organizes, and conducts breaching operations.
- Provides barrier planning and installation.
- Conducts demolition operations beyond the ability of other division units.
- Employs assault bridging and other standard bridging systems when augmented with equipment and trained personnel.
- Provides expedient short-span bridges from local material.
- Constructs and maintains limited combat roads and trails in support of division operations (construction and maintenance requirements are limited to those that can be performed with organic equipment and personnel).
- Provide mission-essential, temporary vertical and horizontal construction.
- Provides provisional infantry.

G-7. The battalion requires motor transport support to move the battalion as a unit. However, there are adequate ground transportation assets (medical and logistic support vehicles, including heavy trucks) organic to the battalion to accomplish its primary mission. Most of the battalion's engineer assets require surface transportation; however, some engineer equipment is helicopter transportable.

G-8. CSS capabilities provide:

- Organic supply support.
- Organizational (1st and 2d echelon) maintenance on organic equipment. Intermediate (3d and 4th echelon) maintenance is provided by the maintenance battalion, force service support group (FSSG).
- Organic transportation support required to accomplish its mission.
- Limited general engineering capability.
- Routine and limited emergency medical support to the battalion.
- Administrative, postal, and chaplain support to the battalion.

ENGINEER SUPPORT BATTALION

G-9. An engineer support battalion (ESB) *(Figure G-2, page G-4)* increases the combat effectiveness of the MAGTF by accomplishing general engineering missions of a more deliberate nature. An ESB is organic to the FSSG but supports the entire MAGTF. An ESB contains an H&S company, engineer support company, three combat engineer companies, a bridge company, and a bulk fuel company. An ESB contains EOD capabilities in the H&S company. The engineer support company contains medium bulldozers, road graders, scrapers, compactors, light and medium cranes, light and medium forklifts, mobile electric power generators, water purification units, welders, and limited organic motor transportation—including dump trucks. The ESB typically performs general engineering tasks such as pioneer road construction, expeditionary airfield construction, cantonment or camp planning and construction, and limited vertical and horizontal construction of a temporary nature. It is also capable of EOD, obstacle reduction, and limited power support.

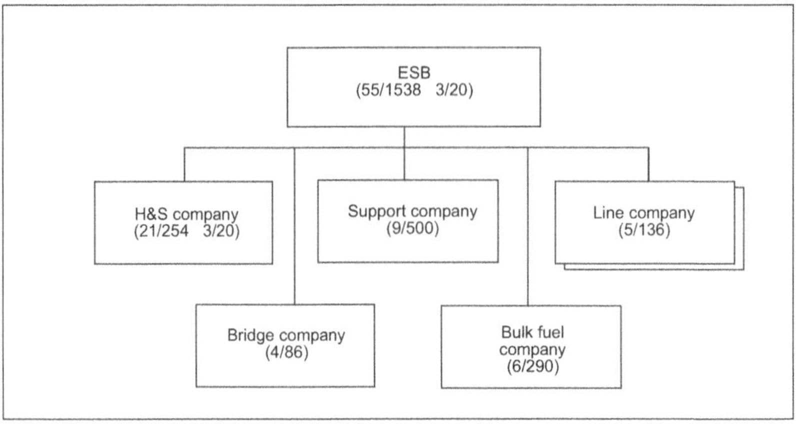

Figure G-2. ESB

G-10. Adequate ground transportation to move essential command and operational elements is organic to the battalion. External motor transport support is required to move all assets (Class III and III [A] bulk fuel), including bulk water supplies, bridging, and heavy engineer equipment.

G-11. CSS capabilities provide—

- Organic supply support.
- Organizational (1st and 2d echelon) maintenance on organic equipment. Intermediate (3d and 4th echelon) maintenance is provided by the maintenance battalion, FSSG.
- Organic transportation support required to accomplish its mission.
- Limited general engineering support.
- Routine and limited emergency medical support to the battalion.
- Administrative, postal, and chaplain support to the battalion.

MARINE WING SUPPORT SQUADRON

G-12. A Marine wing support squadron (MWSS) *(Figure G-3)* provides all essential aviation ground support (AGS) requirements to the Marine aircraft group. MWSSs provide all essential AGS and engineer support requirements to aid components of a Marine Corps FOB. One MWSS supports one Marine aircraft group or the aviation combat element (ACE) of a MEB. A task-organized detachment from the MWSS provides AGS to the ACE of a MEU. The engineers of an MWSS are located in the engineer operations division. MWSS capabilities include providing expeditionary airfield construction; RRR; mobile electric power; water purification and storage; refrigeration; light and medium MHE; light and medium bulldozing; welding, drafting, and survey; and limited vertical construction. The MWSS has a bulk fuel section, under the control of airfield operations, responsible for storing and dispensing aircraft fuel. Limited EOD capabilities provide support for the disposal of UXO, airfield damage assessment, and aircraft emergency landings. Engineer tasks and functions performed by the MWSS are primarily general engineering oriented.

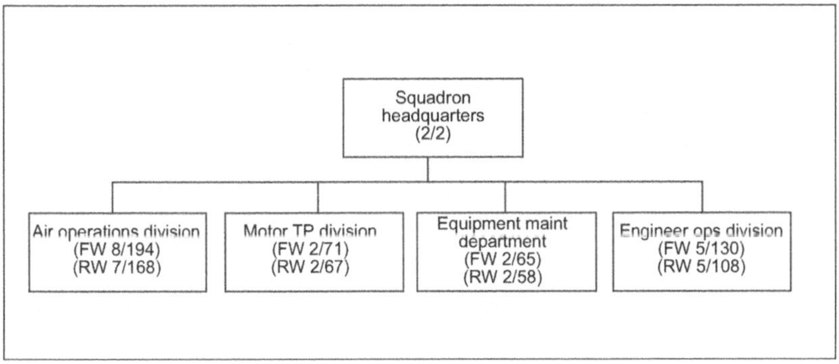

Figure G-3. MWSS

G-13. The engineer operations division of the MWSS organizes to provide limited combat and general engineering support to designated components of the ACE. Engineers located in the airfield operations division receive, store, and dispense aviation and ground fuels from various types of expeditionary fuel systems. The engineer operations division is capable of—

- Providing engineer reconnaissance and survey.
- Repairing, improving, and maintaining existing road networks for the ACE.
- Constructing and maintaining expedient roads.
- Constructing, maintaining, and improving expeditionary airfields, vertical or short takeoff locations, and landing sites.
- Constructing and maintaining mission-essential base camp requirements (temporary bunkers, aircraft revetments, and strongbacks).
- Providing technical and equipment assistance for erecting preengineered buildings.
- Providing utilities support (mobile electric power, water, potable water production, bath and laundry facilities, and refrigeration services).
- Developing, improving, and maintaining drainage systems.
- Providing technical assistance to support camouflage requirements.
- Assessing bomb damage and providing RRR.
- Providing materials and equipment handling services.
- Providing for EOD.

G-14. Organizational vehicles of the squadron provide sufficient transportation for command, control, and routine support activities. The squadron requires external transportation support to displace as a unit.

G-15. CSS capabilities of the squadron provide—

- Organic ground supply support (except for Navy-funded equipment that requires support from a designated aviation supply element).
- Organizational (1st and 2d echelon) maintenance on organic equipment; intermediate (3d and 4th echelon) maintenance by the maintenance battalion, and FSSG.
- Sufficient motor transport equipment to accomplish its mission.
- Limited general engineering capability, as required to support the squadron's mission.
- Service company, H&S battalion, and FSSG service support.

NAVAL CONSTRUCTION FORCE

G-16. The NCF may be assigned as a major subordinate element within the MAGTF to sustain MAGTF operations and maximize naval civil engineering capabilities. Elements of the NCF construct and maintain base facilities; repair battle-damaged facilities; conduct limited defensive operations, as needed; and accomplish disaster control and recovery efforts when required. NCF units contain highly skilled specialists capable of projects of a more permanent nature than those normally accomplished by Marine Corps engineers. *See Appendix E* for a more in-depth discussion of these forces.

MARINE AIR-GROUND TASK FORCE ENGINEERS

G-17. Engineers are normally provided to each corresponding MAGTF as follows:

- **MEF.** A CEB, an ESB, a Marine wing support group, and an NCR.
- **MEB.** A CEB, an ESB, an MWSS, and an NMCB.
- **MEU.** A reinforced platoon from a CEB, an engineer detachment from a MWSS (if a task-organized unit), a detachment from an ESB, and civil engineer support from an NMCB, if required.

G-18. While these are typical support relationships, the actual level of combat engineer support depends on the situation and mission analysis.

ARMY AND MARINE CORPS ENGINEER CONSIDERATIONS

G-19. Marine division CEBs, as part of MAGTF operations, are usually the initial USMC engineering capabilities available in theater during force projection operations. ESBs are also deployed quickly to force projection operations to construct necessary Marine facilities. The Army engineer staff should consider the following when coordinating joint engineer plans and operations with the Marine Corps:

- Requesting the latest engineer intelligence data from deployed or deploying Marine CEB and ESB elements to assist in identifying force projection theater Army engineer requirements and enemy engineer capabilities. These requirements include threat mine and obstacle data, soils data, the availability of construction materials, and HN construction support.

- Establishing engineer staff links between the MAGTF, MARFOR, and ARFOR engineer staff and HQ.
- Providing necessary Army engineer LNO support.
- Developing the joint task organization relationships that enhance Marine engineer capabilities following the deployment of Army engineer units.
- Assessing the need for CEB and ESB support following the arrival of Army combat and construction units in theater.
- Determining if ESB units need augmentation from Army combat and construction units.
- Developing procedures for Army engineer units to be able to acquire Class IV construction materials from ESBs.

G-20. For additional information on Marine Corps engineer units and their doctrine and supporting TTPs, *refer to Marine Corps Warfare Publication (MCWP) 3-17* and *FMs 13-7 and 13-8.*

Appendix H

Multinational, Interagency, Non-Governmental Organizations, and Host Nation Considerations

Military engineers may need to coordinate their activities with multinational, US government, NGO, UN, and HN agencies according to the operational mandate or military objective. In all cases, authority must exist for direct coordination. Military engineers must establish interagency relationships through negotiation. The specific agency will vary, depending on who has federal or state proponency for the situation (for example, disaster relief versus a fire fighting mission). Agreements should be written as memorandums of understanding or terms of reference to ensure understanding and avoid confusion. Most agreements are made at the unified command or JTF level and normally place serious legal restrictions on using military personnel and equipment. These agencies and organizations may have unique engineer capabilities that could be used as part of the overall operational effort. However, these agencies and organizations often request extensive engineer support of their activities and programs. It is critical that an effective engineer liaison is established with the force HQ CMOC to coordinate and execute any engineer support exchange with these agencies.

UNITS AND ORGANIZATIONS

SECTION I - MULTINATIONAL CONSIDERATIONS

H-1. When military operations are considered, the US seeks to develop coalitions rather than conduct unilateral operations. The operation in the Dominican Republic in 1965, for example, was under the aegis of the Organization of American States (OAS). More recently, Operation Provide Comfort (Northern Iraq) and Operation Joint Endeavor (Bosnia) included a coalition of forces from other nations, as well as the US, under the sponsorship of the UN. The US may participate in a US-led coalition such as Operation Restore Hope (Somalia), or a non-US-led coalition, such as Operation Able Sentry (former Yugoslav Republic of Macedonia). The agencies involved in each of these operations are both consumers and possible resources of engineer activity. Army engineer units may be subordinate to, collocated and working alongside, or directing engineer activities and providing oversight or support for the missions assigned to these organizations. The engineer forces' effectiveness to operate within the varied framework surrounding a collective international enterprise can be greatly enhanced by respecting the multinational partners; their construction and engineering techniques; and their ideas, culture, religion, and customs. Equally important and parallel to operating within a US-only joint environment is understanding multinational unit or organization capabilities and training. This understanding ensures the assignment of appropriate missions and avoids the risk of offending national honor or prestige by allocating unsuitable tasks to partners in the multinational endeavor.

MULTINATIONAL ENGINEERS

H-2. The engineer organizations available from deployed national armies are generally a mix of combat and/or construction engineers in company- and battalion-sized units. The training and experience levels and equipment fielding varies among these units. National engineers from Britain, Canada, and Australia have been involved in numerous missions outside their territorial boundaries. The political impact of these missions is important to understand. When German engineers deployed into Somalia in 1992, it took a national legislative amendment to their constitution to allow them to participate in operations off German soil. This was their first experience in multinational efforts outside of NATO. Smaller countries have more regional restrictions on their involvement, and their experience is correspondingly narrow. However, they are also more likely to be attuned to the special circumstances that are relevant to the AO.

MULTINATIONAL ENGINEER CAPABILITIES

H-3. NATO and ABCA engineer capabilities are well known, and data about them is readily available. Standardization agreements (STANAGs) between national armies facilitate engineer interoperability and cooperation. The capabilities of other nation's engineers are normally available through intelligence channels or formal links with the nations concerned. Several nations have engineers that are experts in specific combat engineering tasks, such as mine detection and removal. Other national engineers are focused on specific missions, such as disaster relief. Engineers must have an appreciation for the engineer capabilities and limitations of other nations. Allied Joint Publication (AJP) 3.12 provides the keystone NATO engineer doctrine for working in this allied environment. Engineers must ensure they understand multinational doctrine, terms, and symbols in order to effectively operate in this environment. For example, AJP 3.12 defines two terms that are similar to US Engineer doctrine but the differences must be understood to effectively operate in the multinational environment:

- (NATO)—An Allied term that encompasses those military engineer tasks associated with direct support to current or imminent operations. It is conducted by the military engineers of any service or component to support land, air, maritime, and information operations. With the emphasis on speed of execution, it frequently relies on prefabricated equipment solutions, may involve a degree of improvisation and is likely to fulfill a short-term tactical requirement. It is closely aligned with the Engineer Battlespace function of Combat Engineering.(AJP 3.12)
- (NATO)—An Allied term that describes the deliberate, longer-term preparation for, and indirect support to ongoing or future operations

as well as those military engineering tasks associated with sustaining the joint force throughout all stages of an operation. It will be the predominant engineer focus for precombat and post combat and for rear operations during conflict. It is an area which may involve a greater degree of cross-compartment support and the engineer tasks will usually be more enduring, relying more on purpose designed and built solutions. It is likely to fulfill a longer term, operational requirement. It is closely aligned with the engineer Battlespace function of General Engineering. (AJP 3.12)

MULTINATIONAL ENGINEER COMMAND AND CONTROL

H-4. Depending on the multinational force arrangement in theater, Army engineers may control or work closely with engineers from other nations. Multinational engineer C2 relationships are established to foster cooperation and share information. Critical to this process is providing adequate US engineer LO support, including linguist support, communications equipment, and transportation.

MULTINATIONAL ENGINEER CONSIDERATIONS

H-5. During force projection operations, the initial engineers in theater most likely provide the HN engineering capabilities. As Army engineers deploy into a theater, they may be joined by allied and coalition engineers. The Army engineer staff should consider including the following when coordinating multinational engineer plans and operations:

- Requesting the latest intelligence information concerning the HN, allied and coalition engineer structures and logistics requirements.
- Requesting the latest engineer intelligence data from the HN or deploying allied and coalition engineer elements to help identify force projection theater Army engineer requirements and enemy engineer capabilities. The requirements include threat mine and obstacle data, soils data, construction materials availability, and HN construction support.
- Establishing multinational engineer staff links between the Army, HN, allied, and coalition engineer force staff sections through the JTF or theater engineer staff and HQ.
- Executing NATO multinational C2 using the NATO OPORD format and the NATO decision-making process.
- Providing necessary Army engineer LO support.
- Developing multinational task organization relationships that enhance HN, allied, and coalition engineer capabilities following the deployment of Army engineers.
- Assessing the need for HN, allied, and coalition engineer support following the arrival of Army combat and construction units in theater.
- Determining if multinational engineer units need augmentation from Army combat and construction units.

- Developing procedures for Army engineer units to support multinational engineers with additional Class IV construction materials and engineer equipment.

UNITED NATIONS OPERATIONS

H-6. UN-sponsored operations normally employ a force under a single commander. The force commander is appointed by the Secretary General of the UN, with the consent of the UN Security Council. The force commander reports to a special representative of the secretary general or the chief administrative officer (CAO) or directly to the secretary general. While the force commander conducts day-to-day operations with a fairly wide discretionary authority, the commander refers all policy matters to the CAO or the secretary general for resolution. The CAO not only establishes policy but controls the resources and funding expenditures within a given operation. The US commander, designated by national policy, retains command over all assigned US forces. The US chain of command runs from the President through the combatant commander to the lowest commander in the field. Other agencies may negotiate and exercise OPCON over US military forces subject to prior approval by the President and SECDEF. The degree of OPCON exercised over US forces must be coordinated and agreed upon by the multinational force commander and the combatant commander according to the President's criteria. The US commander is, therefore, responsible for mission success to the UN force commander as well as the theater combatant commander.

NON-UNITED NATIONS ORGANIZATIONS

H-7. Although the US is more likely to undertake involvement in multinational operations with the UN, a number of regional organizations may perform this leadership function, including—

- NATO.
- The Organization of African Unity (OAU).
- The OAS.
- The Organization for Security and Cooperation in Europe (OSCE).

SECTION II - INTERAGENCY AND NONGOVERNMENTAL ORGANIZATION CONSIDERATIONS

INTERAGENCY OPERATIONS

H-8. Interagency operations greatly expand the scope and capabilities of any given response team because of the wide variety of expertise and funding resources that can be tapped to perform functions during a crisis response. This is true whether the response is international or within the territorial confines of the US and its protectorates and territories. Interagency operations increase the resources engaged in any given operation, but they also significantly increase and complicate the coordination necessary to achieve victory. Additionally, they generate mechanisms that reduce efficiency, as organizations work at cross-purposes. The SOPs, report formats, information requirements, and intermediate goals and perceptions of each of these organizations vary greatly. Therefore, coordination and a clear understanding of the commander's intent are absolutely crucial when arranging operational efforts involving multiple interagency organizations. The following are some of the interagency organizations that could be involved:

- Federal Emergency Management Agency (FEMA).
- Environmental Protection Agency (EPA).
- Drug Enforcement Administration (DEA).
- National Oceanic and Atmospheric Administration (NOAA).
- United States Geological Survey (USGS).
- Public Health Service.
- Civil Air Patrol.
- United States Department of Agriculture (USDA).
- Department of State (DOS).
- Office of Foreign Disaster Assistance (OFDA).
- United States Department of the Interior, Fish and Wildlife Agency.
- General Accounting Office (GAO).

H-9. *See JPs 3-07* and *3-08* for an in-depth discussion of interagency coordination during joint operations.

NONGOVERNMENTAL ORGANIZATIONS

H-10. Relationships with international and domestic NGOs must be established through negotiation. Most agreements are made at the strategic level (unified command); however, the operational commander may have some latitude delegated to him. All agreements normally have serious legal restrictions on using military personnel and equipment. Some of these agencies may have unique and significant engineer capabilities and intelligence that could be used as a part of the overall operational concept. These capabilities may be a useful source of Class IV material, not only for the agency's own projects but also as a negotiated barter for services rendered in support of its mission. More often than not, however, these agencies and organizations may request extensive engineer support for their activities and programs. As these organizations play an important part in the combatant commander's achievement of strategic objectives, their demands must be coordinated. Therefore, it is critical that an effective engineer liaison be established and maintained with the force HQ CMOC.

H-11. The UN may designate a regional organization with a greater vested interest and appreciation for the forces at work in a given region as its operational agent to exercise control. Each of these organizations has different operational concepts and organizational procedures; US forces are familiar with some of these concepts and procedures, such as those of NATO. However, there are others with which they are not familiar.

SECTION III - HOST NATION CONSIDERATIONS

HN INTERFACE

H-12. In a forward deployed theater, the ASCC, in conjunction with the other component commanders, the combatant commander, the allies, and the HNs, identifies wartime facility and construction requirements for the Army as part of the deliberate war planning effort. Doctrinal construction requirements for the ASCC may be identified using the planning module in the TCMS. Subsequent analyses further refine construction requirements and provide a basis for—

- Force structuring.
- Procurement.
- Leasing provisions and establishing HN agreements.

H-13. The product of these analyses is the CESP. The goal is to reach HNS agreements in peacetime to provide as many of the facilities as possible that are needed within the theater. Advanced planning and the commitment of resources by HNs reduce the early lift requirements needed to support the RSOI and focus force projection assets C2 and engagement systems. Engineering support from the HN usually involves providing—

- Land.
- Facilities.
- Construction support.
- Manpower.
- Equipment.
- Materials.
- Services.
- Hazardous-waste disposal.

H-14. It is highly desirable to secure written agreements with the HNs regarding support items to foster an understanding of the assistance levels and increase the likelihood of fulfillment upon execution.

REAL ESTATE CONSIDERATIONS

H-15. Real estate operations involves acquiring, managing, and disposing of land and facilities to support joint operations. The JFC should establish real estate acquisition policies and programs in support of contingency operations. The JFC determines what real estate is needed to satisfy operational requirements. Acquiring land and facilities not owned by the United States Government (USG) is accomplished through assignment, international agreements such as SOFAs, and memorandums of agreement (MOAs), leases from the HN, or direct leases from the private sector. Within DOD, the secretaries of the military departments are authorized to lease structures and real property relating to structures that are needed for urgent military purposes in foreign countries (*see Title 10 USC,* initiated as contingency plans and developed to identify land and facility requirements for joint operations).

Real estate acquisition requires special contracting procedures be performed by the USACE, the Naval Facilities Engineering Command (NAVFACENGCOM), or a designated executive agent. Early deployment of real estate personnel is essential to acquire land and facilities in a timely manner.

H-16. Real estate authorities throughout the world have been assigned to components along similar lines corresponding to the designation of DOD construction agents (DODD 4270.5). Within regions designated to the Army, the USACE establishes policies for the acquisition, maintenance, and disposal of real estate, to include leased and rent-free facilities. Real estate teams may be assigned to each ASG or centrally controlled at the senior engineer HQ. These teams coordinate with HN agencies and private owners to acquire and dispose of real estate and establish the terms of lease agreements. Real estate planning and surveys must be initiated as campaign plans are developed to provide timely and adequate facilities to sustain the combat force. Local HN officials can help identify available facilities or land that meets military requirements. Thorough documentation of lease agreements, property conditions at the time of the lease, to include environmental baseline survey (EBS) data and expectations of property conditions at the termination of the lease are crucial to expedite a fair and amiable conclusion of lease activities. Civil affairs and real estate personnel may be required to work through HN governments to settle agreements with property owners. Real estate acquisition is more difficult in contingency operations due to the lack of preparation to identify probable sources of and confirm legal ownership. Real estate is required for—

- Air bases.
- Base camps.
- Medical and logistics complexes.
- Training sites.
- Quarry and borrow sites.
- Trailer transfer points.
- Traffic control points.

H-17. Property is generally acquired by requisition, with all transactions documented thoroughly under the provisions of the combatant commander's directives. Procedures are used that provide the property required for missions while protecting the property owner's legal rights. Using rent-free facilities provided by the host government or a host agency require the same legal responsibilities as using facilities leased from private owners. Real estate policies and procedures are discussed in more detail in *FM 5-104.*

HOST NATION SUPPORT

H-18. Wartime HNS agreements in forward-presence theaters (Europe and Korea) have been negotiated to provide HN construction support, such as facility modifications, LOC maintenance and repair, and utility services. In Southwest Asia, the agreements are less formal and lack the practiced application that accompanies the full-time presence of US forces in Europe and Korea. However, these agreements are no less critical to mission success in the event of an MTW in this region. Such HNS is used whenever possible to free US engineer units for critical missions where HNS alternatives are not viable. Support agreements are negotiated in peacetime on an asset basis. Assets may be facilities, contracts, or equipment. Again, this support is particularly critical during the initial stages of a contingency when RSOI requirements are high and engineer assets are limited.

H-19. Prepositioning equipment within the region reduces the US response time into a particular theater by allowing military forces to deploy by air and fall in on war stocks within the region. These prepositioning locations are a critical element of US force projection national strategy and represent a significant contribution of HNS. The HN's commitment for space, facilities, services, and utility support for these complexes demonstrate the HN's interface with US forces and the partnership of the US and UN in the defense and stability within the region. Beyond direct HNS, allied nations directly or indirectly involved in the crisis may provide other support. Other nations sympathetic to the cause may be limited in their direct participation because of constitutional restrictions or political sensitivities. However, these nations may provide nonlethal equipment or funding much like that provided by the Japanese during the Gulf War.

H-20. During a conflict, the HN may provide construction organizations to repair or construct facilities, usually within the rear area. Construction materials such as cement, asphalt, aggregate, timber, and steel, as well as contract labor, may also be available. HN assets may also be available for local security and the transportation of construction materials and equipment. Third country nationals (TCNs) may also be available, by request through the HN or through direct contact with nationals, to support engineer activities within the rear areas. Engineer reconnaissance and assessment teams engaged in deliberate planning during peacetime or dispatched early in contingency operations are the key to identifying and accessing available HN assets.

Appendix I

Contract Construction Agents

Use of construction contracting and engineering support can play an important role in the support of joint operations. Civilian construction contractors and HN engineering support provide the JFC with a significant engineering capability that becomes a force multiplier when combined with joint force military engineering units. CCAs provide the ability to harness and direct this means of support.

CONSTRUCTION CONTRACTING AND ENGINEERING SUPPORT

I-1. The DOD CCAs are the USACE, the NAVFACENGCOM, or other approved DOD activity *(see DODD 4270.5)*. These organizations and their contractors are a powerful force multiplier, allowing military engineers to concentrate on engineering missions in high-threat areas. The USACE and the NAVFACENGCOM also provide the JFC with a significant engineering capability to be leveraged in joint operations. They are DOD's principal organizations to plan, design, construct, and acquire (lease or buy) facilities and real estate. Inherent in their mission support capabilities is a planning and engineering capability for theater advanced base and infrastructure development. These organizations also maintain in-depth engineering expertise in their operating field organizations and laboratories.

I-2. The responsibilities of the DOD CCAs include designing, awarding, and managing the construction contracts for projects associated with the peacetime military construction program. Overseas, the USACE, the NAVFACENGCOM, and the Air Force are assigned specific geographical areas under DODD 4270.5 *(see Figure I-1, page I-2)*. Related to these responsibilities is the leasing of real estate.

I-3. The combatant commander may also use the USACE and the NAVFACENGCOM as contingency CCAs for designing, awarding, and managing construction contracts in support of military operations. For geographical areas where there is no designated DOD construction agent, the combatant commander usually designates a CCA for support during a contingency. The USACE and the NAVFACENGCOM also provide base development master planning, contract administration, and technical engineering support to JFCs (advanced base master planning, geospatial [topographic] engineering, infrastructure security and force protection engineering, and cold-weather mobility assessments). The Air Force also maintains a limited capability in contract construction in contingencies and facilities and real estate acquisition in England, Turkey, Spain, and Israel.

Designated Geographical Areas of Department of Defense Construction Agents

United States Army Corps of Engineers

Panama Canal Zone, Canada (excluding Newfoundland), Egypt, Europe (excluding Spain, Portugal, Italy, Greece), Greenland, Israel, Japan (including the Ryukyu Islands (Okinawa)), Korea, Taiwan, Marshall Islands, Middle East (including the Saudi Arabian Peninsula), Southern Asia (from Iran to Burma), and Turkey

Naval Facilities Engineering Command

Atlantic Ocean area, Australia, New Zealand, Caribbean Sea area, Greece, Iceland, Indian Ocean area, Italy, Newfoundland, North Africa (including Somalia and Kenya but excluding Egypt) Pacific Ocean area (excluding the Marshall Islands), Portugal (including the Azores), Republic of the Philippines, Southeast Asia, and Spain

United States Air Force

British Isles

Figure I-1. Designated Geographical Areas of Department of Defense Construction Agents

UNITED STATES ARMY CORPS OF ENGINEERS

The word 'miracle' has often been used when reporting the exploits of the American Corps of Engineers during the last war.... In fact there has been no miracl.... there has been only the logical result of a peacetime engineer organization unique in the world, which participates actively... in the development and execution of great public works of National interest.

General Robert J.L. Pinson, 1948, Chief of Engineers, French Army

I-4. USACE is the Army's major command assigned responsibility to execute the following Army and DOD mission areas:

- Engineering and design for the Army and Air Force military construction programs.
- Contract construction.
- Real estate acquisition.
- Technical assistance.
- Geospatial (topographic) engineering support.
- Army's civil works program.
- Public works and engineering for the Federal Response Plan (FRP).

I-5. The USACE's subordinate commands are organized geographically and functionally. It's C2 contains three major organizational structures, including—

- Divisions. The division is the major subordinate C2 organization for the USACE. The division commander provides executive direction to and management of the subordinate district commands. The division's orientation is regional and provides broad interface with regional interests and management of division-wide programs. Division commanders are responsible for contingency support in their directed areas of responsibility.
- Districts. The district command is the operating arm of the division. All CONUS USACE districts in the US have civil works responsibilities. In the US, their boundaries are delineated along major watershed basins, and their work lines are set on state boundaries *(see Figure I-2, page I-4)*. Additionally, some of the districts have military execution responsibilities. The districts maintain in-house capabilities in planning, engineering, construction, operations, project management, and contract administration. USACE provides support through subordinate commands to the following geographic combatant commands and subunified commands:
  - United States Forces Korea by the Pacific Ocean Division through the Far East District—Seoul, South Korea.
  - United States Forces Japan by the Pacific Division through the Japan District—Camp Zama, Japan.
  - Pacific Command (PACOM) by the Pacific Ocean Division through the Honolulu District—Fort Shafter, Hawaii.
  - Southern Command (SOUTHCOM) by the South Atlantic Division through the Mobile District—Mobile, Alabama.
  - EUCOM by the North Atlantic Division through the European District—Wiesbaden, Germany.
  - CENTCOM by Transatlantic Programs Center—Winchester, Virginia. For contingencies, the Transatlantic Programs Center is in support of the Southwestern Division for support to CENTCOM.
- USACE laboratories. The ERDC is the USACE's distributed research and development command. Headquartered in Vicksburg, Mississippi, the ERDC consists of seven unique laboratories that conduct research and development in support of the Army and the other services. By virtue of their engineering expertise, lab personnel provide operational support to the rest of the USACE and the DOD through the aforementioned subordinate commands. Additionally, the USACE maintains several specialized centers of expertise at its districts that provide additional technical engineering services (infrastructure security and force protection, airfield design, electronic systems security, physical security, base camp design, water flow data, and electronic systems security).
  - The Coastal and Hydraulics Laboratory, Vicksburg, Mississippi—provides services in vehicle mobility and trafficability; coastal engineering; structural hardening against weapons effects;

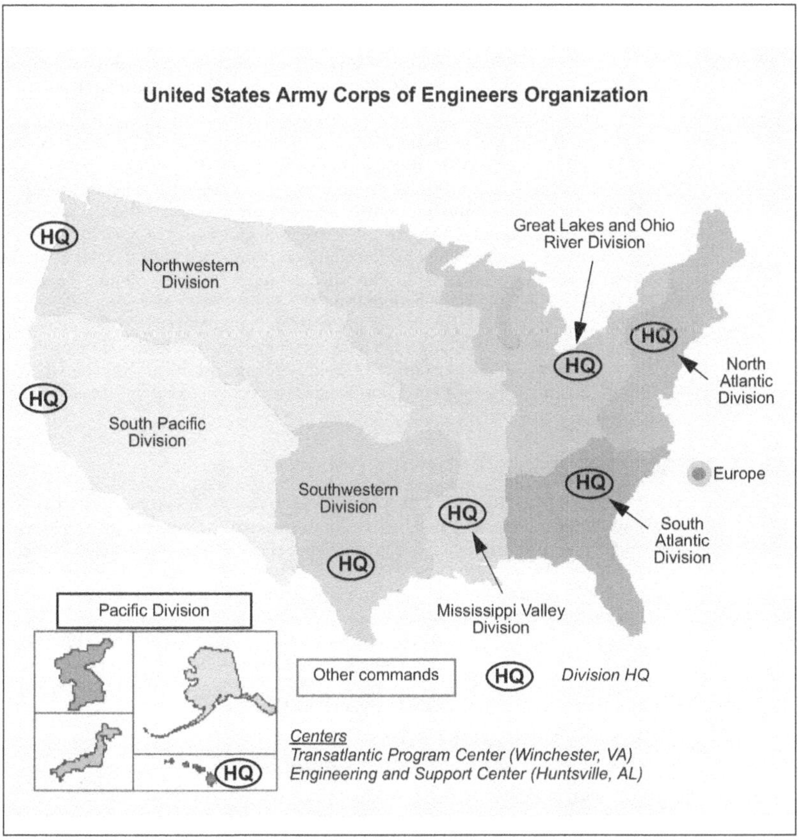

Figure I-2. United States Army Corps of Engineers

facilities camouflage, concealment, and deception technology; geotechnical expertise; and pavement and surface treatment technologies.

- The Cold-Regions Research and Engineering Laboratory (CRREL), Hanover, New Hampshire—provides services in all aspects of engineering research with respect to the conditions unique to cold weather.
- CERL, Champaign, Illinois—provides services in construction management technology and facilities engineering.
- The Environmental Laboratory, Vicksburg, Mississippi.

- The Geotechnical and Structures Laboratory, Vicksburg, Mississippi.
- The Information Technology Laboratory, Vicksburg, Mississippi.
- The TEC, Alexandria, Virginia—specializes in engineer mapping products, analysis of remote sensing data, and terrain analysis.

I-6. The USACE designs and constructs military facilities and supports military installations worldwide. The military engineering expertise of the USACE is focused on the engineering required to plan, design, and construct military facilities and the environmental engineering necessary to execute DOD installation environmental restoration projects. USACE maintains specialized expertise in its laboratories and centers for cold-weather engineering, remote sensing and imagery, force protection design, airfield design, weapons effects (for example, support for operational targeting—assessing the target, recommending the appropriate weapon systems, and providing an attack profile), terrain analysis for mobility and countermobility, topographic engineering, security systems engineering, environmental management, and environmental engineering. Additionally, the USACE's 249th Engineer Battalion (Prime Power) can conduct power assessments and install generators to provide emergency power.

NAVAL FACILITIES ENGINEERING COMMAND

I-7. The NAVFACENGCOM and its field divisions directly support the Navy and Marine Corps and DOD shore establishment throughout the world with a wide variety of military and contract construction, real estate acquisition, and public works support. The NAVFACENGCOM's engineering support units provide project management, planning, design engineering, construction, operations, and maintenance, as well as disposal functions for Navy and Marine Corps shore facilities and real estate. They also provide engineering, logistics, doctrine, and policy support and guidance for NCF units.

NAVAL FACILITIES ENGINEERING COMMAND AND CONTROL

I-8. The NAVFACENGCOM and subordinate commands are organized geographically and functionally *(see figure I-3, page I-6)*.

I-9. The NAVFACENGCOM headquarters is located in Washington, DC. Subordinate commands include four engineering field divisions (EFDs), six engineering field activities (EFAs), officers in charge of construction (OICCs), and the Naval Facilities Engineering Service Center (NFESC). Additionally, the NAVFACENGCOM provides technical support to ten regional Navy public works centers, including—

- EFDs and EFAs. The EFD commanders provide executive management and engineering support for all Navy and Marine Corps facilities in designated geographical areas. EFAs provide similar support for smaller CONUS and overseas areas within the greater geographic area of the EFDs. Both EFDs and EFAs have design, engineering, and architecture capabilities and act as technical consultants for all facilities of supported commands. EFDs play a significant role in contingency operations through forward-deployed resident OICCs that can provide construction contracting and real

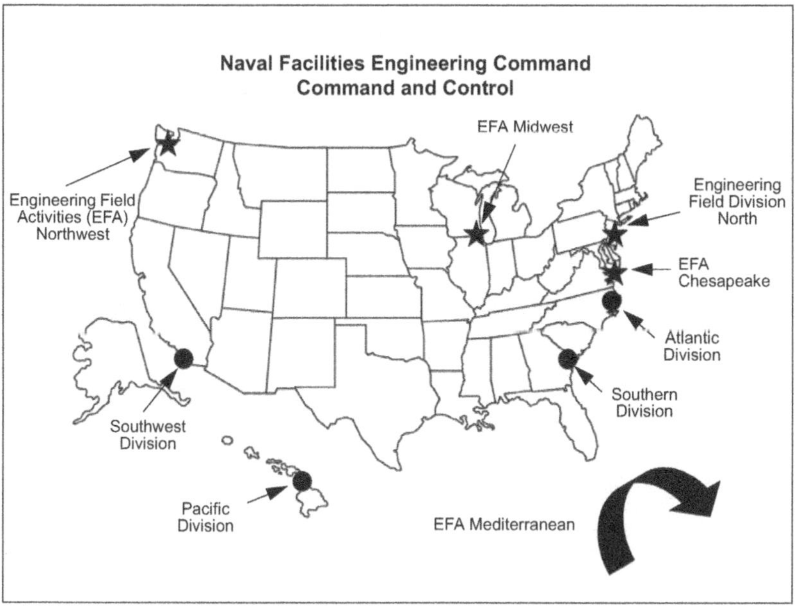

Figure I-3. Naval Facilities Engineering Command, Command and Control

estate support and manage Navy construction capabilities contracts (CONCAPs).

- OICC. The EFD commanders designate OICCs as contracting officers in certain geographical locations where the complexity and contracting tempo warrant. The EFD commander can also assign a resident officer in charge of construction (ROICC) in any geographical location, as required. The ROICC is trained in facilities support, construction, engineering, environmental design, and real estate contract acquisition and management. The ROICC office is especially useful for contingency operations in support of the joint force. The ROICC may also provide construction management support to the Army, Air Force, and DOD, as well as other federal agencies.
- NFESC. The NFESC provides engineering support and expertise to the Navy in undersea and amphibious operations, conventional ammunition storage, mobile utilities and environmental support equipment, and research and consulting engineering services. The NFESC is located in Port Hueneme, California, with a detachment in Washington, DC *(see Figure I-4).*

Figure I-4. Naval Facilities Engineering Command and Naval Facilities Engineering Center

NAVAL FACILITIES ENGINEERING COMMAND SUPPORT FOR MILITARY OPERATIONS

I-10. The NAVFACENGCOM supports Navy and Marine Corps operations, DOD missions, and joint force operations around the world, to include—

- Force projection missions. The Navy accomplishes force projection worldwide by using superior naval, air, and amphibious forces, equipment, tactics, and doctrine. This force projection requires shore logistic platforms, such as advanced logistic support sites and ports to receive supplies, equipment, and personnel to be deployed to ships and advanced bases. The NAVFACENGCOM directly supports these activities by providing engineering, construction forces, contract construction, and facilities management, including the disposition of real estate for these ports and bases.
- Theater missions. The NAVFACENGCOM supports the theater through real estate acquisition, management, and final disposition, as well as through the construction of facilities to support Navy, Marine Corps, combatant commands, subordinate joint forces, and other DOD elements. The NAVFACENGCOM provides this support through the subordinate EFD and EFA and provides technical engineering and construction support to other services and DOD agencies through the

NCF and as a designated CCA. The NAVFACENGCOM commander may provide a forward element dedicated to support the geographic combatant commander or subordinate JFC for military operations in contingencies. The forward element may call on the full capability of the entire NAVFACENGCOM organization to assist in providing this support.

- CONUS contingency missions. The NAVFACENGCOM has responsibility for disaster recovery and other contingency operations at Navy and Marine Corps stations in CONUS. With NCF personnel and equipment, in addition to its contracting capability, the NAVFACENGCOM can quickly mobilize and provide significant assets for disaster recovery.

CONTRACTED SUPPORT

I-11. The US military can contract civilian engineering support, as required, based on the threat situation and available resources. These contracts relieve the workload on US military engineer units in such areas as logistics base construction, real estate and facilities acquisition, RPMA, and demining operations. As discussed earlier, overseas construction and other contracting services are available through the USACE, or the NAVFACENGCOM, or through limited Air Force capability during contract construction in contingencies, as well as facility and real estate acquisition (in England, Turkey, Spain, and Israel).

CIVIL AUGMENTATION PROGRAMS

I-12. Civil augmentation programs such as the LOGCAP, the CONCAP, and the AFCAP can provide worldwide logistics and construction support. Civil augmentation programs can play a significant role in mission accomplishment by providing the JFC and joint force engineer with additional options and flexibility in achieving timely civil engineering and logistic support. Civil augmentation programs are managed by a contract agent and are structured with one contractor responsible for providing support that effectively integrates construction, facility maintenance, and logistic support to the joint force. For example, within an operational area, subcontractors, materiel, and personnel may come from many countries within the region. A single contractor prevents multiple agencies and their contractors from bidding against one another for services and materiel in the operational area. Use of civil augmentation programs requires planning and operational oversight and quality control and assurance to ensure costs are effectively controlled, while providing support consistent with the JFC's concept of operations.

ARMY

I-13. The LOGCAP is an Army program funded in peacetime as a component of Army readiness. The LOGCAP is a broad logistic and engineering contingency support contract that encompasses all Army preplanned contingency contracts and contingency components of contracts. Currently, the LOGCAP is a cost plus award fee contract, managed by the United States Army Materiel Command (USAMC). In peacetime, the LOGCAP contractor maintains an on-call, preplanned, ready capability. The contractor demonstrates readiness through the development of a worldwide plan, supporting plans to OPLANs, specific regional plans, and participation in exercises.

- The USAMC support contract provides both engineering construction and general logistic services. USAMC is supported by the USACE for engineering and construction contract administration and by the Defense Contract Management Agency (DCMA), a component of DLA for logistic services contract administration.
- In the operational area, the USAMC logistics support element, subordinate to the senior Army logistics commander in the operational area, manages the LOGCAP support contract. The USACE provides program management, coordinates the LOGCAP requirements with supported MACOMs, and administers the LOGCAP contract. The MACOM G3, G4, engineer, and comptroller are key players in developing the LOGCAP requirements and ensuring the appropriate mix of contractor and troop support. The contract can support all DOD components and missions and is a cost plus award fee service contract that provides support to the joint force in four major areas:
  - Contractor planning expertise to assist contingency planners.
  - Facilities repair and construction.
  - Base operations and maintenance.
  - Logistics services.

I-14. The LOGCAP is especially suited to support reception, onward movement, and sustainment facilities. The LOGCAP can augment engineer units by operating Class IV supply yards, supplying construction equipment, providing facility engineering support, and supporting theater construction.

NAVY

I-15. The Navy's CONCAP program is a cost-reimbursable contract administered by the Atlantic Division, NAVFACENGCOM. The contractor is usually a large construction firm or joint venture firm, with international capability. The contract offers responsive engineering and construction capabilities for a wide range of construction missions. The CONCAP was used for engineering fieldwork in Bosnia and for disaster response and recovery for Hurricane Bertha at Camp Lejeune, North Carolina. This construction-oriented contract may be used worldwide, including in CONUS. Major capabilities include the—

- Engineering, design, and construction of the following:
  - Airfield and port facilities—piers and dredging.
  - Roads, bridges, ordnance facilities, and landfills.
  - Power plants and utility systems.
  - Communication and supply systems.
  - Medical and EPW facilities.
- Operation of the following:

- Facilities and utilities.
- Billeting, food services, and recreation.
- Waste management—refuse and sanitation.

AIR FORCE

I-16. The AFCAP is a cost plus award fee contract that is centrally administered by the AFCESA at Tyndall Air Force Base, Florida, (through the Air Force's major commands' civil engineers). The contract was designed to augment or relieve specified civil engineering support functions participating in MOOTW. The AFCAP capabilities focus on temporary contingency skills to sustain military forces and to support a 10,000-person force in up to eight different locations for periods of 180 or more days. When AFCAP contract support is initiated for a contingency operation, a cadre of contractor personnel responds on-site, as soon as possible, to begin construction tasks (typically within 30 calendar days). Each task order specifies the required response time and the period of performance.

CONTRACTED CIVILIAN ENGINEER CONSIDERATIONS

I-17. The challenge for engineer planners and executers is to achieve the optimal mix of contractor and military engineer unit capabilities. Construction contractors are best suited for the longer duration and for heavy construction work in stable environments. In turn, contractors leverage local resources (labor and materiel) to minimize costs and impacts on intratheater lift and port facilities. The contractor's presence contributes significantly to local area political and economical stabilization and thereby reduces the need for the presence of US security forces. In turn, the US commander in theater must recognize the need for US military oversight of contract and contractor activities in the areas of project management, financial management, quality assurance, and audit.

I-18. During force projection operations, extensive contracted civilian engineer capabilities are probably available only after 30 days (D+30), due to mobilization and deployment time lines. Civilian engineer contracting may be available sooner when deliberately and properly planned during permissive entry conditions. As Army engineers deploy into the theater, they may be joined by contracted civilian engineers. The Army engineer staff should consider the following when coordinating engineer plans and operations with contracted civilian engineers:

- Requesting the latest engineer intelligence data from any contractors working in the theater to help identify force projection theater Army engineer requirements and enemy engineer capabilities. These requirements should include the availability of real estate, construction materials, and facilities; data on threat mines, obstacles, and soils; and construction support from the HN.
- Establishing engineer staff links between the Army and contracted civilian engineer staff through the JTF, the USACE or, NAVFAC and the theater engineer staff and HQ.
- Providing necessary Army engineer LO support.

- Developing time lines that quickly phase in contracted civilian engineer capabilities to relieve deployed Army engineer units of some responsibilities.
- Assessing the need for additional contracted civilian engineer support following the arrival of Army combat and construction units in theater.
- Determining if contracted civilian engineers need augmentation from Army combat and construction units.
- Developing procedures for Army engineer units to draw on contracted Class IV construction materials and engineer equipment.

Appendix J

Contingency Authorities and Funding

It is especially important that engineers understand contingency authorities and the associated funding. These are the tools that set the conditions for success during contingency operations and provide the basis for legal spending to take care of DOD personnel and activities in support of contingency operations. Contingency operations comprise a very large portion of the operations conducted by Army and engineer forces.

LEGAL PERSONNEL

J-1. Legal personnel can provide invaluable advice and guidance on authorities and sources of funding for civil engineering activities in a variety of situations. From the earliest stages of planning, execution, and redeployment, legal professionals play a vital role in preparing the AO by identifying and assisting in the resolution of legal and political constraints, and provide relevant and responsive readiness programs to the individual civil engineering members.

TYPES OF AUTHORIZATIONS AND SOURCES OF FUNDING

J-2. Services are authorized to use annual operation and maintenance (O&M) funds for construction projects costing less than $500,000 ($1 million to correct a life threatening condition or for new construction and $2 million for maintenance and repair of existing facilities). This is a peacetime provision, applicable during contingencies and emergencies; however, life threatening is generally considered a safety issue vice an emergency in the context of contingency operations. During combat or designated contingency operations, O&M may be used to fund construction projects exceeding these thresholds. The JFC must consult with the Staff Judge Advocate (SJA) before making a determination to use O&M in such a case.

J-3. Several broad authorities have been established under Title 10 USC that enable the JFC to carry out contingency construction, including procuring materials for construction by military forces and funding civilian contracts. *Figures J-1 and J-2, pages J-2 and J-3,* depict decision trees for the contingency construction funding options.

J-4. Section 2803, Emergency Construction, authorizes each service to use $30 million per year of unobligated military construction (MILCON) funds for projects that cannot wait for the normal MILCON submission procedures. Projects must comply with a 21-day congressional notice and wait period before proceeding. Generally, a previously congressionally approved project must be canceled to free the $30 million.

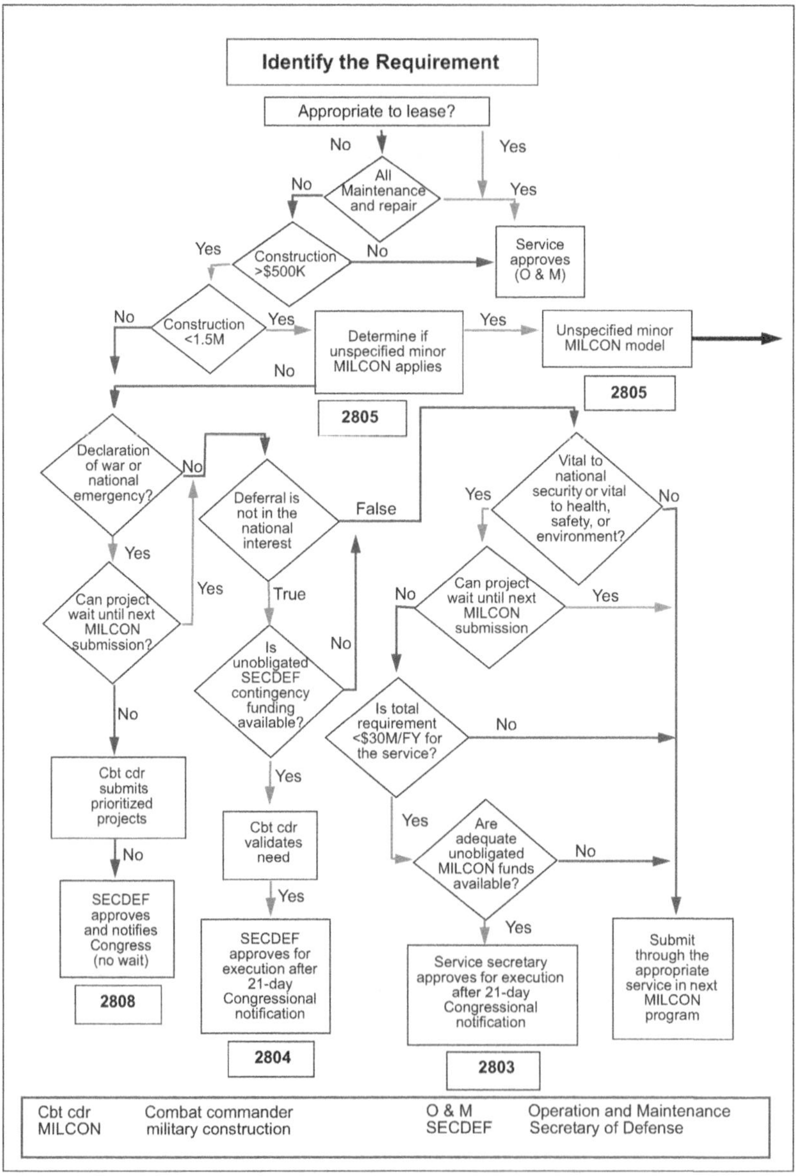

Figure J-1. Contingency Construction Funding Model

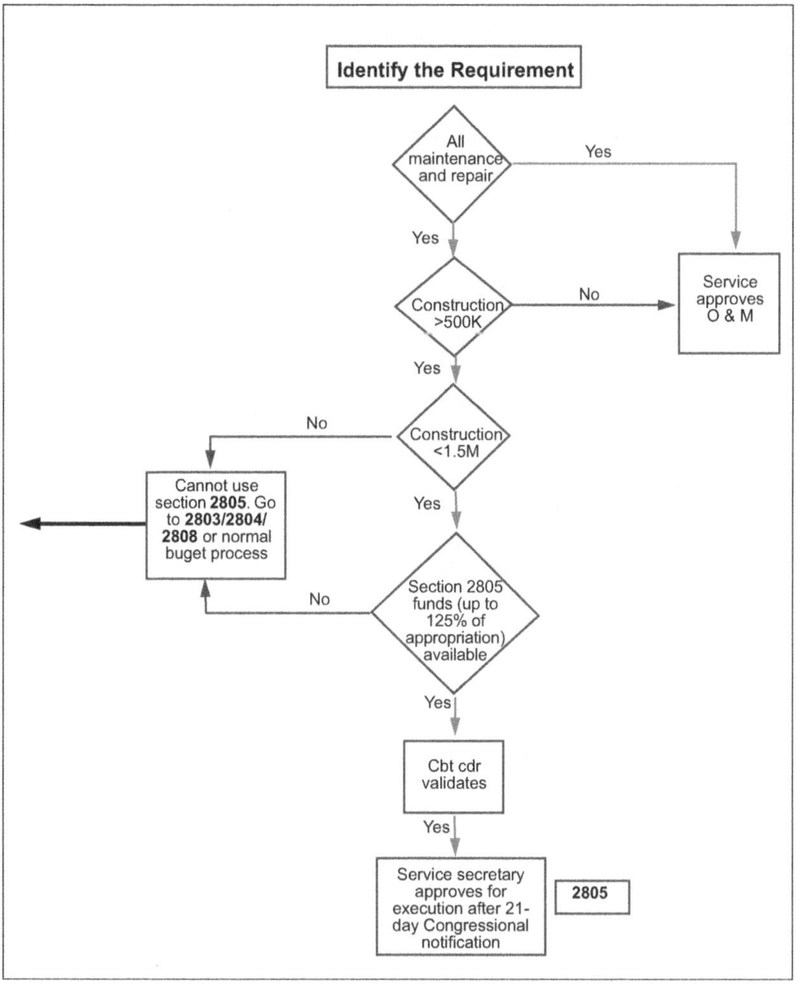

Figure J-2. Contingency Construction Funding Model (Unspecified Minor, $500,000-$1,500,000)

J-5. Section 2804, Contingency Construction, authorizes the SECDEF a specific military construction line item amount for contingency construction projects that cannot wait for the normal military construction program submission process. A project must comply with a 21-day congressional notice and wait period before proceeding. Generally, funding for this section is limited to less than $10 million per year.

J-6. Section 2805, Unspecified Minor Construction, authorizes each service a specific MILCON line item amount that varies annually for unspecified minor construction. Projects must be less than $1.5 million each ($3 million to correct a life threatening safety condition). Projects greater than $500,000 require a 21-day congressional notice and wait period before proceeding.

J-7. Section 2808, Construction Authority requires a Presidential declaration of war or national emergency and authorizes the SECDEF to carry out any military construction project for the war or national emergency within the total amount of unobligated MILCON funds available. Congress must be notified of each project, but there is no wait requirement before the project may begin.

J-8. Combatant commanders do not need specific authority to request projects under Sections 2803 and 2804. To gain approval for a project under either authority, it is necessary to provide the appropriate service secretary or SECDEF with a justification of need, estimated costs, and source of funding.

OTHER AUTHORITIES AND SOURCES OF FUNDING

BURDEN SHARING (TITLE 10 USC 2350.J)

J-9. This regulation provides guidance enabling DOD to accept funds from HN or foreign governments to share the financial burden of construction, supplies, and services.

- Countries must be designated countries which allow DOD to accept cash contributions for these purposes. A listing is held with the Office of the SECDEF (Comptroller).
- If a country is not previously designated, the SECDEF must formally consult with Secretary of State (SECSTATE) for designation of a particular country. The Joint Chiefs of Staff (JCS) would submit a formal request to the SECDEF to initiate this process.
- Contributions can be accepted only for certain costs, including—
  - Compensation for foreign local national employees of DOD.
  - Military construction projects of DOD.
  - Supplies and services for DOD.
- Reporting procedures for military construction projects state that a project is initiated when a report is submitted to congressional committees. There is a 21-day wait period for approval.

SECTION 607A OF THE FOREIGN ASSISTANCE ACT OF 1961 (PL 87-195, AS AMENDED)

J-10. This act provides restoration of HN civil infrastructure. This provision of law allows any USG agency to provide goods and services to friendly countries and NGO agencies on an advance-of-funds or reimbursable basis.

ARMS EXPORT CONTROL ACT OF 1976 (PL 90-629, AS AMENDED)

J-11. HN military facilities may be restored under the foreign military sales provisions of this authority.

ECONOMY IN GOVERNMENT ACT (TITLE 31 USC 1535)

J-12. This act allows USG agencies to support each other provided that the supported agency has the funds and authority to do the work requested.

HUMANITARIAN AND CIVIC ASSISTANCE PROJECTS (TITLE 10 USC 401)

J-13. In HCA facilities projects, the JFC and joint force engineers may work with HN government agencies to repair or improve infrastructure and public facilities. These authorized and funded projects are designed to provide assistance to the HN populace in conjunction with a military operation or exercise. They are usually planned well in advance and are not usually planned in response to disasters, although HCA activities have been executed following disasters. Specific engineer activities for which HCA funds can be used include the construction of rudimentary surface transportation systems, well drilling, construction of basic sanitation facilities, and rudimentary construction and repair of facilities *(refer to Chapter 8).*

FOREIGN HUMANITARIAN ASSISTANCE (TITLE 10 USC 2551)

J-14. In disaster operations, the UN and the DOS, OFDA may generate funded requirements for DOD assistance. FHA programs focus on the use of DOD excess property, emergency transportation support, disaster relief, or other support, as necessary, to alleviate urgent needs caused by some type of disaster or catastrophe in a host country. While other elements of the joint force are focused on immediate humanitarian assistance, civil engineering planning may focus on projects that provide immediate shelter for dislocated civilians. The joint force engineers must work in a close relationship with the representatives of the HN and US country team *(refer to Chapter 8).*

DRAW DOWN OF DEPARTMENT OF DEFENSE ARTICLES AND SERVICES (TITLE 10 USC 506(A)(1), 506(A)(2), AND 552(C)(2)

J-15. Draw-down authority is a means to respond to unforeseen military emergencies or humanitarian relief situations. These recurring authorities have placed annual limitations on the value of articles and services that may be drawn down in any fiscal year.

- Section 506(a)(1). Draw down of DOD stocks and military assistance to countries and NGOs in support of emergencies worldwide. The total is limited to $100 million per year, of which no more than $75 million can come from DOD.
- Section 506(a)(2). Draw down of stocks of any USG agency for counternarcotics, disaster relief, migration, and refugee assistance worldwide. The total is limited to $150 million per year, of which no more than $75 million can come from DOD.

- Section 552(c)(2). Draw down of stocks from any USG agency to countries and international organizations in support of PKOs. The total is limited to $25 million per year.

DEPARTMENT OF DEFENSE DIRECTIVE 5100.46, FOREIGN DISASTER RELIEF

J-16. Normally, DOD components may participate in foreign disaster relief operations only after a determination is made by the DOS. This directive allows the military commander at the scene of a disaster to undertake disaster relief operations without prior approval of the ambassador and/or chief of mission when the emergency is so acute that immediate action is required to save life and property.

Appendix K

Environmental Considerations

"The American people will continue to expect us to win any engagement, but they also expect us to be efficient in protecting lives and resources while accomplishing the mission successfully. Commanders will be expected to reduce the cost and adverse effects of military operations, from environmental disruptions in training to collateral damage in combat." "

Joint Vision 2010

Environmental considerations are the spectrum of environmental media, resources, or programs that may impact on, or are affected by, the planning and executing of military operations. Factors may include, but are not limited to, environmental compliance, pollution prevention, conservation, protection of historical and cultural sites, and protection of flora and fauna (JP 3-34). Military environmental protection is the application and integration of all aspects of environmental considerations as they apply to the conduct of military operations. The purpose of this chapter is to outline the environmental considerations as part of a commander's planning process and specific responsibilities of the engineer/ ENCOORD on the staff.

COMMANDERS AND ENVIRONMENTAL CONSIDERATIONS

K-1. Commanders must be able to identify ways to protect the natural environment while executing their missions by doing the following:

- Considering the natural and the physical environment in planning and decision making in conjunction with other essential considerations of national policy.
- Protecting both the natural and physical environment of the home station and training areas as a means of retaining resources for mission purposes.
- Complying with all federal, state, local, and HN environmental laws and regulations.
- Using environmental risk-assessment and environmental management principles to integrate environmental considerations into mission performance.
- Instilling an environmental ethic in subordinates.

K-2. These actions identify environmental protection as an important undertaking considered in mission planning and execution. The commander will use the staff engineer as the key advisor on all environmental issues. Working with other staff officers, the engineer determines the impact of the operation on the environment and integrates environmental considerations into the decision-making process.

K-3. Commanders, engineers, and other staff officers should refer to *FM 3-100.4* for specific guidance in applying appropriate environmental consideration procedures during military operations.

ENVIRONMENTAL ROLES AND RESPONSIBILITIES

COMBATANT COMMANDER AND SUBORDINATE JOINT FORCE COMMANDER

K-4. The combatant commander and subordinate JFC are responsible for ensuring environmental compliance, within the confines of mission accomplishment, with all domestic environmental laws and international agreements that apply within the joint operations area. JFCs should demonstrate proactive environmental leadership during all phases of joint operations, instill an environmental ethic in subordinate commands, and promote environmental awareness throughout the joint force. JFCs should ensure that environmental considerations are an integral part in the planning and decision-making processes. They should identify specific organizational responsibilities and joint force environmental requirements. These responsibilities should have clearly defined goals, strategies, and measures of success.

COMBATANT COMMAND AND SUBORDINATE JOINT FORCE ENGINEER

K-5. The COCOM and subordinate joint force engineer are responsible for providing guidance to the combatant commander and/or subordinate JFC on environmental considerations in planning and executing joint operations. The COCOM and subordinate joint force engineer and staff develop and assist with implementing the policies, procedures, and practices of the Environmental Considerations annex to an OPLAN and/or OPORD. *See CJCSI 3122.03A* for additional information.

COMBATANT COMMAND AND SUBORDINATE JOINT FORCE STAFF JUDGE ADVOCATE

K-6. The COCOM and subordinate joint force SJA advise the commander on compliance with environmental laws, regulations, treaties, conventions, OEBGDs, and SOFAs and their potential impact on operations. Specifically, the COCOM and subordinate joint force SJA are responsible for legal support in the development of the environmental considerations annex to an OPLAN and/or OPORD to ensure that legal requirements related to environmental considerations are incorporated, as appropriate. The joint force SJA—

- Assists other members of the joint force staff (J-4) and defense agencies in negotiating transit agreements in advance of the actual deployment, to permit the transit of regulated (hazardous) waste and ensure they are disposed of in an environmentally sound manner.
- Helps determine baseline environmental survey requirements and processes civilian claims resulting from environmental damage.

- Participates in the development of baseline environmental survey exemptions.

COMBATANT COMMAND AND SUBORDINATE JOINT FORCE SURGEON

K-7. The COCOM and subordinate joint force surgeon are responsible for health services support (preventive medicine and occupational health) to the joint force. Priorities are on water and wastewater, including water vulnerability assessment support, sanitation, waste disposal (hazardous and infectious waste); health risk assessment (base-camp site selection); environmental health sampling; and surveillance and vector control to protect human health and welfare.

JOINT FORCE PUBLIC AFFAIRS OFFICER

K-8. The joint force public affairs officer (PAO) coordinates with the appropriate staff and commanders to plan and accomplish public relations efforts in support of mission objectives. Special attention should be given to potentially sensitive environmental issues associated with a joint operation. In the US and its territories and possessions, various environmental laws require public involvement. The joint force PAO will be a significant participant in public outreach efforts. He should participate in developing assigned responsibilities in environmental contingency plans.

JOINT FORCE LOGISTICS DIRECTORATE

K-9. The joint force J4 and staff, when augmented by engineer personnel, are responsible for ensuring that waste and effluents from operations materiel and service functions are appropriately controlled. The joint force J4 and staff—

- Are responsible for all aspects of hazardous materials and regulated (hazardous) waste management, to include minimizing the use, storage, transportation, disposition, and redeployment of excess materials.
- Should coordinate closely with the joint force SJA in negotiating transit agreements and establishing procedures for the turn-in of regulated (hazardous) waste for proper treatment and disposal.

JOINT ENVIRONMENTAL MANAGEMENT BOARD

K-10. A JEMB may be established by the combatant commander or subordinate JFC for a joint operation to integrate the environmental-protection efforts of all participating components under a single authority, ensuring unity of effort for environmental protection activities. The JEMB should be chaired by the COCOM or subordinate joint force J4 or joint force engineer and include representatives from each service component and joint force staff representative, as necessary (legal, occupational health, preventive medicine, safety, comptroller, planning, operations, and logistics). The JEMB assists the JFC in establishing the joint force environmental policies, practices, procedures, and priorities and in providing oversight of environmental protection standards and compliance. Establishing a dedicated and appropriately staffed environmental engineering staff, supported by experts from other joint force staff members (legal and medical), may obviate the need for a JEMB in smaller operations.

OTHER FEDERAL AGENCIES AND NGOS

K-11. During operations, such as those involving responses to disasters or support to civilian governmental agencies (cleanup of major oil or hazardous substance spills), the JFC may have to work with other federal agencies or NGOs to ensure successful completion of the operation. Where appropriate, these representatives should be a part of the environmental planning process. The JFC may also consider their participation as ad hoc members of the JEMB.

ARMY FORCES COMMANDER AND STAFF

K-12. Similar to joint-level commanders and staffs, the ARFOR commander and staff have inherent responsibilities within their areas of expertise that require environmental actions. Whether developing the staff estimate, protection levels, or EBS, environmental protection requires active participation from each staff member. Environmental factors may affect or influence a wide range of activities or require a significant expenditure of resources. The unit staff integrates environmental considerations into the planning and execution processes. Common staff duties provide the basis for some environmental responsibilities, and *FM 101-5* provides a basis for others. While some of these responsibilities may depend on the command or commander, all staffs undertake many of them. The commander will use the staff engineer as an integrator of environmental considerations, which is similar to the role of the G2/S2 in the IPB process.

SUBORDINATE COMMANDERS AND STAFFS

K-13. Unit commanders are responsible for complying with the applicable environmental requirements established by the commander in the environmental considerations appendix of the OPLAN and/or OPORD. Subordinate commanders should keep the higher command staff informed of conditions that may result in noncompliance or the potential for noncompliance with this appendix. Unit SOPs at battalion and company levels must incorporate specific responsibilities. Unit commanders should appoint primary and alternate environmental compliance officers (ECOs) to act as liaisons to higher command engineering and environmental staff and communicate applicable issues to the commander and unit personnel in both training and real-world operations.

ENGINEER COORDINATOR

K-14. The ENCOORD is the special staff officer for coordinating engineer assets and operations for the command. As the senior engineer officer in the force, he advises the commander on environmental issues, to include the command environmental program. Working with other staff officers, the ENCOORD determines the impact of operations on the environment and integrates environmental considerations into the decision-making process. He works with the G4 in performing site assessments for installations and facilities. With the SJA, the ENCOORD advises the commander on the necessity for environmental assessments to meet HN or Executive Order (EO) 12114 requirements. He is also responsible for advising the G2 and S2 of significant environmental factors and ensuring these impacts are integrated into the IPB process. *See FM 34-130* for more information. Additionally, *FM 101-5* directs that OPLANs, OPORDs, and CONPLANs contain an appendix to address environmental considerations. Through coordination with the unit staff, the ENCOORD prepares Appendix 2, Environmental Considerations to Annex F, Engineer, for a joint order. *See FM 3-100.4* for the specific appendix format and content.

ENVIRONMENTAL REQUIREMENTS

GENERAL

K-15. Environmental requirements can be divided into overseas requirements and requirements applicable in the US and its territories and possessions, although some US environmental requirements may have extraterritorial application. For example, EO 12114 establishes requirements for the conduct of environmental studies for activities conducted overseas, somewhat similar to the environmental analysis requirements mandated by the National Environmental Policy Act (NEPA) regarding operations conducted within the US. The SJA should be consulted to determine extraterritorial applicability. The activation and incorporation of environmental management systems (EMSs) is critical for all DOD organizations *(see EO 13148)* regardless of whether they operated in CONUS or OCONUS. *See FM 3-100.4* for a more focused discussion on environmental requirements affecting Army operations.

REQUIREMENTS APPLICABLE WITHIN THE UNITED STATES

K-16. All military operations within the US and its territories and possessions are conducted in compliance with applicable federal, state, or local environmental laws and regulations. NEPA requires environmental planning for major federal actions that have the potential for a significant environmental impact. It requires that commanders make environmental considerations an integral part of the mission-planning and decision-making processes. NEPA mandates procedures for environmental planning but does not mandate decisions. Other federal statutes, such as the Clean Water Act (CWA); Clean Air Act (CAA); Resource Conservation and Recovery Act (RCRA); and the Comprehensive Environmental Response, Compensation, and Liability Act (CERCLA), establish environmental requirements that may impact joint operations. During MOOTW, all legal requirements imposed by various jurisdictions during peacetime may still apply. In many cases, US environmental laws allow for national security exemptions for specific activities, but only upon action by the President.

REQUIREMENTS APPLICABLE BEYOND UNITED STATES TERRITORY

K-17. All operations beyond US territory are conducted according to applicable international treaties, conventions, OEBGDs, SOFAs, final governing standards (FGS), and other international agreements. In the absence of definitive environmental guidance within applicable international agreements, geographic combatant commanders and subordinate commanders should establish guidance in the OPLAN and/or OPORD that protect force health, limit adverse public health impacts, consider US liability, and remain consistent with mission goals.

ENVIRONMENTAL PLANNING

THE NEED FOR ENVIRONMENTAL PLANNING

K-18. By considering environmental issues early during the planning process, the commander may continue to achieve operational objectives while minimizing the impact on human health and the environment. Failure to consider the environmental impact of all activities may adversely affect the operation. Potential effects include endangering personnel, delaying operation commencement, limiting future use of exercise or HN areas, and creating an adverse public opinion. Through early assessment of environmental considerations, commanders may become aware of the potential environmental effects or impacts of mission accomplishment while alternatives to address mitigating actions still exist. By planning early, the commander and staff must be aware of the environmental requirements and able to plan more effectively and act to incorporate environmental considerations. Furthermore, careful and visible attention to environmental considerations in military operations can assist in shaping a positive image both internationally and domestically.

K-19. Integrating environmental considerations into planning is very similar to integrating safety and force protection issues. *FM 3-100.4* discusses environmental planning and focuses on how and where the Army integrates environmental considerations into the MDMP. As part of the MDMP, risk management is an effective process to minimize actions that may negatively impact the environment and take appropriate steps to prevent or mitigate damage.

ELEMENTS OF ENVIRONMENTAL PLANNING

K-20. The staff should plan the operation to achieve the mission objectives while minimizing the environmental effects and observing environmental requirements. Although not all elements *(See Figure K-1)* are applicable to all operations (some, such as the identification of alternatives to obtaining objectives, are not required for OCONUS operations), they may prove helpful during planning operations to—

- Identify the operational objectives and the activities that are proposed to obtain these objectives, including the logistics requirements and the identification of HAZMATs that may be used.
- Identify potential alternative means of obtaining the operational objectives. These alternatives may include the use of new technologies to minimize the impact on the environment.
- Identify the environmental requirements applicable to the operational area.
- Identify adverse environmental health and environmental impacts that may result from the operation.

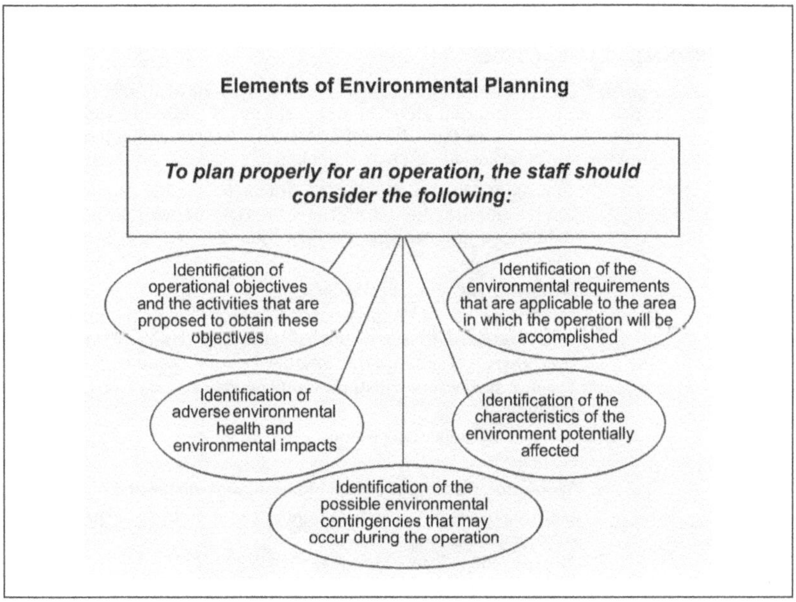

Figure K-1. Elements of Environmental Planning

- Establish formal relationships and coordinate with other disciplines that have roles in environmental planning and operations (medical and legal).
- Identify the environmental characteristics of the potentially affected area.
- Identify possible environmental contingencies that may occur during the operation, such as accidental spills.
- Determine how the environmental contingency would affect the environment in the operational area and how it could be prevented or mitigated should it occur.
- Determine the environmental and operational risk associated with the operation. If risks are unacceptable, identify alternatives that mitigate associated risks.
- Establish standardized identification (signage or markings) of off-limits or high-risk areas.
- Negotiate applicable agreements to allow for the unimpeded transit of HAZMAT or waste by military and contracted assets for environmentally sound treatment or disposal.
- Determine contractor status, to include privileges and immunities in support of the operation.

- Identify environmental resources and reach capabilities.

KEY ENVIRONMENTAL FACTORS

K-21. Commanders should consider environmental and force health protection during each phase of an operation. In planning and conducting military operations, regardless of geographic location, commanders should give appropriate consideration to the following:

- Preexisting environmental conditions impacting site selection and environmental health vulnerabilities, as well as potential US liabilities associated with the operation.
- Air emissions.
- HAZMAT, including pesticides.
- Hazardous waste. Appropriate disposition could include recovery, treatment, or disposal within the operational area or, where necessary, transportation to another location for these purposes.
- Oil and hazardous substance spill prevention, control and response training.
- Medical and infectious waste.
- Solid waste.
- Water and wastewater, to include sanitary wastewater.
- Natural resources, to include endangered or threatened species and marine mammals.
- Historic and cultural resources.
- Noise abatement.
- Resource and energy conservation through pollution prevention practices.
- Camp closure and site cleanup before redeployment.
- Incident reporting and documenting of cleanup actions.
- The transportation of excess material and equipment from the tactical area in an environmentally sound manner.
- Ensuring that contractors and their vehicles have unhindered transit across international borders.

RISK MANAGEMENT

K-22. *FM 101-5* describes risk management as the process of detecting, assessing and controlling risk arising from operational factors and balancing risk with mission benefits. Risk management is an integral part of the MDMP. *FM 100.14* outlines the risk management process and provides the framework for making risk management a routine part of planning, preparing, and executing operational missions and everyday tasks. *FM 3-100.4* further clarifies the elements of risk when focused on environmental considerations. Assessing environmental-related risks is part of the total risk management process. Knowledge of environmental factors is key to planning and decision-making. With this knowledge, leaders quantify risks, detect problem areas, reduce risk of injury or death, reduce property damage, and ensure compliance with environmental laws and regulations. Unit leaders should conduct risk assessments before conducting any training, operations, or logistical activities.

Appendix L

Explosive Ordnance Disposal Organizations and Functions

The mission of EOD is to eliminate or reduce chemical, biological, radiological, nuclear, and CBRNE hazards and protect the commander's combat power. UXO limits battlefield mobility, denies the use of critical assets, and threatens to injure or kill soldiers at levels unprecedented in past wars. UXOs, to include improved conventional munitions (ICM) and IED, have greater emphasis now and in the future because of the potential of significantly reducing the commander's combat power. The continuing development of foreign and US smart and brilliant munitions that disperse hundreds of submunitions and area denial ordnance has lead to the proliferation of UXOs on the battlefield. These munitions are available for a range of weapon systems, including artillery, ballistic and cruise missiles, rockets, and bombs.

INTRODUCTION TO EXPLOSIVE ORDNANCE DISPOSAL

L-1. The EOD force is prepared to deal with the increased quantity, quality, and lethality of UXOs. Sophisticated fuzing and sensors systems developed for the 21st century have the capability to detect, identify, and select specific targets using infrared, proximity, magnetic influence, acoustic, and seismic technologies. Attempts to approach and perform a render-safe procedure (RSP) on these munitions by the traditional 20th century methods will cause detonation of the devices. This is because of their wide kill radius 750 feet (greater than 750 feet [250 meters]) added to the hazards of their antidisturbance, antiremoval, antilift, random-delay, and self-destruct features. UXOs also present access problems from—

- Toxic chemicals from rocket motors and guidance systems.
- Ordnance exposed to fire, especially newer ordnance made of lightweight metals and plastics.
- The location. Destroyed vehicles and aircraft have unique dangers like depleted uranium (DU) and carbon fiber.
- Confined space, bunkers, caves and tunnels.

L-2. EOD missions include—

- Responding to counter terrorism, WMD, and UXO incidents on the sea, in the air, and on land.
- Supporting the United States Secret Service (USSS), DOS, Department of Justice (DOJ), Department of Energy (DOE), and FEMA.
- Advising and assisting civil authorities in the remediation of military ordnance that poses a threat to public safety.

- Providing education on the hazards of UXO.
- Examining, identifying, and reporting new and unusual explosive ordnance for technical intelligence purposes.
- Supporting nuclear and chemical weapons shipments.
- Conducting a range sweep by disposing of UXOs on impact areas.
- Destroying ammunition and routine ammunition stocks and conducting emergency destruction of ammunition to prevent capture by the enemy (including sensitive site exploitation).
- Responding to increased lethality of UXOs, IEDs, and WMD.
- Removing rounds stuck in artillery tubes and other large-caliber weapons.
- Advising and assisting in the instruction of UXOs for humanitarian demining missions.
- Clearing enemy UXOs and booby traps from a captured ASP, airfield, air defense battery, command post, or other key objectives.
- Clearing US dropped UXOs from enemy targets—ASP, airfield, air defense battery, and CP, or other key objectives that have been captured by US forces.
- Responding to an enemy sapper attack clearing satchel charges from targets—aircraft, vehicles, stacks of ammunition, and dead sappers or suicide bombers.
- Responding to terrorist attacks against US facilities with car or truck bombs—American embassies, consulates, and military barracks.
- Retrieving casualties from a minefield and evacuating survivors to medical facilities while assisting graves registration with fatalities.
- Clearing enemy aircraft, armor, artillery, and other materiel identified as having significant intelligence value of booby traps and hazardous ordnance.
- Collecting war souvenirs from departing US forces.

EXPLOSIVE ORDNANCE ORGANIZATION

GROUP HEADQUARTERS

L-3. The EOD group commander serves as the ASCC, G3, and theater EOD special staff officer. The EOD group provides C2 of all Army EOD assets and operations in theater. When directed by the JFC, the EOD group commander becomes the commander of the joint EOD TF and coordinates all EOD assets within theater.

L-4. The EOD group provides C2, mission tasking, theater EOD planning, technical intelligence acquisition and management, and limited administrative and logistics support for two to six EOD battalions. In operations without a fully deployed theater or EOD group a C2 element of an EOD group, will deploy to provide C2 and staff planning for deployed EOD battalions.

BATTALION HEADQUARTERS

L-5. The EOD battalion exercises C2 for three to seven EOD companies in the AO. The commander of the EOD battalion is the EOD officer for the corps. He monitors operations and develops plans to meet the needs of the combatant commander, providing an LO team, as required.

L-6. The EOD battalion provides C2, mission tasking, EOD planning, technical intelligence acquisition and management, and limited administrative and logistics support for up to seven EOD companies.

COMPANY

L-7. The EOD company exercises C2 for two EOD response sections. The primary function of the EOD company is to provide support as directed by the EOD battalion; it does this in a variety of ways. An EOD company provides GS support to assigned AORs and all units within it. Dependent upon METT-TC considerations, this may require the company to perform split-based operations to fully support mission requirements. An EOD company is tasked-organized by the EOD battalion commander and is typically attached to a separate organization for administrative and logistical support. The EOD company commander may further task-organize EOD teams to division or BCT's areas to conduct EOD operations in support of maneuver elements. The commander of an EOD company exercises C2 of the company throughout his assigned AOR, to include split-based operations and fragmented team operations. He must also coordinate and conduct liaison functions with various supported and supporting units, to include civil and HN authorities and other agencies.

L-8. The EOD company provides the ability to eliminate or reduce the hazards of domestic and/or foreign conventional, NBC munitions, and IEDs that threaten personnel, military operations, facilities, and materiel. The EOD company exploits technical intelligence by submitting reports on first-seen ordnance. It provides support to the USSS to protect the President, Vice President, and others, as directed. It also provides support to the FBI and the DOE about CT with emphasis on IEDs.

ENGINEER AND EXPLOSIVE ORDNANCE CONSIDERATIONS

L-9. Army engineers and ordnance (EOD) units have a unique relationship compared to other armies and services around the world. Several allied countries (the United Kingdom, France, Canada, and Australia) teach EOD skills in addition to combat engineer training; we do not. Army engineers and ordnance are two distinct branches and organizations. The role of each function is unique but similar, and this necessitates close coordination when operating in today's operational environment. Engineers are responsible for assuring the mobility of combat forces. The Army's combat engineers approach mine detection and neutralization using three strategies—metal detection, explosive neutralization, and mechanical neutralization. Detection technologies focus on identifying the metal content of mines. Explosive neutralization applies systems such as MICLICs to detonate the mines. The mechanical neutralization applies plows and rollers to push the mines aside or detonate them by pressure. All of these means require the use of intelligence sources and reconnaissance to detect mines and use the appropriate combination of explosive or mechanized means to neutralize and breach conventional minefields. Explosive and mechanized neutralization strategies may not be appropriate in all operations, such as in urban areas. EOD personnel can task-organize directly to a maneuver unit to render safe or neutralize booby traps, UXOs, and IEDs. (Booby traps have increased in sophistication in recent years and may include electronic circuitry—light sensors, motion sensors, and command detonation.) *See FM 3-100.38 (100-38)* and *FM 13-8* for additional multiservice considerations and UXO procedures.

L-10. The engineer commander and/or staff engineer should consider the following when working with EOD units:

- Ensure that you know the EOD LO and work together in the planning process.
- Exchange information on the mines, booby traps, UXOs, and IEDs.
- Work with the S2 and G2 to identify the types of booby traps expected.
- Include the EOD in the planning process and request EOD intelligence of the region.
- Establish and operate a training scenario to teach mine awareness and the most current procedures.
- Use combat engineers to detect, mark, record, and report booby traps.
- Know that EOD is task-organized to evaluate and render safe booby traps, IEDs, and UXOs.
- Know that combat engineers can remove standard structural demolitions that are not booby-trapped.

Glossary

The glossary lists acronyms/abbreviations and terms with Army or joint definitions, and other selected terms. Where Army and joint definitions are different, (Army) follows the term. Terms or acronyms for which *FM 3-34* is the proponent manual (the authority) are marked with an asterisk (*).

1SG	first sergeant
A2C2	Army airspace command and control
A&O	assault and obstacle
AA	assembly area
AA	TOE numbering system variation for a firefighting headquarters
AAP	Allied administrative publication
AAV	assault amphibious vehicle
AB	TOE numbering system variation for a firefighting team
ABCA	American, British, Canadian, and Australia
ABCS	Army Battle Command System
ABE	assistant brigade engineer
ABO	air base operability
ABN	airborne
AC	Active Component
ACB	amphibious construction battalion
ACE	assistant corps engineer
ACofS	assistant chief of staff
ACR	armored cavalry regiment
AD	air defense
ADCOM	administrative command
ACS	assistant chief of staff
ADA	air defense artillery
ADC	area damage control
ADCON	administrative control
ADE	assistant division engineer
ADP	automatic data processing
AECA	Arms Export Control Act
AEF	aerospace expeditionary force
AFB	assault float bridge

AFCAP	Air Force Contract Augmentation Program
AFCESA	Air Force Civil Engineering Support Agency
AFDC	Air Force Doctrine Center
AFDD	Air Force doctrine document
AFFOR	Air Force forces
AFSOC	Air Force Special Operations Command; Air Force special operations component
AGS	aviation ground support
AIGIS	Army imagery and geospatial information systems
AISI	automated integrated survey instrument
AJP	allied joint publication
AL	Alabama
ALSA	Air Land Sea Application (Center)
alt	alternate
ALSS	advanced logistics support site
AMC	Army Materiel Command
AMOPES	Army Mobilization and Operations Planning and Execution System
AO	area of operations
AOI	area of interest
AOR	area of responsibility
AP	antipersonnel
APC	armored personnel carrier
APLA	antipersonnel land mine alternatives
APOD	aerial port of debarkation
APOE	aerial port of embarkation
AR	Army regulation
ARC	American Red Cross
ARCENT	Army Forces Central Command
ARFOR	Army forces
ARNG	Army National Guard
ARSOF	Army Special Operations Forces
ART	Army tactical task
ASAS	All-Source Analysis System
ASAS-RWS	All Source Analysis System—Remote Workstation
ASCC	Army service component commander (FM 101-5-1)
ASG	area support group
ASL	authorized stockage list

ASP ammunition supply (or storage) point

ASR alternate supply route

assault breach A breach tactic used by small unit (company, teams, and platoons) to penetrate an enemy's protective obstacles and seize a foothold within his defense; it is normally a very decentralized operation with SOSRA actions synchronized at the platoon and company level. (FM 3-34.2)

***assured mobility** Actions that give the force commander the ability to maneuver where and when he desires without interruption or delay to achieve the mission. (FM 3-34)

***assured mobility imperatives** There are four imperatives: (1) Develop a mobility COP:—leverage information dominance by using geospatial tools to combine terrain data, and an integrated R&S to gain/attain a level of understanding from friendly and enemy perspectives; (2) select, establish, and maintain operating areas—gain ownership of friendly and identification of enemy; EAs and TAIs, operating areas, and the mobility corridors that are the LOCS connecting those areas; (3) attack the enemy's ability to influence operating areas—proactively allocate combat power and sensors to secure our ability to maneuver) operating areas, LOCs, TAIs and EAs) and or attack the enemy's operating areas and or maneuver to prevent his ability to influence our maneuver; (4) maintain mobility and momentum—synchronize all the BOSs capabilities to protect and sustain our established ability to maneuver when and where we wish, enabling us to maintain pressure and lethality upon the enemy for legacy, interim and future organizations. (FM 3-34)

***assured mobility (fundamentals)** Predict, detect, prevent, avoid, neutralize and protect—these fundamentals support the assured mobility imperatives. (FM 3-34)

ASL authorized stock list

AT antitank

AUTL Army universal task list—a comprehensive listing of Army tactical-level task, missions, and operations; the AUTL complements CJCSM 3500.04B by providing tactical-level Army-specific tasks. (FM 7-15)

AVLB armored vehicle-launched bridge

AWE Army warfighting experiment

bare base (Joint) A base having minimum essential facilities to house, sustain, and support operations to include, if required, a stabilized runway, taxiways, and aircraft parking areas. A bare base must have a source of water that can be made potable. Other requirements to operate under bare base conditions form a necessary part of the force package deployed to the bare base. (JP 3-05.3/JP 1-02)

barrier A coordinated series of obstacles designed or employed to channel, direct, restrict, delay, or stop the movement of an opposing force and to impose additional losses in personnel, time, and equipment on the opposing force; barriers can exist naturally, be man-made, or a combination of both. (JP 3-15/JP 1-02)

barrier, obstacle and mine warfare plan (joint)—A comprehensive, coordinated plan that includes responsibilities; general location of unspecified and specific barriers, obstacles, and minefields; special instructions; limitations; coordination; and completion times; the plan may designate locations of obstacle zones or belts. It is normally prepared as an annex to a campaign plan, OPLAN, or OPORD. (JP 3-15/JP 1-02)

base development (less force beddown) (joint)—Acquisition, development, expansion, improvement, and construction, and/or replacement of the facilities and resources of an area or location to support forces employed in military operations or deployed according to the strategic plans. (JP 4-04/JP 1-02)

baseline environmental survey (joint)—Multidisciplinary site survey conducted before or in the initial stage of a joint operational deployment; the survey documents existing deployment-area environmental conditions determines the potential for present and past site contamination (hazardous substances, petroleum products, and derivatives) and identifies potential vulnerabilities (to include occupational and environmental health risks). Surveys accomplished in conjunction with joint operational deployments that do not involve training or exercises (contingency operations) should be completed to the extent practicable consistent with operational requirements. (JP 4-04/JP 1-02)

BASOPS base operations

battlespace (Joint)—The environment, factors, and conditions that must be understood to successfully apply combat power, protect the force, or complete the mission. Included is the air, land, sea, space, enemy and friendly forces, facilities, weather, terrain, electromagnetic spectrum, and the information environment within the operation areas and AOIs. (JP 1-02)

BB bare base—(joint)—A base having minimum essential facilities to house, sustain, and support operations to include, if required, a stabilized runway, taxiways, and aircraft parking areas. A BB must have a source of water that can be made potable. Other requirements to operate under BB conditions form a necessary part of the force package deployed to the BB. (JP 3-05.3/JP 1-02)

BCAT base camp assistance/assessment team

BCCA Base Camp Coordination Agency

BCOC Base Cluster Operations Center (FM 3-90)

BCT brigade combat team

BDA battle damage assessment

BDOC Base Defense Operations Center

***BDT** base development team—Nondeployable team that can quickly provide base development engineering and planning and facilities design for staging bases, base camps, FOBs, displaced persons camps, and any similar requirement. (FM 3-34)

BDP base development plan

BDR battle damage repair

BEEF base engineer emergency force

BEFV Bradley engineer fighting vehicle

BFT battle-focused training

BFV Bradley fighting vehicle

BOM bill of materials

BOS battlefield operating system—The physical means (soldiers, organizations, and equipment) that tactical commanders use to execute operations and accomplish missions assigned by superior tactical- and operational-level commanders. The seven BOS are (1) intelligence system—the activity to generate knowledge of and products portraying the enemy and the environmental features required by a command planning, preparing, executing, and assessing operations; (2) maneuver system—the movement of forces to achieve a position of advantage with respect to enemy forces. This system includes employing forces in combination with direct fire or fire potential. It also includes the conduct of tactical tasks associated with force protection; (3) fire support system—collective and coordinated use of target-acquisition; data, indirect-fire weapons, fixed-wing aircraft, offensive information operations, and other lethal and nonlethal means against targets located throughout an AO; (4) air defense system—employment of all active measures to nullify or reduce the effectiveness of attack by hostile aircraft and missiles after they are airborne; (5) M/CM/S system—mobility operations preserve the freedom of maneuver of friendly forces, countermobility operations deny mobility to enemy forces, and survivability operations protect friendly forces from the effects of enemy weapon systems; (6) CSS system—the support and services provided to sustain forces during war and MOOTW; (7) C2 system—all collective tasks associated with supporting the exercise of authority and direction by a properly designated commander over assigned and available forces in the accomplishment of the mission. (FM 7-15)

breach A tactical mission task in which the unit employs all available means to break through or secure a passage through an enemy defense, obstacle, minefield, or fortification. (FM 3-90)

breach area Area that is established and fully defined by the higher HQ of the unit conducting breaching operations; it is the area where a breaching operation occurs. (FM 3-34.2)

breach force A combined-arms force that is task-organized with the maneuver and engineer forces necessary to reduce obstacles and create lanes through an obstacle to pass initial assault forces through the lanes. (FM 3-34.2)

breaching fundamentals Suppress, obscure, secure, reduce and assault (SOSRA) are the breaching fundamentals that must be applied when breaching against a defending enemy. These fundamentals will always apply, but they may vary based on the specific battlespace situation. (METT-TC) (FM 3-34.2)

breaching operations Operations conducted to allow maneuver despite the presence of obstacles; breaching is a synchronized combined-arms operation under the control of the maneuver commander. Breaching operations begin when friendly forces detect an obstacle and begin to apply the breaching fundamentals, and they end when battle handover has occurred between follow-on forces and a unit conducting the breaching operation. (FM 3-34.2)

breach tenets Intelligence, breach organization, breach fundamentals, mass, and synchronization. (FM 3-34.2)

BTRA battlefield terrain reasoning and awareness

C2 command and control

C2PC command and control personal computer

C3 command, control, and communications

C4ISR command, control, communications, computers, intelligence, surveillance, and reconnaissance

C&S command and support

CA civil affairs

CAA Clean Air Act

CALL Center for Army Lessons Learned

camouflage (joint)—Use of natural or artificial material on personnel, objects, or tactical positions with the aim of confusing, misleading, or evading the enemy. (JP 1-02/FM 20-3)

CAO chief administrative officer

CAP crisis action plan(ning)

CBMU construction battalion maintenance unit

CBRNE chemical, biological, radiological, nuclear, and high-yield explosive

CBRNE-CM chemical, biological, radiological, nuclear and high-yield explosives consequence management

CBT combating terrorism

cbt combat

CBU construction battalion unit

CCA contract construction agent

CCD camouflage, concealment, and deception

CCIR commander's critical information requirements

CDMP combat decision making process

cdr commander

CEB combat engineer battalion

***CELL** Center for Engineer Lessons Learned

CENTCOM Central Command

CERL Construction Engineering Research Laboratory

CERCLA Comprehensive Environmental Response, Compensation, and Liability Act

CESE civilian engineer support equipment

CESP civil engineering support plan—(joint)—An appendix to the logistics annex or separate annex of an OPLAN that identifies the minimum essential engineering services and construction requirements required to support the commitment of military forces. (JP 4-04/JP 1-02)

CESPG civil engineering support plan group; civil engineering support planning generator

CG commanding general

cgo cargo

civil augmentation program (joint)—Sanding, long-term contracts designed to augment service logistic capabilities with contract support in both preplanned and short notice contingencies; examples include US Army LOGCAP, US Air Force contract augmentation program, and US Navy Construction Capabilities Contract. (JP 4-07)

civil engineering (joint)—Those CS and CSS activities that identify, design, construct, lease, or provide facilities and which operate, maintain, and perform war damage repair and other engineering functions in support of military operations. (JP 4-04/JP 1-02)

CJCS Chairman of the Joint Chief of Staff

CJCSI Chairman of the Joint Chiefs of Staff, Instruction

CJCSM Chairman of the Joint Chiefs of Staff manual

CJTF commander, joint task force

***clear** (1) The total elimination or neutralization of an obstacle that is usually performed by follow-on engineers and is not done under fire (FM 3-34/FM 101-5-1); (2) A tactical mission task that requires the commander to remove all enemy forces and eliminate organized resistance in an assigned area. (FM 3-90/FM 101-5-1)

clearing operation (joint)—Operation designed to clear or neutralize all mines and obstacles from a route or area. (FM 3-34.2/FM 101-5-1/JP 3-15/JP 1-02)

CM consequence management—(joint)—Those measures taken to protect public health and safety, restore essential government services, and provide emergency relief to governments, businesses, and individuals affected by the consequences of an NBC and/or high-yield explosive situation; for domestic consequence management, the primary authority rest with the States to respond and the Federal Government to provide assistance as required. (JP 3-0/JP 1-02)

cmd command

CMO civil-military operations

CMOC civil-military operations center

co company

COA course of action

COB collocated operating base

COCOM combatant command

COE contemporary operational environment—(joint)—Automation services that support the development of the common reusable software modules which enable interoperability across multiple CS application; includes segmentation of common software modules from existing applications, integration of commercial products, development of a common architecture, and development of common tools for application developers. (JP 1-02)

***combat engineering** (joint)—Those engineering task that assist the tactical and/or operational commander to "shape" the battlespace by enhancing mobility, creating the space or time necessary to generate mass and speed while protecting the force, and denying mobility and key terrain to the enemy; these tasks include breaching, bridging and emplacement of obstacles to deny mobility to the enemy. (JP 3-34/JP 1-02); one of the engineer battlespace functions that includes mobility, countermobility and survivability. (FM 3-34)

combat support engineering (NATO)—An Allied term that encompasses those military engineer tasks associated with direct support to current or imminent operations. It is conducted by the military engineers of any service or component to support land, air, maritime, and information operations. With the emphasis on speed of execution, it frequently relies on prefabricated equipment solutions, may involve a degree of improvisation and is likely to fulfill a short-term tactical requirement. It is closely aligned with the Engineer Battlespace function of Combat Engineering. (AJP 3.12)

comdt commandant

commo communications

COMMZ communications zone

COMPT comptroller

COMSEC communications-security

CON construction

CONCAP construction capabilities contract (Navy)

CONPLAN concept or contingency plan

contingency engineering management organization (joint)—An organization that may be formed by the combatant commander, or subordinate joint force commander, to augment the engineering expertise to support both deliberate and crisis action planning and to provide construction management in contingency and wartime operations; the combatant commander may form a theater contingency engineering management cell, and similar organizations may be formed at subordinated levels of command (RCEM cell and/or JTFCEM cell). These organizations should be staffed with expertise in combat engineering, general engineering and topographic (geospatial) engineering. (JP 3-34)

const construction

CONUS continental United States

COP common operational picture—(joint)—An operational picture tailored to the user's requirements, based on common data and information shared by more that one command. (JP 3-0/JP 1-02)

COR contracting officer's representative

COS chief of staff

COSCOM corps support command

countermine (NATO, joint)—To explode the main charge in a mine by the shock of a nearby explosion of another mine or independent explosive charge; the explosion of the main charge may be caused either by sympathetic detonation or through the explosive train and/or firing mechanism of the mine. (JP 1-02); (Army)—the actions taken to detect, bypass, breach, mark, report, record, and eliminate enemy mines or minefields. (FM 20-32)

countermine operations (joint)—In land warfare, an operation to reduce or eliminate the effects of mines or minefields. (JP 1-02) See CJCSI 3207.01.

countermining (joint)—(1) Land mine warfare—tactics and techniques used to detect, avoid, breach, and/or neutralize enemy mines and the use of available resources to deny the enemy the opportunity to employ mines; (2) naval mine warfare—the detonation of mines by nearby explosions, either accidental or deliberate. (JP 1-02)

***countermobility** A component of combat engineering and one of the five engineer battlespace functions. It augments natural terrain with obstacle systems according to the commander's concept. This adds depth to the battle in space and time by attacking the enemy's ability to maneuver his forces. (FM 3-34/FM 101-5-1)

***countermobility operations** (joint)—The construction of obstacles and emplacement of minefields to delay, disrupt, and destroy the enemy by reinforcement of the terrain; the primary purpose of countermobility operations is to slow or divert the enemy, to increase time for target acquisition, and to increase weapons effectiveness. (FM 3-34/FM 101-5-1/JP 3-34/JP 1-02)

covert breach A breach tactic used when the force must reduce lanes through enemy tactical or protective obstacles undetected; in the covert breach, suppression from the support force is a be-prepared task upon detection of the breach force, or an on-order once the breach is completed and the assault is initiated. (FM 3-34.2/FM 101-5-1)

CP command post

CPW Center for Public Works

CRREL Cold Regions Research and Engineering Laboratory

***CREST** contingency real estate support team (FM 3-34)

CS civil support—DOD support to US civil authorities for domestic emergencies and for designated law enforcement and other activities (FM 3-07/JP 3-07.7); combat support.

CSA Chief of Staff, United States Army

CSE combat support equipment

CSS combat service support

CSR controlled supply rate

CSS-BOS The support and service provided to sustain forces during war and MOOTW; contains the subordinate Army Task (ART) of Provide General Engineering Support. (FM 7-15)

CSSE combat service support element—(joint)—Those elements whose primary missions are to provide service support to combat forces and which are a part, or prepared to become a part, of a theater, command, or task force formed for combat operations; the core element of a MAGTF that is task-organized to provide the CSS necessary to accomplish the MAGTF mission. The CSSE varies in size from a small detachment to one or more force service support groups. It provides supply, maintenance, transportation, general engineering, health services, and a variety of other services to the MAGTF. The CSSE itself is not a formal command. (FM 101-5-1/JP 1-02)

CT counterterrorism

CTC Combat Training Center

CTIS Combat Terrain Information System

CUL common-user logistics

cultural resource Monuments, nationally identifiable or distinctive buildings and structures, archives and libraries, ancient artifacts and structures, archaeologically important sites, historically import sites or structures, mosques, cathedrals, temples, other churches or sacred structures, sacred sites or areas, museums, and works of art. (FM 101-5-1)

CWA Clean Water Act

CWD civil works directorate

CZ combat zone

DA Department of the Army

datum (NATO, Joint)—Any numerical or geometrical quantity or set of such quantities which may serve as reference or base for other quantities. Where the concept is geometric, the plural form is "datums" in contrast to the normal plural "data". (JP 1-02)

datum (geodetic) (joint)—(1) A reference surface consisting of five quantities: the latitude and longitude of an initial point, the azimuth of a line from that point, and the parameters of the reference ellipsoid; (2) the mathematical model of the earth used to calculate the coordinates on any map; different nations use different datums for printing coordinates on their maps. The datum is usually referenced in the marginal information of each map. (JP 1-02)

DCD directorate of combat developments

DCMA Defense Contract Management Agency

DCSENG Deputy Chief of Staff for Engineering

DCSINT deputy chief of staff for intelligence

DEA Drug Enforcement Administration

defensive minefield (joint)—In land mine warfare, a minefield laid in accordance with an established plan to prevent a penetration between positions and to strengthen the defense of the positions themselves. (JP 1-02)

delegation of authority (joint)—Action by which a commander assigns part of his authority commensurate with the assigned task to a subordinate commander; while ultimate responsibility cannot be relinquished, delegation of authority carries with it the imposition of a measure of responsibility; the extent of the authority delegated must be clearly stated. (JP 1-02)

deliberate breaching (NATO, joint)—Creation of a lane through a minefield or a clear route through a barrier or fortification, which is systematically planned and carried out. (FM 3-34.2/FM 101-5-1/JP 1-02) (At the tactical level, simply referred to as a breach.) (FM 3-34.2)

deliberate crossing (NATO, joint)—Crossing of an inland water obstacle that requires extensive planning and detailed preparations. (FM 90-13/FM 101-5-1/JP 1-02)

***demining** Activities to remove the hazard of all mines and other unexploded explosive munitions for a defined area. (FM 3-34)

demolition belt (joint)—A selected land area sown with explosive charges, mines, and other available obstacles to deny use of the land to enemy operations, and as a protection to friendly troops; there are two types of demolition belts: (1) primary continuous series of obstacles across the whole front, selected by the division or higher commander. The preparation of such a belt is normally a priority engineer task. (2) subsidiary—a supplement to the primary belt to give depth in front or behind, or to protect the flanks. (JP 1-02)

DHHS Department of Health and Human Services

directed obstacle An obstacle directed by a higher commander as a specified task to a subordinate unit. (FM 3-34.2/FM 101-5-1)

***directed target** A target directed by the responsible commander to be prepared for demolition or destroyed to support his intent. (FM 3-34/FM 101-5-1)

DISCOM Division Support Command

div division

DIVENG division engineer

DLA Defense Logistics Agency

DLEA Drug Law Enforcement Agency

DMA Defense Mapping Agency

DMC distribution management center

doctrine Fundamental principles by which the military forces or elements thereof guide their actions in support of national objectives; authoritative but requires judgment in application. (FM 3-0/FM 101-5-1/JP 1-02)

DOD Department of Defense

DOD construction agent (joint)—USAES, NAVFACENGCOM, or other such approved DOD activity that is assigned design or execution responsibilities associated with military construction programs, facilities support, or civil engineering support to the combatant commanders in contingency operations. (JP 3-34/JP 1-02)

DODD Department of Defense directive

DODI Department of Defense instruction

DOE Department of Energy

DOJ Department of Justice

DOMS Directorate of Military Support

DON Department of the Navy

DOS Department of State

DOT Department of Transportation

DOTMLPF doctrine, organization, training, materiel, leadership and education, personnel, and facilities

DPG defense planning guidance

DPPDB	digital point-positioning database
DPW	Directorate of public works
DS	direct support
DSES	division staff engineer section
dsgn	design
DSO	domestic support operations
DST	decision support template
DTED	Digital Terrain Elevation Data
DTG	date-time group
DTSS	digital topographic support system
DTSS-B	digital topographic support system-base
DTSS-D	digital topographic support system-deployable
DTSS-H	digital topographic support system-heavy
DTSS-L	digital topographic support system-light
DU	depleted uranium
EA	engagement area
EAB	echelons above brigade
EAC	echelons above corps
EAD	echelons above division
EBA	engineer battlefield assessment
EBS	environmental baseline survey
ECO	environmental compliance officer
ECOA	enemy course of action
ECR	Environmental conditions report concise summary of environmental conditions at a base camp site, based on the environmental baseline survey, supported by maps and backup documents, prepared by base camp commanders for each base camp; the ECR documents conditions at the site it claims or other legal challenges arise against the government. (FM 3-100.4)
EEFI	essential elements of friendly information
EEM	early entry module
EFA	engineering field activity
EFD	engineering field division
ELCAS	elevated causeways system
EMS	Environmental Management System

***EMST** essential mobility/survivability task—Specified or implied BOS-specific tasks that are critical to mission success; although ultimately executed by a combined-arms element, the staff (typically MANSPT elements such as engineer, chemical, MP, EOD) identifies the EMSTs. A fully developed EMST has four components including effect (the qualitative or qualitative impact desired by the higher commander on the friendly or adversary caused by the completion of the task); purpose (the desired or intended result of the task stated in terms relating to the purpose of the supported unit); tasks (one or more clearly defined and measurable tasks accomplished by individuals or organizations required to achieve the desired effects); and method (how the task and purpose will be achieved, described by a quantifiable use of assets or capabilities and communicates their priority). (FM 3-34)

ENCOM engineer command

***ENCOORD** engineer coordinator—Special staff officer who coordinates engineer assets and operations for the command, usually the senior engineer officer in the force. (FM 6-0 and FM 3-34)

enabling mine/ countermeasures (joint)—Countermeasures designed to counter mines once they have been laid. This includes both passive and active mine countermeasures. (JP 3-15/JP 1-02)

***engineer battlespace functions** The three battlespace functions include combat (M/CM/S grouped under combat engineering), geospatial and general engineering. (FM 3-34)

***Engineer Regiment** All AC and RC engineer organizations (as well as the DOD civilians and affiliated contractors and agencies within the civilian community) with a diverse range of capabilities that are all focused toward supporting the Army and its warfighting mission. (FM 3-34)

engr engineer

ENSIT enemy situation

environmental area of interest An assessment or study done on an area of interest (a property) to define the environmental state or condition of that property prior to use by US forces; survey is used to determine the environmental impact of property use by US forces and the level of environmental restoration needed before returning the property upon US departure. (FM 3-100.4)

environmental base line survey An assessment or study done on an area of interest (a property) in order to define the environmental state or condition of that property prior to use by US forces. Used to determine the environmental impact of property use by US forces, and the level of environmental restoration needed prior to returning the property upon US departure. (FM 3-100.4)

environmental compliance The unconditional obeying of international, foreign nation, federal, state and local environmental rules, regulation and guidelines that affect current operations. (FM 3-100.4)

environmental considerations (joint)—The spectrum of environmental media, resources or programs that may impact on, or are affected by, the planning and execution of military operations; factors may include, but are not limited to, environmental compliance, pollution prevention, conservation, protection of historical and cultural sites and protection of flora and fauna. (JP 3-34/JP 1-02)

environmental hazard All activities that may pollute, create negative noise related effects, degrade archaeological/cultural resources, or negatively affect threatened or endangered species habitats; also included are environmental health-related hazards. (FM 3-100.4)

environmental protection The application of human ingenuity and resources, through the disciplines of science and engineering, as required by environmental protection laws, regulations, and policies to protect the natural environment. (FM 3-100.4)

environmental protection level The varying level of environmental protection that can reasonably be afforded at any particular time during warfare of battlefield conditions, given the absolute requirement that such a diversion of resources away from the mission at hand does not adversely affect that mission, or any friendly personnel, or indigenous or refugee populations. (FM 3-100.4)

environmental reconnaissance The systematic observation and recording of site or area data collected by visual or physical means, dealing specifically with environmental conditions as they exist, and identifying areas that are environmentally sensitive or of relative environmental concern, for information and decision-making purposes. (FM 3-100.4/FM 101-5-1)

environmental services (joint)—Various combinations of scientific, technical and advisory activities (including modification processes, that is, the influence of man-made and natural factors) required to acquire, produce, and supply information on the past, present, and future states of space, atmospheric, oceanographic, and terrestrial surrounding for use in military planning and decisionmaking processes, or to modify those surroundings to enhance military operations. (JP 1-02)

environmental stewardship (joint)—The integration and application of environmental values into the military mission to sustain readiness, improve quality of life, strengthen civil relations, and preserve valuable natural resources. (JP 1-02)

ENVST environmental support team

EO executive order

EOD Explosive ordnance disposal(Joint)detection, identification, on-site evaluation, rendering safe, recovery, and final disposal of UXO; may also include explosive ordnance which has become hazardous by damage or deterioration. (FM 3-100.38/FM 101-5-1/JP 1-02)

EPA Environmental Protection Agency

EPW enemy prisoner of war

equip equipment

ERC exercise related construction

ERDC Engineer Research and Development Center

***ERP** Engineer regulating point checkpoint to ensure that vehicles do not exceed the capacity of the crossing means and to give drivers final instructions on site-specific procedures and information, such as speed and vehicle interval. (FM 90-13)

ERT engineer reconnaissance team

ESB engineer support battalion

ESF emergency support function

ESP engineer supply points

ESV engineer squad vehicle

EUCOM European Command

EUSA Eighth United States Army

EW electronic warfare

***EWL** Engineer work line coordinated boundary or phase line used to compartmentalize an operational area to indicate where specific engineer units have primary responsibility for the engineer effort; it may be used at division level to discriminate between a sector supported by division engineer assets and a sector supported by DS or GS corps engineer units. (FM 3-34)

executive agent (joint)—Term used to indicate a delegation of authority by the SECDEF to a subordinate to act on the Secretary's behalf; an agreement between equals does not create an executive agent. For example, a service cannot become a DOD executive agent for a particular matter with simply the agreement of the other services; such authority must be delegated by the SECDEF. Designation as executive agent, in and of itself, confers no authority. The exact nature and scope of the authority delegated must be stated in the document designating the executive agent. An executive agent may be limited to providing only administration and support or coordinating common functions, or it may be delegated authority, direction, and control over specified resources for specified purposes. (JP 0-2/JP 1-02)

explosive ordnance (joint)—All munitions containing explosives, nuclear fission or fusion materials, and biological and chemical agents; this includes bombs and warheads; guided and ballistic missiles; artillery, mortar, rocket, and small arms ammunition; all mine, torpedoes, and depth charges; demotion charges; pyrotechnics; clusters and dispensers; cartridge and propellant actuated devices; electro-explosive devices; clandestine and improvised explosive devices; and all similar or related items or components explosive in nature. (JP 1-02)

explosive ordnance disposal unit (joint)—Personnel with special training and equipment who render explosive ordnance safe (such as bombs, mines, projectiles, and booby traps), make intelligence reports on such ordnance, and supervise the safe removal thereof. (JP 1-02)

FAA Foreign Assistance Act

fac facility

FACE forward aviation combat engineering—(joint)—A mobility operation in which engineers perform tasks in support of forward aviation ground facilities; tasks include reconnaissance; construction of low altitude parachute extraction zones, landing strips, and airstrips; and providing berms, revetments, and trenches for forward arming and refueling points. (JP 3-34/JP 1-02)

FAO foreign area officer

FARP forward area refueling point

FASCAM family of scatterable mines—A grouping of munitions that dispenses SCATMINEs by artillery, helicopter, fixed wing, or ground launchers; there are AP and AT mines. (FM 20-37)

fascines Large cylindrical bundles of material, usually wooden poles or plastic or metal pipe loosely bound together, which are dropped into ditches or gaps to create crossings. (FM 3-34.2/FM 101-5-1)

FBCB2 Force XXI Battle Command Brigade and Below

FBI Federal Bureau of Investigation

FE facilities engineer(ing)

FEC facility engineer center

FED facilities engineer detachment

FEG facility engineer group

FEMA Federal Emergency Management Agency

***FEST** Forward engineer support teamdeployable USACE organization that executes the USACE mission in the AO; it is usually subordinate to the senior engineer commander in the AO. (FM 3-34)

***FEST-A** forward engineer support team-advance (FM 3-34)

***FEST-M** forward engineer support team-main (FM 3-34)

FET facility engineer team

FF fire fighting

***FFE** field force engineering—The application of all of the Engineer Regiment's capabilities (to include TeleEngineering) across the range of engineer battlespace functions (although primarily general engineering intensive) and in all phases and types of operations (offense, defense, stability, support) through both reach and forward presence.) (FM 3-34); flame field expedients(Joint)simple, hand-made devices used to produce flame or illumination. (JP 3-15/JP 1-02)

FFIR friendly forces information requirements

FGS final governing standards

FHA foreign humanitarian assistance

FID foreign internal defense

field fortifications (joint)—An emplacement of shelter of a temporary nature which can be constructed with reasonable facility by units requiring no more than minor engineer supervisory and equipment participation. (FM 5-102/FM 101-5-1/JP 1-02)

FLOT forward line of own troops

flt flat

FM field manual; frequency modulation

FMTV family of tactical vehicles

FOB forward operating base

FOL forward operating location

force beddown The provision of expedient facilities for troop support to provide a platform for the projection of force; these facilities may include modular or kit-type facility substitutes. (JP 4-04/JP 1-02)

force support engineering (NATO)—An Allied term that describes the deliberate, longer-term preparation for, and indirect support to ongoing or future operations as well as those military engineering tasks associated with sustaining the joint force throughout all stages of an operation. It will be the predominant engineer focus for precombat and post combat and for rear operations during conflict. It is an area which may involve a greater degree of cross-compartment support and the engineer tasks will usually be more enduring, relying more on purpose designed and built solutions. It is likely to fulfill a longer term, operational requirement. It is closely aligned with the engineer Battlespace function of General Engineering. (AJP 3.12)

FORSCOM United States Army Forces Command

ford A shallow par of a body of water or wet gap that can be crossed without bridging, boats, ferries, or rafts. It is a location in a water barrier where the physical characteristics of current, bottom, and approaches permit the passage of personnel, vehicles, and other equipment where the wheels or track remain in contact with the bottom at all times. (FM 90-13/FM 101-5-1)

FP force protection(joint)—Security program designed to protect service members, civilian employees, family members, facilities, and equipment, in all locations and situations, accomplished through planned and integrated application of combating terrorism, physical security, operations security, personal protective service, and support by intelligence, counterintelligence, and other security programs. (Army)those actions taken to prevent or mitigate hostile actions against DOD personnel (to include family members), resources, facilities, and critical information; these actions conserve the forces's fighting potential so it can be applied at a decisive time and place and incorporates the coordinated and synchronized offensive and defensive measures to enable the effective employment of the joint force while degrading opportunities for the enemy. Force protection does not include actions to defeat the enemy or protect against accidents, weather, or disease. (FM 3-0/FM 101-5-1/JP 3-0/JP 1-02)

FPOL forward passage of lines

fps foot(feet) per second

FRAGO(RD) fragmentary order

FRP Federal Response Plan

FS fire support

FSB forward support base

FSCOORD fire support coordinator

FSSG force service support group (USMC)

***FTEE** forward theater engineer element (FM 3-34)

functions (joint)—The appropriate or assigned duties, responsibilities, missions or tasks of an individual, office or organization; as defined in National Security Act of 1947, as amended, the term function includes functions, powers and duties. (JP 1-02)

fwd forward

FXXI Force XXI

G1 Assistant Chief of Staff, Personnel, General Staff (Division or higher staff)

G2 Assistant Chief of Staff, G2 (intelligence)

G3 Assistant Chief of Staff, G3 (operations and plans)

G4 Assistant Chief of Staff, G4 (logistics)

G5 Assistant Chief of Staff, G5 (Civil Affairs)

G6 Assistant Chief of Staff, G6 (command, control, communications and computer systems)

GAO General Accounting Office

gal gallon

***GE** general engineering—(joint)—Encompasses the construction and repair of LOCs, MSRs, airfields, and logistic facilities to support joint military operations and may be performed in DS of combat operations, such as battle damage repair; these operations include both horizontal and vertical construction, and may include use of both expedient repair methods and more deliberate construction methods characterized by the application of design criteria, advanced planning, and preparation, depending on the mission requirements. (JP 3-34/JP 1-02); one of the engineer battlespace functions. (FM 3-34)

***geospatial engineering** Provides commanders with terrain visualization, operational and tactical terrain analysis, digitized terrain products, nonstandard map products, and baseline survey data; one of the engineer battlespace functions. (FM 3-34)

geospatial information Foundation information upon which all other battlespace information is referenced to form the COP. (FM 5-105/FM 101-5-1)

(GI&S) geospatial information and services—(joint)—The concept for collection; information extraction; storage; and dissemination and exploitation of geodetic, geomagnetic, imagery (both commercial and national source), gravimetric, aeronautical, topographic, hydrographic, littoral, cultural, and toponymic data accurately referenced to a precise location on the earth's surface; these data are used for military planning, training, and operations, including navigation, mission planning, mission rehearsal, modeling, simulation, and precise targeting. Geospatial information provides the basic framework for battlespace visualization. It is information produced by multiple sources to common interoperable data standards. It may be presented in the form of printed maps, charts, and publications; in digital simulation and modeling databases; in photographic form; or in the form of digitized maps and charts or attributed centerline data. Geospatial services include tools that enable users to access and manipulate data, and also includes instruction, training, laboratory support, and guidance for the use of geospatial data. (FM 5-105/FM 101-5-1/JP 2-03/JP 1-02)

GPH gallons per hour

GPR general support reinforcing

GPS global positioning system

GS general support

GSA General Services Administration

Harvest Eagle A US-owned facility substitute of flexible tempered tents with insulation and environmental control units; the structures can be used to house more than 500 people and include a dining facility, showers and latrines.

Harvest Falcon A US-owned facility substitute used to accommodate USAF personnel and required operational facilities when deployed in off-shore environments; this system includes living accommodations, aircraft flight line facilities, air-contingency hospital operations, specialized clinics, and administrative support functions.

hasty breaching (joint)—Rapid creation of a route through a minefield, barrier, or fortification by any expedient method. (JP 1-02)

hasty breaching (land mine warfare) (joint)—Creation of lanes through enemy minefields by expedient methods such as blasting with demolitions, pushing rollers or disabled vehicles through the minefields when the time factor does not permit detailed reconnaissance, deliberate breaching, or bypassing the obstacle. (JP 1-02)

hasty crossing (NATO, joint)—Crossing of an inland water obstacle using the crossing means at hand, or those readily available, and made without pausing for elaborate preparations. (FM 3-34.13/ FM 101-5-1/JP 1-02)

hazard (DOD)—Condition with the potential to cause injury, illness, or death of personnel; or damage to or loss of equipment or property; or mission degradation. (FM 3-100.4/FM 100-14)

HAZMAT hazardous material

HCA humanitarian and civic assistance

HD humanitarian demining—(joint)—DOD and DOS program to promote the foreign policy interests of the US by assisting other nations in protecting their populations from landmines and clearing land of the threat posed by landmines remaining after conflict has ended; the HD program includes training of HN deminers, establishment of national demining organizations, provision of demining equipment, mine awareness training, and research development. (JP 3-07/JP 1-02)

HDO humanitarian demining operation

HET heavy equipment transporter

HHC headquarters and headquarters company

HHD headquarters and headquarters detachment

HLC helicopter landing zones

HLD homeland defense—The protection of US territory, sovereignty, domestic population, and critical infrastructure against external threats and aggression. (JP 3-26)

HLS homeland security—The preparation for, prevention of, deterrence of, preemption of, defense against, and response to threats and aggression directed toward US infrastructure; as well as crisis management, consequence management, and other domestic civil support. Homeland Defense and Civil Support are sub elements of HLS. (JP 3-26)

HMMWV high mobility, multipurpose wheeled vehicle

HN host nation

HNS host nation support—(joint)—Civil and/or military assistance rendered by a nation to foreign forces within its territory during peacetime, crisis or emergencies or war, based upon agreements mutually concluded between nations. (FM 3-07/JP 4-0/JP 1-02)

holding line In retrograde river crossing operations, the outer limit of the area established between the enemy and water obstacle to preclude direct and observed indirect fires in the crossing areas. (FM 90-13/FM 101-5-1)

horizontal action mine (joint)—In land mine warfare, a mine designed to produce a destructive effect in a plane approximately parallel to the ground. (JP 1-02)

HPTL high-payoff target list

HQ headquarters

H&S headquarters & service

HSC headquarters support company

humanitarian demining (HD) (joint)—Department of Defense and Department of State program to promote the foreign policy interests of the United States by assisting other nations in protecting their populations from landmines and clearing land of the threat posed by landmines remaining after conflict has ended. The humanitarian demining program includes training of host nation deminers, establishment of national demining organizations, provision of demining equipment, mine awareness training, and research development. Also called HD. (JP 3-07/JP 1-02)

HUMINT human intelligence; human resources intelligence

HVAC heating, ventilation, and air conditioning

HVT high-value target

hvy heavy

***IAT** infrastructure assessment team—nondeployable team that provides engineering infrastructure assessments for military deployments and civil military operations in forward areas; focus areas for the IAT are infrastructure-related to USACE missions and aspects of the AO impacting contract construction, to include roads, utilities, water resources, and HN support. (FM 3-34)

ICM improved conventional munitions

IDAD internal defense and development

IEB interentity boundary

IED improvised explosive device—(joint)—A device placed or fabricated in an improvised manner incorporating destructive, lethal, noxious, pyrotechnic, or incendiary chemicals and designed to destroy , incapacitate, harass, or distract; it may incorporate military stores, but is normally devised from nonmilitary components. (JP 1-02)

IG inspector general

info information

information operations Use of offensive and defensive information means to degrade, destroy, and exploit an adversary's information-based process while protecting one's own. The actions taken affect adversary and influence others' decision making processes, information, and information systems while protecting one's own information and information systems. (FM 3-0)

infrastructure (joint)—All building and permanent installations necessary for the support, redeployment, and military forces operations (for example., barracks, HQ, airfields, communications, facilities, stores, port installations, and maintenance stations). (JP 4-01.8/ JP 1-02)

intel intelligence

IPDS inland petroleum distribution system

intelligence (joint)—Product that results from collecting, processing, integrating, analyzing, evaluating, and interpreting available information concerning foreign countries or areas. (JP 1-02)

intelligence BOS Activity that generates knowledge of and products which portray the enemy and the environmental features required by a command planning, preparing, executing, and assessing operations. (FM 7-15)

intelligence requirement (joint)—(1) Any subject, general or specific, for which there is a need to collect information or produce intelligence; (2) a requirement for intelligence to fill a gap in the command's knowledge or understanding of the battlespace or threat forces. (JP 1-02)

interchangeability (joint)—Condition that exists when two or more items possess such functional and physical characteristics as to be equivalent in performance and durability and are capable of being exchanged one for the other without alteration of the items themselves or of adjoining items, except for adjustment, and without selection for fit and performance. (JP 1-02)

interoperability (joint)—The ability of systems, units, or forces to provide services to and accept services from other systems, units, or forces and to use the services so exchanged to enable them to operate effectively together. (JP 1-02)

IO information operations; international organization

IPB intelligence preparation of the battlefield/battlespace—An analytical methodology employed as part of intelligence planning to reduce uncertainties concerning the enemy, environment, and terrain or all types of operation; IPB is conducted during mission planning to support the commander's decision making and to form the basis for the direction of intelligence operation in support of current and future missions. It utilizes existing databases and identifies gaps in intelligence needed to determine the impact of the enemy, environment, and terrain on operations and presents this in an appropriate form to facilitate operational planning. It forms the basis for situation development. (FM 2-01.3); (Joint) an analytical methodology employed to reduce uncertainties concerning the enemy, environment, and terrain for all types of operations; IPB builds an extensive database for each potential area in which a unit may be required to operate. The database is then analyzed in detail to determine the impact of the enemy, environment, and terrain on operations and presents it in graphic form. IPB is a continuing process. (JP 2-02/JP 1-02/FM 2-0)

IPDS inland petroleum distribution system

IR information requirement

IRR individual ready reserve

IS intelligence and surveillance

ISB intermediate staging base

ISR Intelligence, surveillance, and reconnaissance integrating and synchronizing of all BOS to collect and process information about the enemy and the environment to produce relevant information to facilitate the commander's decision making. (FM101-5-1)

ISR assets Those organizations, systems, sensors, and equipment dedicated to or directed toward the collection of information and the analysis and production of intelligence from the collected information in response to the commander's PIRs. (FM 101-5-1)

ISR plan Integrated plan for collecting information from all available sources and analysis of that information to produce intelligence to meet requirements; specifically, a logical plan for transforming PIRs into orders or requests to reconnaissance and surveillance assets to collect pertinent information within a required time. (FM 101-5-1)

ISYSCON integrated systems control

J2 Intelligence Directorate

J3 Operations Directorate

J4 Logistics Directorate

J7 development directorate, joint staff

JAG Judge Advocate General

JCMEB Joint Civil-Military Engineering Board

JCS Joint Chiefs of Staff

JDWP joint doctrine working party—(joint)—A forum to include representatives of the services, combatant command, and joint staff (represented by the operational plans and joint force development directorate, J7) which meets semiannually to address and vote on project proposals; discuss key joint doctrinal or operational issues; keep up to date on the status of the joint publication projects and emerging publications; and keep abreast of other initiatives of interest to the members. The JDWP meets under the sponsorship of the J7. (JP 1-01/JP 1-02)

JEMB Joint Environmental Management Board

JEPES Joint Engineer Planning and Execution System

JFACC joint force air component commander—(joint)—Derives authority from the JFC, who has the authority to exercise OPCON, assign missions, direct coordination among subordinate commanders, redirect and organize forces to ensure unity of effort in the accomplishment of the overall mission; normally designated by the JFC; responsibilities are assigned by the joint force commander (normally these include, but are not limited to, planning, coordination, allocation, and tasking, based on the JFC's apportionment decision); using the JFC's guidance and authority, and in coordination with other service component commanders and other assigned or supporting commanders, recommends to the JFC apportionment of air sorties to various missions or geographic areas. (JP 3-0/JP 1-02)

JFC joint force commander

JFCOM Joint Forces Command

JFLCC joint force land component commander—(joint)—The commander within a unified command, subordinate unified command, or joint task force who is responsible to the establishing commander for making recommendations on the proper employment of land forces, planning and coordinating land operations, or accomplishing such operational missions as may be assigned; JFLCC is given the authority necessary to accomplish missions and tasks assigned by the establishing commander. JFLCC is normally the commander with the preponderance of land forces and the requisite C2 capabilities. (JP 3-0/JP 1-02)

JFLSC Joint Force Logistic Support Command

JFUB Joint Facilities Utilization Board—(joint)—Joint board that evaluates and reconciles component requests for real estate use of existing facilities, interservice support, and construction to ensure compliance with JCMEB priorities. (JP 4-04/JP 1-02)

JIC joint implementation committee

JIM joint, interagency and multinational

JIMP Joint Vision Implementation Master Plan

JLOTS joint logistics over-the-shore

JIP joint implementation plan

JLP joint logistics plan

JLSP joint logistics support plan

JMC joint movement center

JOA joint operations area

joint doctrine (joint)—Fundamental principles that guide the employment of forces of two or more military departments in coordinated action toward a common objective, it is authoritative; as such. Joint doctrine will be followed except when, in the judgment of the commander, exceptional circumstances dictate otherwise. It will be promulgated by or for the Chairman of the Joint Chiefs of Staff, in coordination with the combatant commands and services. (JP 1-01/1-02)

JOPES Joint Operation Planning and Execution System

JP joint publication

JRSOI joint reception, staging, onward movement, and integration

JTF joint task force

JTFCEM joint task force contingency engineering management

K-span automatic building machine

kW kilowatt

lane A route through an enemy or friendly obstacle that provides a passing force safe passage; the route may be reduced and proofed as part of a breach operation or constructed as part of a friendly obstacle. It is a clear route that moves all the way through an obstacle. (FM 3-34.2)

LCC land component commander

LE light equipment

LFT lead federal agency

LMTV light, medium-tracked vehicle

LNO liaison officer

LO logistics order

LOC line of communications

LOCTDA line of communication tactical decision aid

LOGCAP Logistics Contract Augmentation Program

logistics preparation of the battlefield All actions taken by combat service support to maximize the means of supporting the commander's plans. (FM 4-0)

LOGPAC logistics package

LOS line of sight

LOTS logistics over the shore operations

LSA life support area

LSE logistics support element

lt light

LZ landing zone

MACA military assistance to civil authorities

MACOM major Army command

MAGTF Marine air-ground task force

maint maintenance

maintenance engineering (joint)—Application of technique, engineering skills, and effort, organized to ensure that the design and development of weapon systems and equipment provide adequately for their effective and economical maintenance. (JP 1-02)

MAN maneuver

***MANSPT** maneuver support—Includes all of the functions identified in the M/CM/S BOS. (FM 3-34)

MARFOR Marine Corps forces

MBA main battle area

MC multicomponent

MCC movement control center

MCM mine countermeasures—(joint)—All methods for preventing or reducing damage or danger from mines. (JP 3-15/JP 1-02)

MCMOPS mine countermeasures operations

M/CM/S mobility, countermobility, and survivability (both for the collective components of combat engineering and the M/CM/S BOS).

M/CM/S BOS mobility operations preserve the freedom of maneuver for friendly forces; These missions include breaching obstacles, increasing battlefield circulation, improving or building roads, providing bridge and raft support, and identifying routes around contaminated areas. Countermobility denies mobility to enemy forces. Survivability operations protect friendly forces from the effects of enemy weapon systems and from natural occurrences. (FM 7-15)

MCO major contingency operation

MCOO modified combined-obstacle overlay—(joint)—A joint IPB product used to portray the effects of each battlespace dimension on military operations; it normally depicts militarily significant aspects of the battlespace environment, such as obstacles restricting military movement, key geography, and military objectives. (JP 2-01.3/JP 1-02)

MCS maneuver control system; modular causeway system

MCS-ENG maneuver control system-engineer

MCWP Marine Corps warfare publication

MDMP military decision-making process

MEB Marine expeditionary brigade

MEDEVAC medical evacuation

MEDCOM medical command

MEF Marine expeditionary force

MEP mobile electric power; Military environmental protection application and integration of all aspects of natural environmental considerations, as they apply to the conduct of military operations. (FM 3-34.500)

METL mission-essential task list—Compilation of collective mission-essential tasks an organization must perform successfully to accomplish its wartime mission(s). (FM 7-0)

METT-TC mission, enemy, terrain and weather, troops and support available, time available, and civil considerations (FM 101-5-1)

MEU Marine expeditionary unit

MFO multinational force and observers

MFPF minefield planning folder

MGB medium girder bridge

mgt management

MHE materials handling equipment

MICC mine information coordination cell

MICLIC mine clearing line charge

MILCON military construction—(joint)—Any construction, alteration, development, conversion, or extension of any kind carried out with respect to a military installation. (JP 4-04/JP 1-02)

mine (NATO, joint)—(1) In land mine warfare, an explosive or material, normally encased, designed to destroy or damage ground vehicles, boats, or aircraft or designed to wound, kill, or otherwise incapacitate personnel; it may be detonated by the action of its victim, by the passage of time, or by controlled means; (2) In naval mine warfare, an explosive device laid in the water with the intention of damaging or sinking ships or of deterring shipping from entering an area; the term does not include devices attached to the bottoms of ships or to harbor installations by personnel operating underwater, nor does it include devices which explode immediately on expiration of a predetermined tie after laying. (FM 20-32/FM 101-5-1/JP 3-15/JP 1-02)

minefield (NATO, joint)—In land warfare, an area of ground containing mines emplaced with or without a pattern. (FM 20-32/JP 3-15/JP 1-02)

minefield breaching (NATO, joint)—In land mine warfare, the process of clearing a lane through a minefield under tactical conditions. (FM 20-32/JP 1-02)

minehunting (joint)—Employment of sensor and neutralization systems, whether air, surface, or subsurface, to locate and dispose of individual mines; minehunting is conducted to eliminate mines in a known field when sweeping is not feasible of desirable, or to verify the presence or absence of mines in a given area. (JP 3-15/JP 1-02)

MINEOPS joint minelaying operations

minesweeping (joint)—Technique of clearing mines using either mechanical, explosive, or influence seep equipment; mechanical sweeping removes, disturbs, or otherwise neutralizes the mine. Explosive sweeping causes sympathetic detonation in, damages, or displaces the mine; influence sweeping produces either the acoustic and/or magnetic influence required to detonate the mine. (JP 3-15/JP 1 02)

MISCAP mission capability (Air Force)

MISP multispectral imagery processor

MIW mine warfare—(joint)—Strategic, operational, and tactical use of mines and mine countermeasures; mine warfare is divided into two basic subdivisions: the laying of the mines to degrade the enemy's capabilities to wage land, air, and maritime warfare; and the countering of enemy-laid mines to permit friendly maneuver or use selected land or sea areas. (FM 20-32/JP 3-15/JP 1-02)

MKT mobile kitchen trailer

MLC military load classification—(NATO, joint)—Standard system in which a route, bridge, or raft is assigned class number(s) representing the load it can carry; vehicles are also assigned number(s) indicating the minimum class of route, bridge, or raft they are authorized to use. (FM 101-5-1)

MMC Materiel Management Center

MOA memorandum of agreement

MOB military operating base; mobility

mobility (NATO, joint)—Quality or capability of military forces which permits them to move from place to place while retaining the ability to fulfill their primary mission; (Army)—those activities that enable a force to move personnel and equipment on the battlefield without delays due to terrain or obstacles. (JP 1-02)

mobility analysis (joint)—In-depth examination of all aspects of transportation planning in support of OPLAN and OPORD development. (JP 1-02)

***mobility operations** Obstacle reduction by maneuvering and engineer units to reduce or negate the effects of existing or reinforcing obstacles; the objectives are to maintain freedom of movement for maneuver units, weapon systems and critical supplies. (FM 3-34)

MOMAT mobility matting

MOOTW military operations other than war

MOPMS modular pack mine system

MOS military occupational specialty

MOUT military operations on urbanized terrain (being replaced by UO)

MP military police

MPF maritime pre-positioning force

MPS maritime pre-positioning ship

MPSC mobilization planning support cell

MPSPRON maritime pre-positioning force squadrons

MRB multirole bridge

MRE mission rehearsal exercise

MSDS mission specific data sets—(joint)—Further densification of global geospatial foundation data; information created to support specific operations, OPLANs, training or system development. Information conforms to established DOD data specifications. (JP 2-03/JP 1-02)

MSR main supply route

MS Mississippi

MSB main support battalion

MSC major subordinate command

MRBC multi-role bridge company

MTC movement to contact

MTOE modified table of organization and equipment

MTT mobile training team

MTW major theater war

MULTICOMPO multicomponent

MWRS morale, welfare, recreation and services

MWSS Marine wing support squadron

NAI named area of interest

NATO North Atlantic Treaty Organization

natural environment The human ecosystem, including the physical and biological systems that provide resources (clean air, clean water, healthy surroundings, sufficient food) necessary to sustain productive human life; included in the natural environment are man-made structures, such as water and wastewater treatment facilities and natural/cultural resources. (FM 3-100.4)

NAVAID navigational aid

NAVFAC naval facilities

NAVFACENGCOM Naval Facilities Engineering Command

NAVFOR Navy forces

NAVSTAR navigation satellite timing and ranging

NBC nuclear, biological, and chemical

NBG naval beach group

NCB naval construction brigade

NCF naval construction force

NCFSU naval construction force support unit

NCR naval construction regiment

NCS National Communications System

NEO noncombatant evacuation operation

NEPA National Environmental Policy Act

neutralization (NATO, joint)—In mine warfare, a mine is said to be neutralized when it has been rendered, by external means, incapable of firing on passage of a target, although it may remain dangerous to handle. (FM 20-32/JP 1-02)

neutralize (joint)—(1) As pertains to military operations, to render ineffective or unusable; (2) a tactical mission task that results in rendering enemy personnel or material incapable of interfering with a particular operation (FM 3-90); (3) to render safe mines, bombs, missiles, and booby traps; (4) to render harmless. (FM 20-32/JP 1-02)

NFESC Naval Facilities Engineering Service Center

NGO nongovernmental organization

NIMA National Imagery and Mapping Agency

NMCB naval mobile construction battalion

NOAA National Oceanic and Atmospheric Administration

NORTHCOM Northern Command

NSE Navy support element

NWP naval warfare publication

OAKOC observation, avenues of approach, key terrain, obstacles and cover and concealment

OAS offensive air support; Organization of American States

OASD Office of the Assistant Secretary of Defense

OAU Organization of African Unity

obstacle (joint)—Any obstruction designed or employed to disrupt, fix, turn, or block the movement of an opposing force and to impose additional losses in personnel, time, and equipment on the opposing force; obstacles can be natural, man-made, or a combination of both. (FM 90-7/FM 101-5-1/JP 3-15/JP 1-02)

obstacle belt (joint)—Brigade-level C2 measure, normally given graphically, to show where within an obstacle zone and ground tactical commander plans to limit friendly obstacle employment and focus the defense; it assigns an intent to the obstacle plan and provides the necessary guidance on the overall effect of obstacles within a belt; (Army)—graphic control measure that corps, division, and brigade echelon commanders use to constrain tactical obstacle employment. (FM 3-90/FM 20-32/JP 3-15/JP 1-02)

obstacle clearing (joint)—The total elimination or neutralization of obstacles. (JP 1-02)

obstacle control measure Specific measures that simplify the granting of obstacle-emplacing authority while providing obstacle control. (FM 3-90/FM 101-5-1/FM 20-32)

obstacle effect block The block effect integrates fire planning and obstacle effort to stop an attacker along a specific AA or prevent him from passing through an EA. (FM 90-7/FM 101-5-1)

obstacle groups One or more individual obstacles grouped to provide a specific obstacle effect. (FM 3-90/FM 101-5-1)

obstacle line A conceptual control measure used at battalion or brigade level to show placement intent without specifying a particular type of linear obstacle. (FM 90-7)

obstacle plan That part of an operation plan or order that is concerned with the use of obstacles to enhance friendly fires or to affect the movement of an enemy. Obstacle plans are used at corps level and below. (FM 90-7)

obstacle restricted areas (joint)—C2 measure used to limit the type or number of obstacles with an area. (FM 90-7/JP 3-15/JP 1-02)

obstacle zone (joint)—Division-level C2 measure, normally done graphically, to designate specific land areas where lower echelons are allowed to employ tactical obstacles; (Army)—a graphic control measure that corps and division commanders use to grant obstacle-emplacement authority to brigades (including armored cavalry regiments and other major subordinate units). FM 3-90/FM 90-7/JP 3-15/JP 1-02)

OBSTINTEL obstacle intelligence—The first breach tenet; an IR prior to conducting combined-arms breaching operations. Obstacle intelligence includes a broad range of information to include location, composition, and orientation of exiting obstacles. (FM 3-34.2)

OCONUS outside the continental United States

ODSS offense, defense, stability, and support

OEBGD overseas environmental baseline guidance document—(joint)—Set of objective criteria and management practices developed by DOD to protect human health and the environment. (JP 4-04/JP 1-02)

OE operational environment

OFDA Office of Foreign Disaster Assistance

OICC officer in charge of construction

O&M operation and maintenance

OODA observe, orient, decide and act

OPCOM (NATO)—operational command

OPCON operational control

OPDS Offshore Petroleum Discharge System

OOTW operations other than war

operational area Subareas defined by the commander and designated within the AO that bounds the specific areas the commander will maneuver through and therefore can focus ISR capabilities to achieve the assured mobility imperative to select, establish, and maintain operating areas. (JP 1-02)

operational environment (joint)—A composite of all conditions, circumstances, and influences which affect the employment of military forces and bear on the decisions of the unit commander. (JP 1-02)

operational countermobility To delay, channel, or stop offensive air, land, and sea movement by enemy operational formations to help create positional advantage for friendly joint and multinational operational forces; operational countermobility exposes enemy decisive points and centers of gravity or high-payoff targets to destruction in conformance with the joint force commander's plans and intent. It includes the execution of sanctions, embargoes, blockades, and no-fly zones. (CJCSM 3500.04B/JP 3-15) (OP 1.4 in CJCSM [UJTL])

operational mobility To facilitate the movement of joint and multinational operational formations in a campaign or major operation without delays due to operationally significant terrain or obstacles. (CJCSM 3500.04B) (OP 1.3 in CJCSM [UJTL])

OPLAN operation plan

OPNAVINST operational naval instruction

opns operations

OPORD operation order

OPP offload preparation party

ops operations

Ops Det operations detachment

OPSEC operations security

OSCE Organization for Security and Cooperation in Europe

OSD Office of the Secretary of Defense

OTSG Office of the Surgeon General

OVE on-vehicle equipment

P&A production and analysis

PACOM Pacific Command

PAO public affairs office(r)

PARC principal assistant for contracting

PBO property book officer

PE peace enforcement

PEO peace enforcement operations

pers personnel

PIR priority intelligence requirements

PKO peacekeeping operation

PL phase line; public law

plan planning

PLL prescribed load list

PLS palletized load system

PME peacetime military engagement

PO peace operations

POD port of debarkation

POE port of embarkation

point obstacle Any obstruction designed or employed to deny ease of movement on a road, lane, or bridge that is normally part of or completes a larger man-made or natural obstacle that has a designed effect on the attacking enemy force. (FM 90-7/FM 101-5-1)

POL petroleum, oils, and lubricants

prestock point Location where mines, explosives, and materials for use in preparation of a planned obstacle are stored. (FM 20-32/FM 101-5-1)

proactive mine countermeasures (joint)—Measures intended to prevent the enemy from successfully laying mines. (JP 3-15/JP 1-02)

protective minefield (NATO, joint)—In land mine warfare, a minefield employed to assist a unit in its local, close-in protection. (FM 20-32/FM 101-5-1/JP 1-02)

protective obstacles Obstacles employed to assist a unit in its local, close-in protection. (FM 90-7)

PSS personnel service support

PSYOP psychological operations

pub public

PW public works

QC quality control

QDR Quadrennial Defense Review

QOL quality of life

R reinforcing

RAA Redeployment Assembly Area

RC Reserve Component

RCEM regional-contingency engineering manager

RCRA Resource Conservation and Recovery Act

RDL Reimer Digital Library

***reach** To communicate and access the Engineer Regiment's capabilities via means such as TeleEngineering; reach may occur from forward units to the rear or laterally. (FM 3-34) (The Air Force and Navy have similar capabilities in their services.)

real property (joint)—Lands, buildings, structures, utilities systems, improvements, and appurtenances; includes equipment attached to and made part of buildings and structures (such as heating systems) but not movable equipment (such as plant equipment). (JP 1-02)

rearming (joint)—(1) An operation that replenishes the prescribed stores of ammunition, bombs, and other armament items for an aircraft, naval ship, tank, or armored vehicle (including replacement of defective ordnance equipment) to make it ready for combat service; (2) resetting the fuze on a bomb, or on an artillery, mortar, or rocket projectile so that it will detonate at the desired time. (FM 101-5-1/JP 1-02)

RED HORSE Rapid Engineers Deployable Heavy Operations Repair Squadron, Engineer—(joint)—Air Force units wartime-structured to provide a heavy engineer capability; they have a responsibility across the operational area, are not tied to a specific base, and are not responsible for base operation and maintenance. These units are mobile, rapidly deployable, and largely self-sufficient for limited periods of time. (JP 3-34/JP 1-02)

reduce (1) Tactical mission task that involves the destruction of an encircled or bypassed enemy force (FM 3-90); (2) One of the five breaching fundamentals; the creation of a lane through, over, or around an obstacle. In the case of minefields, refers to destroying, neutralizing, removing, or bypassing mines. (FM 3-34.2/FM 101-5-1)

reduction (joint)—Creation of lanes through a minefield or obstacle to allow passage of the attacking ground force. (JP 3-15/JP 1-02)

reinforcing obstacles (joint)—Those obstacles specifically constructed, emplaced, or detonated through military effort and designed to strengthen existing terrain to disrupt, fix, turn, or block enemy movement. (FM 90-7/JP 3-15/JP 1-02)

reserved demolition target (NATO, joint)—Target for demolition, the destruction of which must be controlled at a specific level of command because it plays a vital part in the tactical or strategic plan, or because of the importance of the structure itself, or because the demolition may be executed in the face of the enemy. (FM 90-7/FM 101-5-1/JP 1-02)

reserved obstacles (joint)—Those demolition obstacles that are deemed critical to the plan for which the authority to detonate is reserved by the designating commander; (Army)—obstacles of any type, for which the commander restricts execution authority. (FM 20-37/JP 3-15/JP 1-02)

RHQ regional headquarters

RIBS readiness base support

river crossing Operation required before ground combat power can be projected and sustained across a water obstacle; it is centrally planned and controlled offensive operation that requires the allocation of external crossing means and a force dedicated to the security of the bridgehead. (FM 90-13/FM 101-5-1)

road block (NATO, joint)—Barrier or obstacle (usually covered by fire) used to block or limit the movement of hostile vehicles along a route. (FM 90-7/FM 101-5-1/JP 1-02)

ROE rules of engagement

ROICC resident officer in charge of construction

RO/RO roll-on/roll-off

***route classification** (NATO, joint)—Classification assigned to a route using factors of minimum width and worst route type, least bridge, raft, or culvert MLC; and obstructions to traffic flow. (FM 3-34/FM 101-5-1)

route reconnaissance Reconnaissance that focuses along a specific LOC, such as a road, railway, or cross-country mobility corridor. (FM 3-90)

RPE rapid port enhancement

RPMA real-property maintenance activities

RPOL rearward passage of lines

RRR rapid runway repair

R&S reconnaissance and surveillance

RSOI reception, staging, onward movement, and integration

RSP render-safe procedure

RSTA reconnaissance, surveillance, and target acquisition

RTE/rte route

RSR required supply rate

rupture Task to create a gap in enemy defensive positions quickly. (FM 3-34.2)

S1 Adjutant (U.S. Army brigade and battalion)

S2 Intelligence Officer (U.S. Army)

S3 Operations and Training Officer (U.S. Army)

S4 Supply Officer (U.S. Army)

S5 Civil Affairs Officer (U.S. Army)

S6 Communications Staff Office (U.S. Army)

SA security assistance

SASO stability and support operation

SBCT Stryker brigade combat team

SCATMINE scatterable mine—(NATO, joint)—In land mine warfare, a mine laid without regard to classical pattern and which is designated to be delivered by aircraft, artillery, missile, ground dispenser, or by hand; once laid, it normally has a limited life. (FM 20-32)

SCBA self-contained breathing apparatus

scuba self-contained underwater breathing apparatus

SD self-destruct

SECDEF Secretary of Defense

SECSTATE Secretary of State

secure (NATO, joint)—In an operational context, to gain possession of a position or terrain feature with or without force, and to make such disposition as will prevent, as far as possible, its destruction of loss by enemy action. [Note: Army classifies this as a tactical mission task]. (FM 3-90); (Army)—one of the five breaching fundamental; those actions which eliminate the enemy's ability to interfere with the reduction and passage of combat power through a lane. Secure may be accomplished by maneuver or by fires. (JP 1-02)

SEE small emplacement excavator

SES staff engineer section

SIC survey information center

SIGINT signal intelligence

situational obstacle An obstacle that a unit plans and possibly prepares before starting an operation, but does not execute unless specific criteria are met; it provides the commander flexibility for emplacing tactical obstacles based on battlefield development. (FM 90-7)

SITEMP situational template

SJA Staff Judge Advocate

SLRP survey, liaison, and reconnaissance party

SMART special medical augmentation response team

SME subject matter expert

SOEO scheme of engineer operations

SOF special operations forces

SOFA Status of Forces Agreement

SOP standing operating procedure

SOSRA suppress, obscure, secure, reduce and assault

SOUTHCOM Southern Command

SPACECOM Space Command

SPCE survey planning and coordination element

SPOD seaport of debarkation

SPOE seaport of embarkation

spt support

SSC smaller-scale contingency

stab stability

STAMIDS Standoff Mine Detection System

STANAG standardization agreement

STRATCOM Strategic Command

strong point (NATO, joint)—Key point in a defensive position, usually strongly fortified and heavily armed with automatic weapons, around which other positions are grouped for its protection (JP 1-02); (Army)—a heavily fortified battle position tied to a natural or reinforcing obstacle to create an anchor for the defense or to deny the enemy decisive or key terrain. (FM 3-90)

SU situational understanding

SUPCOM support command

support force That force in a breaching operation whose mission is to eliminate enemy interference with the breach through suppressive, direct, and indirect fires; the support force is one of the three breach organizations used in breaching operations. (FM 3-34.2)

suppress (1) A tactical mission task that results in temporary degradation of the performance of a force or weapons system below the level needed to accomplish the mission (FM 3-90) (2) One of the five breaching fundamentals; the focus of all fires on enemy personnel, weapons, or equipment to prevent effective fires on friendly forces. The purpose of suppression is to protect forces reducing and maneuvering through the obstacle and to soften the initial foothold. (FM 3-34.2)

***survivability** (joint)—Concept which includes all aspects of protecting personnel, weapons, and supplies while simultaneously deceiving the enemy; survivability tactics include building a good defense; employing frequent movement; using concealment, deception, and camouflage; and constructing fighting and protective positions for both individuals and equipment. (JP 3-34/ JP 1-02) [Note: Army adds, "Encompasses planning and locating position sites, designing adequate overhead cover, analyzing terrain conditions and construction materials, selecting excavation methods, and countering the effects of direct and indirect fire weapons."] (FM 3-34)

survivability operations Development and construction of protective positions, such as earth berms, dug-in positions, overhead protection, and countersurveillance means, to reduce the effectiveness of enemy weapon systems. (FM 5-103/FM 101-5-1)

sustaining operations Operations at any echelon that enable shaping and decisive operations by providing CSS, rear area and base security, movement control, terrain management, and infrastructure development. (FM 3-0)

TA theater Army

TAA tactical assembly areas

TAACOM Theater Army Area Command

TAC Transatlantic Programs Center

TACCP tactical command post

TACON tactical control

tactical minefield (NATO, Joint)—Minefield that is employed to directly attack enemy maneuver as part of a formation obstacle plan and is laid to delay, channel, or break up an enemy advance, giving the defending element a positional advantage over the attacker. (FM 20-32/JP 1-02)

tactical mobility the ability to move rapidly from one part of the battlefield to another, relative to the enemy. (FM 3-90)

tactical obstacles (joint)—Those obstacles employed to disrupt enemy formations, to turn them into a desired area, to fix them in position under direct and indirect fires, and to block enemy penetrations. (FM 90-7/JP 3-15/JP 1-02)

TAI targeted area of interest

TC training circular

TCEM theater contingency engineering management

TCF tactical combat force

TCMS theater construction management system

TCN third country national

TDA table of distribution and allowance

TEC Topographic Engineering Center

***TeleEngineering** Reflects an engineering telepresence to the force under the proponency of USAES; this capability is focused on assisting engineers and the commanders they support in planning and executing their operational and tactical missions. It is a piece of the capabilities inherent in FFE. The overarching concept is the exploitation of the Army's C3 architectures to provide a linkage between engineers and the appropriate nondeployed SMEs for resolution of engineer challenges. (FM 3-34)

tempo the rate of military action (FM 3-0/FM 101-5-1)

TEOC TeleEngineering Emergency Operations Center

terrain analysis (NATO, joint)—Collection, analysis, evaluation, and interpretation of geographic information on the natural and man-made features of the terrain, combined with other relevant factors, to predict the effect of the terrain on military operations. (FM 5-105/JP 1-02)

***terrain expert** Person who demonstrates skills and knowledge in rendering geospatial engineering to the tactical and operational situation to leverage the battlespace; experts understand the limits/ capabilities of GI&S and can integrate them into MDMP. (FM 3-34)

terrain intelligence (joint)—Intelligence on the military significance of natural and man-made characteristics of an area. (JP 1-02)

***terrain reinforcement** The development of terrain using obstacles to degrade enemy mobility or to enhance friendly survivability through the construction of fighting positions and cover. (FM 3-34)

terrain study (joint)—An analysis and interpretation of natural and man-made features of an area, their effects on military operations, and the effect of weather and climate on these features. (JP 1-02)

TF task force

theater support contractors (joint)—Contract personnel hire in, and operating in, a specific operational are (JP 4-07/JP 1-02)

TIH toxic industrial hazard—(NATO)—The hazards resulting from the release, by any means, of toxic industrial materials, resulting in contamination or irradiation of personnel or the environment area or any particular object. (FM 3-34.500)

TLP troop leading procedure

TO theater of operation

TOA table of allowances

TOC tactical operations center

TOE table of organization and equipment

topo topographic

***topographic engineering** (joint)—Those engineering tasks that provide geospatial information and services to commanders and staffs across the range of military operations; these tasks include terrain analyses, terrain visualization, digitized terrain products, nonstandard map products, and baseline survey data. See also GI&S. (JP 3-34/JP 1-02) (Replaced in Army doctrine by geospatial engineering [FM 3-34])

topographic map (joint)—A map that presents the vertical position of features in measurable form, as well as their horizontal position. (FM 3-25.26/ FM 101-5-1/JP 1-02)

topography (joint)—The configuration of the earth's physical surface, to include its relief and natural and man-made features; topography addresses both dry land and the sea floor (underwater topography). It also refers to mapping or describing these features. (JP 4-01/JP 1-02)

TP transportation

TPFDD time-phased force and deployment data

tph ton(s) per hour

TRADOC United States Army Training and Doctrine Command

trl trailer

TRANSCOM Transportation Command

trk truck

TSC theater support command

TTP tactics, techniques and procedures

TUAV tactical unmanned aerial vehicle

UAV unmanned aerial vehicle

UCT underwater-construction team

UIC unit identification code

UJTL universal joint task list—(joint)—A menu of capabilities (mission-derived tasks with associated conditions and standards, such as the tools) that may be selected by a joint force commander to accomplish the assigned mission; once identified as essential to mission accomplishment, the tasks are reflected with the command joint METL. (JP 1-02)

UMT unit maintenance team

UN United Nations

UO urban operations—Offense, defense, stability, and support operations conducted in a topographical complex and adjacent natural terrain where man-made construction and high population density are the dominant features. (FM 3-0)

US United States

USACE United States Army Corps of Engineers

USACOM United States Atlantic Command

USAEHSC United States Army Engineering and Housing Support Center

USAES United States Army Engineer School

USAFEG United States Army Facility Engineer Group

USAID United States Agency for International Development

USAMC United States Army Materiel Command

USAMEDCOM United States Army Medical Command

USAR United States Army Reserve

USARC United States Army Reserve Center or Command

USAREUR United States Army, Europe

USARPAC United States Army, Pacific

USARSO United States Army Forces Southern Command—The MACOM with Central and South America and the Caribbean as its AOR; it is the Army component of USSOUTHCOM and is one of five components of the command, which when combined, work together as a team to accomplish USSOUTHCOM's theater engagement missions.

USC United States Code

USDA United States Department of Agriculture

USFK United States Forces Korea

USG United States Government

USGS United States Geological Survey

USJFCOM United States Joint Forces Command

USPACOM United States Pacific Command

USSOUTHCOM United States Southern Command

USSS United States Secret Service

USTRANSCOM United States Transportation Command

UTC unit type code

util utility

UXO unexploded ordnance—(NATO, joint)—Explosive ordnance which has been primed, fused, armed, or otherwise prepared for action, and which has been fired, dropped, launched, projected, or placed in such a manner as to constitute a hazard to operations, installations, personnel, or materiel, and remains unexploded either by malfunction or design or for any other cause. (FM 4-30.11/FM 101-5-1/JP 3-15/JP 1-02)

VA Virginia

VDD visualize, describe, direct

WAN wide-area network

WO warning order

whl wheel

WIN Warfighter Information Network

WES Waterways Experiment Station

WMD weapons of mass destruction

WO warning order

wtr water

XO executive officer

ZOC zone of control

ZOP zone of penetration

ZOS zone of seperation

Bibliography

AAP-4. *NATO Standardization Agreements and Allied Publications.* December 1991.

AAP-6 (2002). *NATO Glossary of Terms and Definitions (English and French).* December 2002.

AAP-15 (F). *NATO Glossary of Abbreviations Used in NATO Documents and Publications.* December 2002.

AFDD 2-4. *Combat Support.* 22 November 1999.

AFDD 2-4.4. *Bases, Infrastructure, and Facilities.* 13 November 1999.

AFDD42. *Air Force Doctrine Document 42 Civil Engineering.* 1 December 1994.

AJP 3.12. *Joint Engineering.* This is a new manual that is in draft and fills a doctrinal gap.

American-British-Canadian-Australian Handbook. Coalition Operations Handbook. 01 November 2001.

AR 71-9. *Materiel Requirements.* 30 April 1997.AR 200-1. *Environmental Protection and Enhancement.* 21 February 1997.

AR 310-50. *Authorized Abbreviations, Brevity Codes, and Acronyms.* 15 November 1985.

AR 500-60. *Disaster Relief.* 1 August 1981.

AR 600-8-105. *Military Orders.* 28 October 1994.

AR 700-4. *Logistic Assistance.* 20 September 2002.

AR 715-9. *Contractors Accompanying the Force.* 29 October 1999.

Army Homeland Security (HLS) Strategic Planning Guidance. This soon to be published document is in draft and fills a doctrinal gap.

CJCSI 3010.02A. *Joint Vision Implementation Master Plan (JIMP).* 15 April 2001.

CJCSI 3121.01A. *Standing Rules of Engagement for US Forces.* 1 September 1999.

CJCSI 3122.03A. *Joint Operation Planning and Execution System Volume II; Planning Formats and Guidance.* 27 November 2000.

CJCSI 3207.01. *Military Support to Humanitarian Demining Operations.* 1 March 1999.

CJCSM 3500.04C. *Universal Joint Task List.* 1 July 2002.

DA Memo 10-1. *Executive Agent Responsibilities Assigned to the Secretary of the Army.* 15 January 1997.

DOD Civil Disturbance Plan, Annex C (Concept of Operations), Appendix 8 (Special Instructions), 15 Feb. 1991; modified by Director of Military Support message 161639Z Jul. 96, Subject: Changes to DOD Civil Disturbance Plan--"Garden Plot."

DODD 3025.1. *Military Support to Civil Authorities (MSCA).* 15 January 1993.

DODD 3025.15. *Military Assistance to Civil Authorities.* 18 February 1997.

DODD 4270.34. *Host Nation-Funded Construction Programs.* 19 October 1982.

DODD 4270.36. *DOD Emergency, Contingency, and other Unprogrammed Construction Projects.* 17 May 1997.

DODD 4270.5. *Military Construction Responsibilities.* 2 March 1982.

DODD 5100.1. *Functions of the Department of Defense and Its Major Components.* 1 August 2002.

DODD 5100.46. *Foreign Disaster Relief.* 4 September 1975.

DODD 6050.7. *Environmental Effects Abroad of Major DOD Actions.* 31 March 1979.

DODI 4715.5. *Management of Environmental Compliance at Overseas Installations.* 22 April 1996.

EO 12114. *Environmental Effects Abroad of Major Federal Actions.* 4 January 1979.

EO 13148. *Greening the Government through Leadership in Environmental Management.* 21 April 2000.

FM 1 (100-1). *The Army.* 14 June 2001.

FM 3-0 (100-5). *Operations.* 14 June 2001.

FM 3-06 (90-10). *Urban Operations.* 1 June 2003.

FM 3-06.11 (90-10-1). *Combined Arms Operations in Urban Terrain.* 28 February 2002.

FM 3-07 (100-20). *Stability Operations and Support Operations.* 20 February 2003.

FM 3-07.2. *Force Protection.* This is a new manual that is in draft and fills a doctrinal gap.

FM 3-21.31 (7-32). *The Stryker Brigade Combat Team.* 13 March 2003.

FM 3-34.2 (90-13-1). *Combined-Arms Breaching Operations.* 31 August 2000. With change 3 dated 11 October 2002.

FM 3-34.221. *Engineer Operations: Stryker Brigade Combat Team.* This is a new manual that is in draft and fills a doctrinal gap.

FM 3-34.230 (5-105). *Topographic Operations.* 3 August 2000.

FM 3-34.331 (5-232). *Topographic Surveying.* 16 January 2001.

FM 3-34.343 (5-446). *Military Nonstandard Fixed Bridging.* 12 February 2002.

FM 3-34.471 (5-420). *Plumbing, Pipefitting, and Sewerage.* 31 August 2001.

FM 3-50. *Smoke Operations*. 04 December 1990. With change 1 dated 11 September 1996. Will be revised as FM 3-11.50.

FM 3-55. *Intelligence, Surveillance, and Reconnaissance Operations*. This is a new manual that is in draft and fills a doctrinal gap.

FM 3-90. *Tactics*. 4 July 2001.

FM 3-100.4. *Environmental Considerations in Military Operations*. 15 June 2000. With change 1 dated 11 May 2001. This is in the process of revision to FM 3-34.500.

FM 3-100.21 (100-21). *Contractors on the Battlefield*. 3 January 2003.

FM 3-100.38 (100-38). *UXO Multiservice Procedures for Unexploded Ordnance Operations*. 23 August 2001.

FM 3-100.81. *Multinational Operations Handbook*. This is a new manual that is in draft and fills a doctrinal gap.

FM 4-0 (100-10). *Combat Service Support*. 29 August 2003

FM 4-01.30 (55-10). *Movement Control*. 1 September 2003.

FM 4-93.4 (63-4). *Theater Support Command*. 15 April 2003.

FM 5-7-30. *Brigade Engineer Combat Operations (Airborne, Air Assault, Light)*. 28 December 1994. Will be revised as FM 3-34.223.

FM 5-10. *The Engineer Platoon*. 3 October 1995. Will be revised as FM 3-34.220.

FM 5-33. *Terrain Analysis*. 11 July 1990. With change 1 dated 8 September 1992. Will be revised as FM 3-34.330.

FM 5-71-2. *Armored Task-Force Engineer Combat Operations*. 28 June 1996. With change 2 dated 4 September 1997. Will be revised as FM 3-34.222.

FM 5-71-3. *Brigade Engineer Combat Operations (Armored)*. 03 October 1995. With change 1 dated 13 November 1997. Will be revised as FM 3-34.224.

FM 5-71-100. *Division Engineer Combat Operations*. 22 April 1993. Will be revised as FM 3-34.225.

FM 5-100-15. *Corps Engineer Operations*. 6 June 1995. Will be revised as FM 3-34.226.

FM 5-102. *Countermobility*. 15 March 1985. Will be integrated into FM 3-34.2.

FM 5-103. *Survivability*. 10 June 1985. Will be revised as FM 3-34.112.

FM 5-104. *General Engineering*. 12 November 1986. Will be revised as FM 3-34.250.

FM 5-116. *Engineer Operations: Echelons Above Corps*. 09 February 1999. Will be revised as FM 3-34.211.

FM 5-125. *Rigging Techniques, Procedures and Applications*. 3 October 1995. With change 1 dated 23 February 2001. Will be revised as FM 3-34.311.

FM 5-134. *Pile Construction*. 18 April 1985. Will be revised as FM 3-34.334.

FM 5-170. *Engineer Reconnaissance*. 5 May 1998. With change 1 dated 13 July 1998. Will be revised as FM 3-34.212.

FM 5-212. *Medium Girder Bridge.* 07 February 1989. Will be revised as FM 3-34.342.

FM 5-233. *Construction Surveying.* 4 January 1985. Will be revised as FM 3-34.352.

FM 5-410. *Military Soils Engineering.* 23 December 1992. With change 1 dated 4 June 1997. Will be revised as FM 3-34.353.

FM 5-412. *Project Management.* 13 June 1994. Will be revised as FM 3-34.350.

FM 5-415. *Firefighting Operations.* 09 February 1999. Will be revised as FM 3-34.480.

FM 5-422. *Engineer Prime Power Operations.* 7 May 1993. Will be revised as FM 3-34.483.

FM 5-424. *Theater of Operations Electrical Systems.* 25 June 1997. Will be revised as FM 3-34.472.

FM 5-426. *Carpentry.* 3 October 1995. Will be revised as FM 3-34.473.

FM 5-428. *Concrete and Masonry.* 18 June 1998. Will be revised as FM 3-34.474.

FM 5-430-00-1. *Planning and Design of Roads, Airfields, and Heliports in the Theater of Operations – Road Design. Volume I.* 26 August 1994. Will be revised as FM 3-34.461.

FM 5-430-00-2. *Planning and Design of Roads, Airfields, and Heliports in the Theater of Operations – Airfield and Heliport Design. Volume II.* 29 September 1994. Will be revised as FM 3-34.462.

FM 5-434. *Earthmoving Operations.* 15 June 2000. Will be revised as FM 3-34.464.

FM 5-436. *Paving and Surfacing Operations.* 28 April 2000. Will be revised as FM 3-34.466.

FM 5-472. *Materials Testing.* 27 October 1999. With change 2 dated 1 July 2001. Will be revised as FM 3-34.354.

FM 5-480. *Port Construction and Repair.* 12 December 1990. Will be revised as FM 3-34.484.

FM 5-482. *Military Petroleum Pipeline Systems.* 26 August 1994. Will be revised as FM 3-34.485.

FM 5-484. *Multiservice Procedures for Well-Drilling Operations.* 8 March 1994. Will be revised as FM 3-34.481.

FM 5-490. *Engineer Diving Operations.* 31 March 1992. Will be revised as FM 3-34.216.

FM 5-499. *Hydraulics.* 1 August 1997. With change 1 dated 12 December 2001. Will be revised as FM 3-34.467.

FM 5-553. *General Drafting.* 6 January 1984. Will be revised as FM 3-34.351.

FM 6-0 (101-5). *Mission Command: Command and Control of Army Forces.* 11 August 2003.

FM 7-0 (25-100). *Training the Force.* 22 October 2002.

FM 7-1 (25-101). *Battle Focused Training.* 15 September 2003.

FM 7-15. *The Army Universal Task List.* 17 December 2002.

FM 20-3. *Camouflage, Concealment and Decoys.* 30 August 1999. Will be revised as FM 3-58.1.

FM 20-11. *Military Diving.* 20 January 1999. With change 1 dated 14 April 2000. Will be revised as FM 3-34.12.

FM 20-32. *Mine/Countermine Operations.* 29 May 1998. With change 3 dated 01 October 2003. Will be revised as FM 3-34.32.

FM 21-10. *Field Hygiene and Sanitation.* 21 June 2000. Will be revised as FM 4-25.10.

FM 21-31. *Topographic Symbols.* 19 June 1961. With change 1 dated 31 December 1968. Will be revised as FM 3-34.31.

FM 34-1. *Intelligence and Electronic Warfare Operations.* 27 September 1994. Will be revised as FM 2-0.

FM 34-130. *Intelligence Preparation of the Battlefield.* 8 July 1994. This is in the process of revision to FM 2-01.3.

FM 90-7. *Combined-Arms Obstacle Integration.* 29 September 1994. With change 1 dated 10 April 2003. Will be revised as FM 3-34.1, incorporating FM 5-102.

FM 90-13. *River-Crossing Operations.* 26 January 1998. Will be revised as FM 3-34.13.

FM 100-6. *Information Operations.* 27 August1966. This is in the process of revision to FM 3-13.

FM 100-7. *Decisive Force: The Army in Theater Operations.* 31 May 1991. This is in the process of revision to FM 3-93, Strategic Army.

FM 100-8. *The Army in Multinational Operations.* 24 November 1994. This is in the process of revision to FM 3-16.

FM 100-10-2. *Contracting Support on the Battlefield.* 4 August 1999. Will be revised as FM 4-100.2.

FM 100-11. *Force Integration.* 15 January 1998. Will be revised as FM 3-100.11.

FM 100-14. *Risk Management.* 23 April 1998. Will be revised as FM 5-19.

FM 100-17. *Mobilization, Deployment, Redeployment, Demobilization.* 28 October 1992. This is in the process of revision to FM 3-35.

FM 100-17-3. Recep*tion, Staging, Onward Movement, and Integration.* 17 March 1999. Will be revised as FM 4-01.8.

FM 100-19. *Domestic Support Operations.* 1 July 1993. Will be revised as FM 3-07.7.

FM 100-23-1. *HA—Multiservice Procedures for Humanitarian Assistance Operations.* 31 October 1994. Will be revised as FM 3-07.6.

FM 101-5. *Staff Organization and Operations.* 31 May 1997. Will be revised as

FM 5-0, Army Planning and Orders Production.

FM 101-5-1. *Operational Terms and Graphics.* 30 September 1977. This is in the process of revision to FM 1-02.

FM 101-5-2. *US Army Report and Message Formats.* 29 June 1999. Will be revised as FM 6-99.2.

FMFM 13-7. *MAGTF Breaching Operations.* 13 June 1995. Currently under revision and will be renumbered as MCWP 3-17.3.

FMFM 13-8. *MAGTF Explosive Ordnance Disposal.* 20 December 1993. Currently under revision and will be renumbered as MCWP 3-17.2.

JP 0-2. *Unified Action Armed Forces (UNAAF).* 10 July 2001.

JP 1. *Joint Warfare of the Armed Forces of the United States.* 14 November 2000.

JP 1-02. *Department of Defense Dictionary of Military and Associated Terms.* 12 April 2001.

JP 2-03. *Joint Tactics, Techniques, and Procedures for Geospatial Information and Services Support to Joint Operations.* 31 March 1999.

JP 3-0. *Doctrine for Joint Operations.* 10 September 2001.

JP 3-07. *Joint Doctrine for Military Operations Other Than War.* 16 June 1995.

JP 3-07.1. *Joint Tactics, Techniques, and Procedures for Foreign Internal Defense (FID).* 26 June 1996. Currently under revision.

JP 3-07.3. *Joint Tactics, Techniques, and Procedures for Peace Operations.* 12 February 1999.

JP 3-07.5. *Joint Tactics, Techniques, and Procedures for Noncombatant Evacuation Operations.* 30 September 1997.

JP 3-07.7. *Doctrine for Civil Support.* This is a new manual that is in draft and fills a doctrinal gap.

JP 3-08. *Interagency Coordination During Joint Operations*, 2 volumes. 9 October 2002. Currently under revision.

JP 3-10.1. *Joint Tactics, Techniques, and Procedures for Base Defense.* 23 July 1996.

JP 3-15. *Joint Doctrine for Barriers, Obstacles, and Mine Warfare.* 24 February 1999.

JP 3-16. *Joint Doctrine for Multinational Operations.* 5 April 2000.

JP 3-26. *Joint Doctrine for Homeland Defense.* This is a new manual that is in draft and fills a doctrinal gap.

JP 3-33. *Joint Force Capabilities.* 13 October 1999.

JP 3-34. *Engineer Doctrine for Joint Operations.* 5 July 2000.

JP 3-35. *Joint Doctrine for Deployment and Redeployment Operations.* 7 September 1999.

JP 3-55. *Joint Doctrine for Intelligence, Surveillance, and Reconnaissance (ISR) Operations.* This is a new manual that is in draft and fills a doctrinal gap.

JP 4-0. *Doctrine for Logistic Support of Joint Operations.* 6 April 2000.

JP 4-01.3. *Joint Tactics, Techniques, and Procedures for Movement Control.* 9 April 2002.

JP 4-01.8. *Joint Tactics, Techniques, and Procedures for Joint Reception, Staging, Onward Movement, and Integration.* 13 June 2000.

JP 4-04. *Joint Doctrine for Civil Engineering Support.* 27 September 2001.

JP 4-05. *Joint Doctrine for Mobilization Planning.* 22 June 1995.

JP 4-05.1. *Joint Tactics, Techniques, and Procedures for Manpower Mobilization and Demobilizations Operations: Reserve Component (RC) Callup.* 11 November 1998.

JP 5-0. *Doctrine for Planning Joint Operations.* 13 April 1995. Currently under revision.

JP 5-00.1. *Joint Doctrine for Campaign Planning.* 25 January 2002.

JP 5-00.2. *Joint Task Force Planning Guidance and Procedures.* 13 January 1999.

MCWP 3-17. *Engineering Operations.* 14 February 2000.

MCWP 3-17B. *Engineer Forms and Reports.* 3 October 1997.

MCWP 3-35.3. *Military Operations on Urbanized Terrain (MOUT).* 26 April 1998.

NWP 4-11. *Environmental Protection.* 1 March 1999.

OPNAVINST 3501.93D, *Projected Operational Environmental (POE) and Required Operational Capabilities (ROC) for Naval Beach Groups and Their Elements.* 19 June 2001.

OPNAVINST 5450.46K, *Naval Construction Force Policy.* 25 May 1999.

STANAG 2014/QSTAG 506. *Operations Orders, Warning Orders, and Administrative/Logistics Orders.* 27 April 1994.

STANAG 2020/QSTAG 510. *Operational Situation Reports.* 14 July 1995.

STANAG 2437. *Allied Joint Operations Doctrine* (AJP-1). 15 December 1994.

STANAG 3680. *NATO Glossary of Terms and Definitions* (AAP-6). 3 October 1986.

TC 3-34.489. *The Soldier and the Environment.* 8 May 2001. With change 1 dated 26 October 2001.

TC 5-150. *Engineer Qualification Tables.* 16 June 1998.

TC 5-210. *Military Float Bridging Equipment.* 27 December 1988. Will be revised as FM 3-34.340.

TC 5-230. *Army Geospatial Guide for Commanders and Planners.* This is a new manual that is in draft and fills a doctrinal gap.

TC 5-340. *Air Base Damage Repair (Pavement Repair).* 27 December 1988. Will be revised as FM 3-34.463.

Title 10. *Armed Forces.*

Title 10. *Burden sharing.*

Title 32. *National Guard.*

TM 5-235. *Special Surveys.* 18 September 1964. Will be revised as FM 3-34.335.

TM 5-240. *Compilation and Color Separation of Topographic Maps.* 15 June 1971. Will be revised and combined with FM 5-243 as FM 3-34.332, Cartographic Photography.

TM 5-243. *Cartographic Aerial Photography.* 2 January 1970. Will be revised and combined with TM 5-240 as FM 3-34.332, Cartographic Photography.

TM 5-245. *Offset Photolithography and Map Reproduction.* 21 July 1970. Will be revised as FM 3-34.333.

TM 5-323. *Military Standardization Handbook: Generator Sets, Electrical Measurement.* 30 August 1962. Will be revised as FM 3-34.476.

TM 5-349. *Arctic Construction.* 19 February 1962. Will be revised as FM 3-34.486.

TM 5-432. *Pits and Quarries.* 15 December 1967. Will be revised as FM 3-34.465.

TM 5-545. *Geology.* 8 July 1971. Will be revised as FM 3-34.360.

TRADOC PAM 525-3-25. *The United States Army Objective Force Maneuver Support Operational Concept.* This yet to be published document is critical to understanding the future Army and the impact on doctrine.

Quadrennial Defense Review. 30 September 2001.

White Paper. *Offshore Force Projection Operations and Enabling Systems: Deployment of the IBCT.* 1 June 2001.

Index

N

O

U

V

W

X

Made in United States
Cleveland, OH
10 November 2026

48115152R00243